Modern Trends in Structural and Solid Mechanics 2

Series Editor
Noël Challamel

Modern Trends in Structural and Solid Mechanics 2

Vibrations

Edited by

Noël Challamel
Julius Kaplunov
Izuru Takewaki

ISTE

WILEY

First published 2021 in Great Britain and the United States by ISTE Ltd and John Wiley & Sons, Inc.

ISTE Ltd
27-37 St George's Road
London SW19 4EU
UK

www.iste.co.uk

John Wiley & Sons, Inc.
111 River Street
Hoboken, NJ 07030
USA

www.wiley.com

Library of Congress Control Number: 2021932076

British Library Cataloguing-in-Publication Data
A CIP record for this book is available from the British Library
ISBN 978-1-78630-715-6

MIX
Paper from responsible sources
FSC® C013604

Contents

Preface

Short Bibliographical Presentation of Prof. Isaac Elishakoff

This book is dedicated to Prof. Isaac Elishakoff by his colleagues, friends and former students, on the occasion of his seventy-fifth birthday.

Figure P.1. *Prof. Isaac Elishakoff*

For a color version of all the figures in this chapter, see www.iste.co.uk/challamel/mechanics2.zip.

Prof. Isaac Elishakoff is an international leading authority across a broad area of structural mechanics, including dynamics and stability, optimization and anti-optimization, probabilistic methods, analysis of structures with uncertainty, refined theories, functionally graded material structures, and nanostructures. He was born in Kutaisi, Republic of Georgia, on February 9, 1944.

Figure P.2. *Elishakoff in middle school in the city of Sukhumi, Georgia*

Elishakoff holds a PhD in Dynamics and Strength of Machines from the Power Engineering Institute and Technical University in Moscow, Russia (Figure P.4 depicts the PhD defense of Prof. Isaac Elishakoff).

Figure P.3. *Elishakoff just before acceptance to university. Photo taken in Sukhumi, Georgia*

Figure P.4. *Public PhD defense, Moscow Power Engineering Institute and State University; topic "Vibrational and Acoustical Fields in the Circular Cylindrical Shells Excited by Random Loadings", and dedicated to the evaluation of noise levels in TU-144 supersonic aircraft*

His supervisor was Prof. V. V. Bolotin (1926–2008), a member of the Russian Academy of Sciences (Figure P.5 shows Elishakoff with Bolotin some years later).

Figure P.5. *Elishakoff with Bolotin (middle), member of the Russian Academy of Sciences, and Prof. Yukweng (Mike) Lin (left), member of the US National Academy of Engineering. Photo taken at Florida Atlantic University during a visit from Bolotin*

Currently, Elishakoff is a Distinguished Research Professor in the Department of Ocean and Mechanical Engineering at Florida Atlantic University. Before joining the university, he taught for one year at Abkhazian University, Sukhumi in the

Republic of Georgia, and 18 years at the Technion – Israel Institute of Technology in Haifa, where he became the youngest full professor at the time of his promotion (Figure P.6 shows Elishakoff presenting a book to Prof. Josef Singer, Technion's former president).

Figure P.6. *Prof. Elishakoff presenting a book to Prof. J. Singer, Technion's President; right: Prof. A. Libai, Aerospace Engineering Department, Technion*

Elishakoff has lectured at about 200 meetings and seminars, including about 60 invited, plenary or keynote lectures, across Europe, North and South America, the Middle East and the Far East.

Prof. Elishakoff has made vital and outstanding contributions in a number of areas in structural mechanics. In particular, he has analyzed random vibrations of homogeneous and composite beams, plates and shells, with special emphasis on the effects of refinements in structural theories and cross-correlations. Free structural vibrations have been tackled using a non-trivial generalization of Bolotin's dynamic edge effect method. Nonlinear buckling has been investigated using a novel method, incorporating experimental analysis of imperfections. As a result, the fundamental concept of closing the gap – spanning the entire 20th century – between theory and practice in imperfection-sensitive structures has been proposed. Novel methods of evaluating structural reliability have been proposed, taking into account the error associated with various low-order approximations, as well as human error; innovative generalization of the stochastic linearization method has been advanced.

A non-probabilistic theory for treating uncertainty in structural mechanics has been established. Dynamic stability of elastic and viscoelastic structures with imperfections has been studied. An improved, non-perturbative stochastic finite element method for structures has been developed. The list of Elishakoff's remarkable research achievements goes on.

His research has been acknowledged by many awards and prizes. He is a member of the European Academy of Sciences and Arts, a Fellow of the American Academy of Mechanics and ASME, and a Foreign Member of the Georgian National Academy of Sciences. Elishakoff is also a recipient of the Bathsheva de Rothschild prize (1973) and the Worcester Reed Warner Medal of the American Society of Mechanical Engineers (2016).

Figure P.7. *Elishakoff having received the William B. Johnson Inter- Professional Founders Award*

Elishakoff is directly involved in numerous editorial activities. He serves as the book review editor of the "Journal of Shock and Vibration" and is currently, or has previously been an associate editor of the *International Journal of Mechanics of Machines and Structures, Applied Mechanics Reviews*, and *Chaos, Solitons & Fractals*. In addition, he is or has been on the editorial boards of numerous journals, for example *Journal of Sound and Vibration*, *International Journal of Structural Stability and Dynamics*, *International Applied Mechanics* and *Computers & Structures*. He also acts as a book series editor for Elsevier, Springer and Wiley.

Figure P.8. *Inauguration as the Frank Freimann Visiting Professor of Aerospace and Mechanical Engineering; left: Rev. Theodore M. Hesburgh, President of the University of Notre Dame; right: Prof. Timothy O'Meara, Provost*

Prof. Elishakoff has held prestigious visiting positions at top universities all over the world. Among them are Stanford University (S. P. Timoshenko Scholar); University of Notre Dame, USA (Frank M. Freimann Chair Professorship of Aerospace and Mechanical Engineering and Henry J. Massman, Jr. Chair Professorship of Civil Engineering); University of Palermo, Italy (Visiting Castigliano Distinguished Professor); Delft University of Technology, Netherlands (multiple appointments, including the W. T. Koiter Chair Professorship of the Mechanical Engineering Department – see Figure P.9); Universities of Tokyo and Kyoto, Japan (Fellow of the Japan Society for the Promotion of Science); Beijing University of Aeronautics and Astronautics, People's Republic of China (Visiting Eminent Scholar); Technion, Haifa, Israel (Visiting Distinguished Professor); University of Southampton, UK (Distinguished Visiting Fellow of the Royal Academy of Engineering).

Figure P.9. *Prof. Elishakoff with Prof. Warner Tjardus Koiter, Delft University of Technology (center), and Dr. V. Grishchak, of Ukraine (right)*

Figure P.10. *Elishakoff and his colleagues during the AIAA SDM Conference at Palm Springs, California in 2004; Standing, from right to left, are Prof. Elishakoff, the late Prof. Josef Singer and Dr. Giora Maymon of RAFAEL. Sitting is the late Prof. Avinoam Libai*

Elishakoff has made a substantial contribution to conference organization. In particular, he participated in the organization of the Euro-Mech Colloquium on "Refined Dynamical Theories of Beams, Plates and Shells, and Their Applications" in Kassel, Germany (1986); the *Second International Conference on Stochastic Structural Dynamics*, in Boca Raton, USA (1990); "International Conference on Uncertain Structures" in Miami, USA and Western Caribbean (1996). He also coordinated four special courses at the International Centre for Mechanical Sciences (CISM), in Udine, Italy (1997, 2001, 2005, 2011).

Prof. Elishakoff has published over 540 original papers in leading journals and conference proceedings. He championed authoring, co-authoring or editing of 31 influential and extremely well-received books and edited volumes.

Here follows some praise of his work and books:

– "It was not until 1979, when Elishakoff published his reliability study … that a method has been proposed, which made it possible to introduce the results of imperfection surveys … into the analysis …" (Prof. Johann Arbocz, Delft University of Technology, The Netherlands, *Zeitschrift für Flugwissenschaften und Weltraumforschung*).

– "He has achieved world renown … His research is characterized by its originality and a combination of mathematical maturity and physical understanding which is reminiscent of von Kármán …" (Prof. Charles W. Bert, University of Oklahoma).

– "It is clear that Elishakoff is a world leader in his field … His outstanding reputation is very well deserved …" (Prof. Bernard Budiansky, Harvard University).

– "Professor Isaac Elishakoff … is subject-wise very much an all-round vibrationalist" (P. E. Doak, Editor in Chief, *Journal of Sound and Vibration*, University of Southampton, UK).

– "This is a beautiful book …" (Dr. Stephen H. Crandall, Ford Professor of Engineering, M.I.T.).

– "Das Buch ist in seiner Aufmachunghervorragendgestaltet und kannalsäusserstwertvolleErganzung … wäzmstensempfohlenwerden …" [The book's appearance is perfectly designed and can be highly recommended as a valuable addition.] (Prof. Horst Försching, Institute of Aeroelasticity, Federal Republic of Germany, *Zeitschrift für Flugwissenschaften und Weltraumforschung*).

– "Because of you, Notre Dame is an even better place, a more distinguished University" (Prof. Rev. Theodore M. Hesburgh, President, University of Notre Dame).

– "It is an impressive volume …" (Prof. Warner T. Koiter, Delft University of Technology, The Netherlands).

– "This extremely well-written text, authored by one of the leaders in the field, incorporates many of these new applications … Professor Elishakoff's techniques for developing the material are accomplished in a way that illustrates his deep insight into the topic as well as his expertise as an educator … Clearly, the second half of the text provides the basis for an excellent graduate course in random vibrations and buckling … Professor Elishakoff has presented us with an outstanding instrument for teaching" (Prof. Frank Kozin, Polytechnic Institute of New York, *American Institute of Aeronautics and Astronautics Journal*).

– "By far the best book on the market today …" (Prof. Niels C. Lind, University of Waterloo, Canada).

– "The book develops a novel idea … Elegant, exhaustive discussion … The study can be an inspiration for further research, and provides excellent applications in design …" (Prof. G. A. Nariboli, *Applied Mechanics Reviews*).

– "This volume is regarded as an advanced encyclopedia on random vibration and serves aeronautical, civil and mechanical engineers …" (Prof. Rauf Ibrahim, Wayne State University, *Shock and Vibration Digest*).

– "The book deals with a fundamental problem in Applied Mechanics and in Engineering Sciences: How the uncertainties of the data of a problem influence its solution. The authors follow a novel approach for the treatment of these problems … The book is written with clarity and contains original and important results for the engineering sciences …" (Prof. P. D. Panagiotopoulos, University of Thessaloniki, Greece and University of Aachen, Germany, *SIAM Review*).

– "The content should be of great interest to all engineers involved with vibration problems, placing the book well and truly in the category of an essential reference book …" (Prof. I. Pole, *Journal of the British Society for Strain Measurement*).

– "A good book; a different book … It is hoped that the success of this book will encourage the author to provide a sequel in due course …" (Prof. John D. Robson, University of Glasgow, Scotland, UK, *Journal of Sound and Vibration*).

– "The book certainly satisfies the need that now exists for a readable textbook and reference book ..." (Prof. Masanobu Shinozuka, Columbia University).

– "[the] author ties together reliability, random vibration and random buckling ... Well written ... useful book ..." (Dr. H. Saunders, *Shock and Vibration Digest*).

– "A very useful text that includes a broad spectrum of theory and application" (*Mechanical Vibration*, Prof. Haym Benaroya, Rutgers University).

– "A treatise on random vibration and buckling ... The reviewer wishes to compliment the author for the completion of a difficult task in preparing this book on a subject matter, which is still developing on many fronts ..." (Prof. James T. P. Yao, Texas A&M University, *Journal of Applied Mechanics*).

– "It seems to me a hard work with great result ..." (Prof. Hans G. Natke, University of Hannover, Federal Republic of Germany).

– "The approach is novel and could dominate the future practice of engineering" (*The Structural Engineer*).

– "An excellent presentation ... well written ... all readers, students, and certainly reviewers should read this preface for its excellent presentation of the philosophy and raison d'être for this book. It is well written, with the material presented in an informational fashion as well as to raise questions related to unresolved ... challenges; in the vernacular of film critics, 'thumbs up'" (Dr. R. L. Sierakowski, U.S. Air Force Research Laboratory, *AIAA Journal*).

– "This substantial and attractive volume is a well-organized and superbly written one that should be warmly welcomed by both theorists and practitioners ... Prof. Elishakoff, Li, and Starnes, Jr. have given us a jewel of a book, one done with care and understanding of a complex and essential subject and one that seems to have ably filled a gap existing in the present-day literature and practice" (*Current Engineering Practice*).

– "Most of the subjects covered in this outstanding book have never been discussed exclusively in the existing treatises ... (*Ocean Engineering*).

– "The treatment is scholarly, having about 900 items in the bibliography and additional contributors in the writing of almost every chapter ... This reviewer believes that *Non-Classical Problems in the Theory of Elastic Stability* should be a useful reference for researchers, engineers, and graduate students in aeronautical, mechanical, civil, nuclear, and marine engineering, and in applied mechanics" (*Applied Mechanics Reviews*).

– "What more can be said about this monumental work, other than to express admiration? … The study is of great academic interest, and is clearly a labor of love. The author is to be congratulated on this work …" (Prof. H. D. Conway, Department of Theoretical and Applied Mechanics, Cornell University).

– "This book … is prepared by Isaac Elishakoff, one of the eminent solid mechanics experts of the 20th century and the present one, and his distinguished coauthors, will be of enormous use to researchers, graduate students and professionals in the fields of ocean, naval, aerospace and mechanical engineers as well as other fields" (Prof. Patricio A. A. Laura, Prof. Carlos A. Rossit, Prof. Diana V. Bambill, Universidad Nacional del Sur, Argentina, *Ocean Engineering*).

– "This book is an outstanding research monograph … extremely well written, informative, highly original … great scholarly contribution …. There is no comparable book discussing the combination of optimization and anti-optimization … magnificent monograph …. This book, which certainly is written with love and passion, is the first of its kind in applied mechanics literature, and has the potential of having a revolutionary impact on both uncertainty analysis and optimization" (Prof. Izuru Takewaki, Kyoto University, *Engineering Structures*).

– "This book is a collection of a surprisingly large number of closed form solutions, by the author and by others, involving the buckling of columns and beams, and the vibration of rods, beams and circular plates. The structures are, in general, inhomogeneous. Many solutions are published here for the first time. The text starts with an instructive review of direct, semi-inverse, and inverse eigenvalue problems. Unusual closed form solutions of column buckling are presented first, followed by closed form solutions of the vibrations of rods. Unusual closed form solutions for vibrating beams follow. The influence of boundary conditions on eigenvalues is discussed. An entire chapter is devoted to boundary conditions involving guided ends. Effects of axial loads and of elastic foundations are presented in two separate chapters. The closed form solutions of circular plates concentrate on axisymmetric vibrations. The scholarly effort that produced this book is remarkable" (Prof. Werner Soedel, then Editor-in-Chief of *Journal Sound and Vibration*).

– "The field has been brilliantly presented in book form …" (Prof. Luis A. Godoy *et al.*, Institute of Advanced Studies in Engineering and Technology, Science Research Council of Argentina and National University of Cordoba, Argentina, *Thin-Walled Structures*).

– "Elishakoff is one of the pioneers in the use of the probabilistic approach for studying imperfection-sensitive structures" (Prof. Chiara Bisagni and Dr. Michela Alfano, Delft University of Technology; *AIAA Journal*).

– "Recently, Elishakoff *et al.* presented an excellent literature review on the historical development of Timoshenko's beam theory" (Prof. Zhenlei Chen *et al.*, *Journal of Building Engineering*).

Professor Isaac Elishakoff is the author or co-author of an impressive list of seminal books in the field of deterministic and non-deterministic mechanics, presented below.

Books by Elishakoff

Ben-Haim, Y. and Elishakoff, I. (1990). *Convex Models of Uncertainty in Applied Mechanics*. Elsevier, Amsterdam.

Cederbaum, G., Elishakoff, I., Aboudi, J., Librescu, L. (n.d.). *Random Vibration and Reliability of Composite Structures*. Technomic, Lancaster.

Elishakoff, I. (1983). *Probabilistic Methods in the Theory of Structures*. Wiley, New York.

Elishakoff, I. (1999). *Probabilistic Theory of Structures*. Dover Publications, New York.

Elishakoff, I. (2004). *Safety Factors and Reliability: Friends or Foes?* Kluwer Academic Publishers, Dordrecht.

Elishakoff, I. (2005). *Eigenvalues of Inhomogeneous Structures: Unusual Closed-Form Solutions of Semi-Inverse Problems*. CRC Press, Boca Raton.

Elishakoff, I. (2014). *Resolution of the Twentieth Century Conundrum in Elastic Stability*. World Scientific/Imperial College Press, Singapore.

Elishakoff, I. (2017). *Probabilistic Methods in the Theory of Structures: Random Strength of Materials, Random Vibration, and Buckling*. World Scientific, Singapore.

Elishakoff, I. (2018). *Probabilistic Methods in the Theory of Structures: Solution Manual to Accompany Probabilistic Methods in the Theory of Structures: Problems with Complete, Worked Through Solutions*. World Scientific, Singapore.

Elishakoff, I. (2020). *Dramatic Effect of Cross-Correlations in Random Vibrations of Discrete Systems, Beams, Plates, and Shells*. Springer Nature, Switzerland.

Elishakoff, I. (2020). *Handbook on Timoshenko-Ehrenfest Beam and Uflyand-Mindlin Plate Theories*. World Scientific, Singapore.

Elishakoff, I. and Ohsaki, M. (2010). *Optimization and Anti-Optimization of Structures under Uncertainty*. Imperial College Press, London.

Elishakoff, I. and Ren, Y. (2003). *Finite Element Methods for Structures with Large Stochastic Variations*. Oxford University Press, Oxford.

Elishakoff, I., Lin, Y.K., Zhu, L.P. (1994). *Probabilistic and Convex Modeling of Acoustically Excited Structures*. Elsevier, Amsterdam.

Elishakoff, I., Li, Y., Starnes Jr., J.H. (2001). *Non-Classical Problems in the Theory of Elastic Stability*. Cambridge University Press, Cambridge.

Elishakoff, I., Pentaras, D., Dujat, K., Versaci, C., Muscolino, G., Storch, J., Bucas, S., Challamel, N., Natsuki, T., Zhang, Y., Ming Wang, C., Ghyselinck, G. (2012). *Carbon Nanotubes and Nano Sensors: Vibrations, Buckling, and Ballistic Impact*. ISTE Ltd, London, and John Wiley & Sons, New York.

Elishakoff, I., Pentaras, D., Gentilini, C., Cristina, G. (2015). *Mechanics of Functionally Graded Material Structures*. World Scientific/Imperial College Press, Singapore.

Books edited or co-edited by Elishakoff

Ariaratnam, S.T., Schuëller, G.I., Elishakoff, I. (1988). *Stochastic Structural Dynamics – Progress in Theory and Applications*. Elsevier, London.

Casciati, F., Elishakoff, I., Roberts, J.B. (1990). *Nonlinear Structural Systems under Random Conditions*. Elsevier, Amsterdam.

Chuh, M., Wolfe, H.F., Elishakoff, I. (1989). *Vibration and Behavior of Composite Structures*. ASME Press, New York.

David, H. and Elishakoff, I. (1990). *Impact and Buckling of Structures*. ASME Press, New York.

Elishakoff, I. (1999). *Whys and Hows in Uncertainty Modeling*. Springer, Vienna.

Elishakoff, I. (2007). *Mechanical Vibration: Where Do We Stand?* Springer, Vienna.

Elishakoff, I. and Horst, I. (1987). *Refined Dynamical Theories of Beams, Plates and Shells and Their Applications*. Springer Verlag, Berlin.

Elishakoff, I. and Lin, Y.K. (1991). *Stochastic Structural Dynamics 2 – New Applications*. Springer, Berlin.

Elishakoff, I. and Lyon, R.H. (1986), *Random Vibration-Status and Recent Developments*. Elsevier, Amsterdam.

Elishakoff, I. and Seyranian, A.P. (2002). *Modern Problems of Structural Stability*. Springer, Vienna.

Elishakoff, I. and Soize, C. (2012). *Non-Deterministic Mechanics*. Springer, Vienna.

Elishakoff, I., Arbocz, J., Babcock Jr., C.D., Libai, A. (1988). *Buckling of Structures: Theory and Experiment*. Elsevier, Amsterdam.

Lin, Y.K. and Elishakoff, I. (1991). *Stochastic Structural Dynamics 1 – New Theoretical Developments*. Springer, Berlin.

Noor, A.K., Elishakoff, I., Hulbert, G. (1990). *Symbolic Computations and Their Impact on Mechanics*. ASME Press, New York.

Figure P.11. *Elishakoff with his wife, Esther Elisha, M.D., during an ASME awards ceremony*

On behalf of all the authors of this book, including those friends who were unable to contribute, we wish Prof. Isaac Elishakoff many more decades of fruitful works and collaborations for the benefit of world mechanics, in particular.

Modern Trends in Structural and Solid Mechanics 1 – the first of three separate volumes that comprise this book – presents recent developments and research discoveries in structural and solid mechanics, with a focus on the statics and stability of solid and structural members.

The book is centered around theoretical analysis and numerical phenomena and has broad scope, covering topics such as: buckling of discrete systems (elastic chains, lattices with short and long range interactions, and discrete arches), buckling of continuous structural elements including beams, arches and plates, static investigation of composite plates, exact solutions of plate problems, elastic and inelastic buckling, dynamic buckling under impulsive loading, buckling and post-buckling investigations, buckling of conservative and non-conservative systems, buckling of micro and macro-systems. The engineering applications concern both small-scale phenomena with micro and nano-buckling up to large-scale structures, including the buckling of drillstring systems.

Each of the three volumes is intended for graduate students and researchers in the field of theoretical and applied mechanics.

Prof. Noël CHALLAMEL
Lorient, France
Prof. Julius KAPLUNOV
Keele, UK
Prof. Izuru TAKEWAKI
Kyoto, Japan
February 2021

1

Bolotin's Dynamic Edge Effect Method Revisited (Review)

A comprehensive review of Bolotin's edge effect method is presented. This chapter begins with a toy problem and is concluded by nonlinear considerations that have not been developed by Bolotin himself. Various generalizations and modifications of the method are described, along with a variety of solved problems for which a wide list of references is provided. Attempts are also made to frame the method among the known methods for finding rapidly oscillating solutions.

1.1. Introduction

Professor Isaac E. Elishakoff was a doctoral student of the world-renowned scientist V.V. Bolotin (March 29, 1926 to May 28, 2008) (Bolotin 2006). The first research works of I. Elishakoff and his PhD thesis were devoted to the application and development of the dynamic edge effect (EE) method proposed by V.V. Bolotin. After moving from the Soviet Union to the Western world, Prof. Elishakoff made great efforts to popularize the dynamic EE method in the Western scientific community (Elishakoff 1974, 1976; Elishakoff and Wiener 1976).

Therefore, the appearance of a review of papers related to Bolotin's method in the volume devoted to Prof. Elishakoff's 75th birthday seems quite reasonable. Moreover, the previous comprehensive reviews of the subject were published in 1976 (Elishakoff 1976) and 1984 (Bolotin 1984).

In the early 1960s, V.V. Bolotin put forward an asymptotic method for studying natural oscillations of plates and shells, which used the inverse of the dimensionless

---

Chapter written by Igor V. ANDRIANOV and Lelya A. KHAJIYEVA.

vibration frequency as a small parameter (Bolotin 1960a, 1960b). In a more general formulation, it is a method for solving self-adjoint eigenvalue problems defined in a rectangular domain, called the boundary value problems with quasi-separable variables, according to Bolotin's terminology. For this reason, the method is referred to as Bolotin's method or the dynamic edge effect method (DEEM). And despite the fact that 60 years have passed since the method creation, it is still relevant. The purpose of this review is describing various generalizations and modifications of DEEM, the problems solved with the use of this method and also trying to determine the place of DEEM among the known methods for finding rapidly oscillating solutions. Thus, we demonstrate that DEEM can be broadly applied for solving modern problems.

1.2. Toy problem: natural beam oscillations

Demonstrate the main idea of DEEM on a spatially 1D problem, which can be reduced to a transcendental equation and solved numerically with any degree of accuracy (Weaver *et al.* 1990). Consider the natural oscillations of a beam of length L, described by the following PDE:

$$\frac{\partial^4 w}{\partial x^4} + a^2 \frac{\partial^2 w}{\partial t^2} = 0, \quad a^2 = \frac{\rho F}{EI}. \qquad [1.1]$$

Here, w is the normal displacement, E is the Young modulus, F is the cross-sectional area of the beam, I is the axial inertia moment of the beam cross-section, and ρ is the density of the beam material.

Let us compare two versions of boundary conditions:

$$w = 0, \quad \frac{\partial^2 w}{\partial x^2} = 0 \quad \text{at } x = 0, L, \qquad [1.2]$$

$$w = 0, \quad \frac{\partial w}{\partial x} = 0 \quad \text{at } x = 0, L. \qquad [1.3]$$

We use the following ansatz:

$$w(x,t) = W(x)\exp(i\omega t),$$

where ω is the eigenfrequency and $W(x)$ is the eigenfunction.

The equation for eigenfunction $W(x)$ has the form

$$\frac{d^4W}{dx^4} - a^2\omega^2 W = 0. \qquad [1.4]$$

The solution of the eigenvalue problem [1.4], [1.2] is given by

$$W = \sin\frac{m\pi}{L}x \quad , \quad m = 1, 2, \ldots ; \qquad [1.5]$$

$$\omega_m = \frac{1}{a}\left(\frac{m\pi}{L}\right)^2. \qquad [1.6]$$

The eigenvalue problem [1.4], [1.3] does not allow separation of variables. However, if the eigenfunction oscillates rapidly along x (i.e. a rather high form of oscillations is considered), then we can hope that in this case a solution of the form [1.5] is also valid for the inner domain sufficiently distant from the boundaries (Figure 1.1). Such an expression does not satisfy the boundary conditions. However, if the solution that compensates the residuals at the boundary conditions and decays rapidly, then the approximate expressions for the eigenfunctions and eigenfrequencies can be obtained.

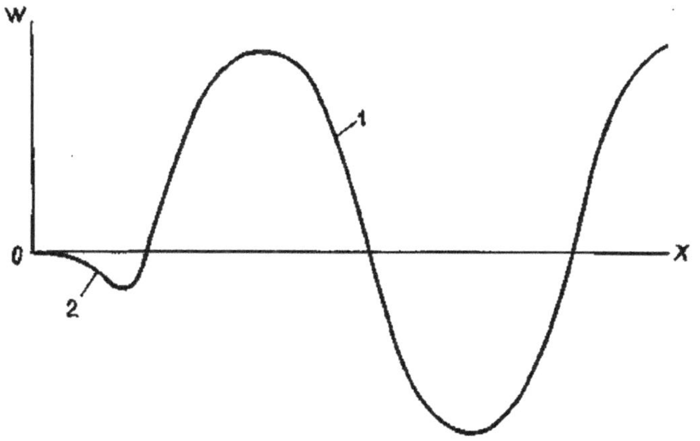

Figure 1.1. *Curve 1 corresponds to the rapidly oscillating solution in the inner domain, and curve 2 corresponds to the sum of DEE and the rapidly oscillating solution*

Let us suppose the solution of equation [1.4] in the form

$$W_0 = \sin\frac{\pi(x - x_0)}{\lambda}. \qquad [1.7]$$

The oscillation frequency ω is

$$\omega = \frac{1}{a}\left(\frac{\pi}{\lambda}\right)^2. \qquad [1.8]$$

Factorization of ODE [1.4] is (Vakhromeev and Kornev 1972)

$$\left(\frac{d^2}{dx^2} + a\omega\right)\left(\frac{d^2}{dx^2} - a\omega\right)W = 0. \qquad [1.9]$$

The general solution of ODE [1.9] is given by

$$W = W_0 + W_1 + W_2,$$

where functions W_0 and $W_{1,2}$ are the general solutions of the following equations:

$$\frac{d^2W_0}{dx^2} + a\omega W_0 = 0, \qquad [1.10]$$

$$\frac{d^2W_{1,2}}{dx^2} - a\omega W_{1,2} = 0. \qquad [1.11]$$

For large frequencies $(a\omega >> 1)$, the following estimates for the derivatives of functions W_0 and $W_{1,2}$ are obtained:

$$\frac{dW_0}{dx} \sim a\omega W_0, \quad \frac{dW_{1,2}}{dx} \sim a\omega W_{1,2}.$$

The behavior of these solution components is different: W_1 is the rapidly oscillating function, and W_2 is the sum of exponentials rapidly decreasing from the edges of the beam.

Therefore, the situation under consideration is fundamentally different from the case when the characteristic equation has small and large modulo roots, which is typical for boundary layer theory. In our case, we are talking about the separation of solutions, one of which oscillates at the same rate as the EE decays (i.e. the characteristic equation has large real and imaginary roots with moduli of the same order). The self-adjoint eigenvalue problem [1.1], [1.2] can be referred to as the boundary value problem with quasi-separable variables (Bolotin 1960a, 1960b, 1961a, 1961b, 1961c; Bolotin *et al.* 1950, 1961).

We proceed to the construction of the EE described by equation [1.11]. Taking into account the expression for the natural frequency [1.8], we obtain the following relations for EEs localized in the vicinity of the edges $x = 0$ and $x = L$, respectively:

$$W_1 = C_1 \exp\left(-\pi\lambda^{-1}x\right), \qquad [1.12]$$

$$W_2 = C_2 \exp\left[-\pi\lambda^{-1}(x-L)\right]. \qquad [1.13]$$

We assume that the beam is so long such that EEs do not affect each other, i.e. $\exp\left(-\pi\lambda^{-1}L\right) << 1$.

Now, it remains to find the quantities x_0, λ and constants C_1, C_2 from the boundary conditions to determine the eigenmodes and eigenfrequencies:

$$W_0 + W_1 = 0, \quad \frac{d}{dx}(W_0 + W_1) = 0 \quad \text{at } x = 0, \qquad [1.14]$$

$$W_0 + W_2 = 0, \quad \frac{d}{dx}(W_0 + W_2) = 0 \quad \text{at } x = L. \qquad [1.15]$$

Note that in original works by Bolotin, a slightly different matching procedure is used. Namely, the matching conditions [1.14], [1.15] are not given at the domain boundaries. They are set at some points, which are then determined from the solution of the system of transcendental equations along with other unknown constants.

Substituting expressions [1.7] and [1.12] into conditions [1.14], and expressions [1.7] and [1.13] into [1.15], we obtain

$$C_1 - \sin\frac{\pi x_0}{\lambda} = 0, \quad C_1 - \cos\frac{\pi x_0}{\lambda} = 0, \qquad [1.16]$$

$$C_2 + \sin\left(\pi\frac{L-x_0}{\lambda}\right) = 0, \quad C_2 + \cos\left(\pi\frac{L-x_0}{\lambda}\right) = 0. \qquad [1.17]$$

For λ and x_0, we have the following expressions:

$$\lambda = \frac{L}{m+0.5}, \quad m = 1,2,\ldots; \quad x_0 = \lambda(0.25+n), \quad n = 1,2,\ldots$$

Finally, the formula for the natural oscillation frequencies of the clamped beam is written as follows:

$$\omega_m = \pi^2 \frac{(m+0.5)^2}{aL^2}, \quad m = 1,2,\ldots \qquad [1.18]$$

The same formula follows from the analysis of the transcendental equation (Weaver *et al.* 1990). Its difference from the numerical solution is less than 1% even for the first natural oscillation frequency.

Using DEEM to solve various problems proved its high accuracy even when calculating the first eigenfrequencies (Bolotin 1960a, 1960b, 1961a, 1961b, 1961c, 1961d; Gavrilov 1961a, 1961b; Bolotin 1963, 1970; Elishakoff and Wiener 1976; Elishakoff and Steinberg 1979; Emmerling 1979; Bolotin 1984; Elishakoff *et al.* 1993, 1994). It confirms “Crighton’s principle” (Crighton 1994):

> All experience suggests that asymptotic solutions are useful numerically far beyond their nominal range of validity, and can often be used directly, at least at a preliminary product design stage, for example, saving the need for accurate computation until the final design stage where many variables have been restricted to narrow ranges.

The problem of natural oscillations was considered above. In the case of forced oscillations, the response is given as a series expansion in terms of normal modes of natural oscillations. It is important that one can use only the rapidly oscillating part of solution W_0, neglecting the dynamic edge effect (DEE) (Bolotin 1961b, 1961c, 1970, 1984).

1.3. Linear problems solved

DEEM in its original form can be effectively used for calculating natural, free and forced oscillations of rectangular plates, shells of revolution and shallow shells with a rectangular base.

In particular, DEEM is widely used for the analysis of rectangular isotropic (Bolotin *et al.* 1958; Bolotin 1961d, 1963; Kudryavtsev 1964; Dickinson and Warburton 1967; Bolotin 1970; King and Lin 1974; Elishakoff 1976; Emmerling 1979; Bolotin 1984; Elishakoff *et al.* 1994) and orthotropic (Dickinson 1971; Elishakoff 1974; Kaza and Ramaiah 1978; Vijaykumar and Ramaiah 1978) plates with various types of boundary conditions, plates on elastic foundations (Gibigaye *et al.* 2016), laminated and stiffened plates and panels (Lin and King 1974; Ueng and Nickels 1978; Meilani 2012, 2015). Oscillations and buckling of stressed plates were studied in Dickinson (1971, 1975), and box structures were studied in Dickinson (1975b). The Timoshenko theory of beams, and Mindlin, Reissner and Ambartsumyan theories of plates were also used in some works (Kudryavtsev 1960; Moskalenko 1961; Nelson 1978).

In the paper by Dickinson and Warburton (1967), DEEM was applied to consider free flexural vibrations of systems built up from rectangular plates.

Many papers are devoted to multi-span plates (Moskalenko and Chen 1965; Moskalenko 1968, 1969; Elishakoff and Steinberg 1979; Elishakoff *et al.* 1993).

Dynamics of cylindrical, conical and shallow spherical shells were analyzed in Bolotin (1960b, 1961c, 1984), Gavrilov (1961a), Zhinzher (1975) and Elishakoff and Wiener (1976).

One of the problems of the DEEM application is the degeneracy of DEE when decaying solutions cannot be constructed for some wave numbers (Bolotin 1984). The resolution of this problem was proposed in Elishakoff (1974) and Elishakoff and Wiener (1976). The solution of the original problem is represented as a sum of solutions of two subproblems. Each of these solutions satisfy the boundary conditions at two opposite boundaries only. The matching conditions described in Bolotin's original papers (Bolotin 1960a, 1960b, 1961a, 1961b, 1961c) are not used, and it gives us the possibility to avoid difficulties caused by the degeneracy of DEEM.

Thereby, we note the following point. Asymptotic methods can be used in two versions (Andrianov *et al.* 2014). From the very beginning, a small parameter can be introduced into the PDEs or ODEs and then asymptotic fractional analysis (Kline 1965) can be used. However, we can use variational approaches (Rayleigh–Ritz, Bubnov–Galerkin, Kantorovich, Trefftz, etc.) to solve the original problem and reduce it to the infinite systems of coupled ODEs or algebraic equations.

Then, a small parameter, caused by the physical nature of the problem or an artificial (homotopy) one, can be introduced into the infinite system to split it into simplified subsystems. The second approach makes it possible to avoid the degeneracy of DEE.

Other generalizations of DEEM are described in the following sections.

1.4. Generalization for the nonlinear case

To describe the generalization of DEEM to the nonlinear case, we use the nonlinear Kirchhoff beam equation (Kauderer 1958):

$$EI\frac{\partial^4 w}{\partial x^4}-\frac{EF}{2L}\left(\int_0^L\left(\frac{\partial w}{\partial x}\right)^2 dx\right)\frac{\partial^2 w}{\partial x^2}+\rho F\frac{\partial^2 w}{\partial t^2}=0. \qquad [1.19]$$

Let the beam be elastically supported:

$$w=0,\quad \frac{\partial^2 w}{\partial x^2}-c^*\frac{\partial w}{\partial x}=0 \quad \text{at } x=0,L, \qquad [1.20]$$

where $c^*=c/EI$, c is the coefficient characterizing elastic support.

We search a generating solution in the form

$$w_0=A\sin\frac{\pi(x-x_0)}{\lambda}\xi(t). \qquad [1.21]$$

Substituting ansatz [1.21] into PDE [1.19], we obtain an ODE for determining the time function $\xi(t)$:

$$\frac{d^2\xi}{dt^2}+\omega^2\left(1+\gamma\xi^2\right)\xi=0, \qquad [1.22]$$

where

$$\omega^2=\frac{EI}{\rho}\left(\frac{\pi}{\lambda}\right)^2,\quad \gamma=0.25(1+\lambda_1)\left(\frac{A}{r}\right)^2,\quad r=\sqrt{I/F},$$

$$\lambda_1=\frac{\lambda}{2\pi L}\left(\sin\frac{2\pi(L-x_0)}{\lambda}+\sin\frac{2\pi x_0}{\lambda}\right).$$

ODE [1.22] with initial conditions

$$\xi = 1, \quad \frac{d\xi}{dt} = 0 \text{ at } t = 0 \qquad [1.23]$$

has the solution

$$\xi(t) = \text{cn}(\sigma t, k), \quad \sigma = \omega\sqrt{1+\gamma}, \qquad [1.24]$$

where $\text{cn}(\sigma t, k)$ is the Jacobi cosine elliptic functions with period $T = 4K$, $K = \int_0^{\pi/2} \left(1 - k^2 \sin^2\varphi\right)^{-1/2} d\varphi$ is the complete elliptic integral of the first kind with modulus $k = \sqrt{0.5\gamma/(1+\gamma)}$ (Abramowitz and Stegun 1965).

The solution to the problem far from the edges is

$$w_0 = W_0(x)\,\text{cn}(\sigma t, k), \qquad [1.25]$$

where $W_0(x) = A \sin\frac{\pi(x - x_0)}{\lambda}$.

Solution [1.25] satisfies the original equation [1.19], but does not satisfy the boundary conditions [1.20]. To construct the states localized near the edges, we represent the solution of the original problem in the form

$$w = w_0 + w_{ee}. \qquad [1.26]$$

Substituting ansatz [1.26] into ODE [1.19], we obtain

$$\frac{\partial^4}{\partial x^4}(w_0 + w_{ee}) - 0.5\left(r^2 L\right)^{-1} \frac{\partial^2}{\partial x^2}(w_0 + w_{ee}) \int_0^L \left(\frac{\partial w_0}{\partial x} + \frac{\partial w_{ee}}{\partial x}\right)^2 dx$$
$$+ \frac{\rho}{EI}\frac{\partial^2}{\partial t^2}(w_0 + w_{ee}) = 0. \qquad [1.27]$$

In contrast to the previously considered linear case, the equations for functions w_0 and w_{ee} are coupled due to the nonlinearity of the problem. At the same time, the solution in the inner domain and EEs differ energetically since the EEs are localized in a small vicinity of the beam ends (Andrianov *et al.* 1979; Awrejcewicz *et al.* 1998;

Andrianov *et al.* 2004, 2014). Let us estimate the orders of the integrand terms in equation [1.27] with respect to $L/\lambda >> 1$:

$$\int_0^L \left(\frac{\partial w_0}{\partial x}\right)^2 dx \sim \left(\frac{L}{\lambda}\right)^2, \ \int_0^L \frac{\partial w_0}{\partial x}\frac{\partial w_{ee}}{\partial x} dx \sim \left(\frac{L}{\lambda}\right), \ \int_0^L \left(\frac{\partial w_{ee}}{\partial x}\right)^2 dx \sim 1. \qquad [1.28]$$

Restricting ourselves to the term of order $(\pi/\lambda)^2 >> 1$ in equation [1.27], we reduce it to the form:

$$\begin{aligned} &\frac{\partial^4 w_0}{\partial x^4} - 0.5\left(r^2 L\right)^{-1} \frac{\partial^2 w_0}{\partial x^2} \int_0^L \left(\frac{\partial w_0}{\partial x}\right)^2 dx + \frac{\rho}{EI}\frac{\partial^2 w_0}{\partial t^2} + \frac{\partial^4 w_{ee}}{\partial x^4} \\ &- 0.5\left(r^2 L\right)^{-1} \frac{\partial^2 w_{ee}}{\partial x^2} \int_0^L \left(\frac{\partial w_0}{\partial x}\right)^2 dx + \frac{\rho}{EI}\frac{\partial^2 w_{ee}}{\partial t^2} = 0. \end{aligned} \qquad [1.29]$$

Substituting function w_0 into equation [1.29], we obtain a PDE for function w_{ee}:

$$\frac{\partial^4 w_{ee}}{\partial x^4} - B\,\mathrm{cn}^2(\sigma t, k)\frac{\partial^2 w_{ee}}{\partial x^2} + \frac{\rho}{EI}\frac{\partial^2 w_{ee}}{\partial t^2} = 0, \qquad [1.30]$$

where $B = \gamma\left(\frac{\pi}{\lambda}\right)^2$.

It is important that PDE [1.30] is linear. The spatial and time variables are not separated exactly; therefore, we apply the Kantorovich variational method (Kantorovich and Krylov 1958) to solve equation [1.30], presenting w_{ee} in the form

$$w_{ee}(x,t) \cong W_{ee}(x)\,\mathrm{cn}(\sigma t, k). \qquad [1.31]$$

On substituting ansatz [1.31] into PDE [1.30] and applying the Kantorovich method (Kantorovich and Krylov 1958), the following ODE is obtained:

$$\frac{d^4 W_{ee}}{dx^4} - B_1 \frac{d^2 W_{ee}}{dx^2} - \left(\frac{\pi}{\lambda}\right)^2 \left[\left(\frac{\pi}{\lambda}\right)^2 + B_1\right] W_{ee} = 0, \qquad [1.32]$$

with

$$B_1 = A\left(\frac{2k^2-1}{2k^2} + \frac{\sqrt{1-k^2}}{2k\arcsin k}\right). \qquad [1.33]$$

Hereinafter, we use the principal value of the arcsin(...) function.

Among the four roots of the characteristic equation for ODE [1.32], two purely imaginary ones correspond to the generating solution W_0. To construct DEE, we should use real roots of the characteristic equation. Then, the DEE solution is

$$W_{ee}(x) = C_1 \exp\left[\left(-\sqrt{\left(\frac{\pi}{\lambda}\right)^2 + B_1}\right)x\right] + C_2 \exp\left[\left(\sqrt{\left(\frac{\pi}{\lambda}\right)^2 + B_1}\right)x\right]. \qquad [1.34]$$

Let us construct DEE near the edge $x = 0$. For a sufficiently long beam, we can suppose

$$C_2 = 0. \qquad [1.35]$$

Then, at $x = 0$, we have from the boundary conditions

$$W_0 + W_{ee} = 0, \quad \frac{d^2W_0}{dx^2} + \frac{d^2W_{ee}}{dx^2} = c^*\left(\frac{dW_0}{dx} + \frac{dW_{ee}}{dx}\right). \qquad [1.36]$$

Using expressions [1.34]–[1.36], we obtain

$$C_1 = A\sin\frac{\pi x_0}{\lambda}, \qquad [1.37]$$

$$x_0 = \frac{\lambda}{\pi}\arctan\frac{\pi}{\lambda\left[\left(2(\pi/\lambda)^2 + B_1\right)/c^* + \sqrt{(\pi/\lambda)^2 + B_1}\right]}. \qquad [1.38]$$

Note that when $c^* \to 0$ and $c^* \to \infty$, formulas [1.34]–[1.38] yield solutions for simply supported and clamped ends of the beam, respectively.

Similarly, we can construct DEE localized at the edge $x = L$.

The modes of natural nonlinear oscillations of the beam can be divided into groups according to the types of symmetry. For the modes that are symmetric relative to the point $x = L/2$, from the condition

$$\frac{dW_0}{dx} = 0 \quad \text{at } x = L/2,$$

we obtain

$$L - 2x_0 = (2m+1)\pi, \quad m = 1,2,... \qquad [1.39]$$

For antisymmetric modes, from the condition

$$W_0 = 0 \quad \text{at } x = L/2,$$

we have

$$L - 2x_0 = 2n\pi, \quad n = 1,2,... \qquad [1.40]$$

Equations [1.39] and [1.40] can be reduced to the following form:

$$L - 2x_0 = m\pi, \quad m = 1,2,..., \qquad [1.41]$$

in which even values of m correspond to antisymmetric modes, and odd values of m to symmetric modes relative to the point $x = L/2$.

Thus, the system of equations [1.37], [1.38] and [1.41] can be applied to determine the constants λ and x_0.

The described technique was used to study nonlinear oscillations of isotropic (Andrianov *et al.* 1979; Zhinzher and Denisov 1983; Awrejcewicz *et al.* 1998; Andrianov *et al.* 2004) and orthotropic (Zhinzher and Khromatov 1984) plates, circular cylindrical and shallow shells (Zhinzher and Denisov 1983; Andrianov and Kholod 1985; Zhinzher and Khromatov 1990; Andrianov and Kholod 1993a, 1993b, 1995).

1.5. DEEM and variational approaches

DEEM, designed to calculate high eigenfrequencies, also gives enough accurate results for lower vibration modes at kinematic boundary conditions. For static conditions, the accuracy of determining the lowest natural frequencies decreases.

Attempts to apply the method to stability problems have shown that the error of determining the buckling load is quite high.

One of the promising ways to improve the DEEM accuracy is its combination with variational approaches. The first works in this direction were the papers (Vijaykumar and Ramaiah 1978a, 1978b), where the Rayleigh–Ritz method (RRM) was applied and the asymptotic expressions for natural modes were used as basis functions (the Rayleigh–Ritz–Bolotin method, RRBM). According to the comparative estimates, this modification grants a much more accurate determination of natural frequencies (see also Krizhevskii 1988, 1989).

As an example, we use RRBM for natural oscillations of a square plate $(0 \le x, y \le a)$ with free contour. The governing equation is

$$D\nabla^4 w + \rho h \frac{\partial^2 w}{\partial t^2} = 0 . \qquad [1.42]$$

Here, $D = \dfrac{Eh^3}{12\left(1-\nu^2\right)}$, h is the plate thickness and ν is Poisson's ratio.

Boundary conditions have the form

$$w_{xx} + \nu w_{yy} = 0, \quad w_{xxx} - 2(1-\nu) w_{xyy} = 0 \quad \text{at} \quad x = 0, x = a, \qquad [1.43]$$

$$w_{yy} + \nu w_{xx} = 0, \quad w_{yyy} - 2(1-\nu) w_{yxx} = 0 \quad \text{at} \quad y = 0, y = a. \qquad [1.44]$$

According to the principle of virtual work,

$$U + V = 0, \qquad [1.45]$$

where U and V are, respectively, the potential and kinetic energy, defined as follows:

$$U = \frac{D}{2} \int_0^a \int_0^a \left(w_{xx}^2 + w_{yy}^2 + 2\nu w_{xx}^2 w_{yy}^2 + 2(1-\nu) w_{xy}^2 \right) dxdy , \qquad [1.46]$$

$$V = \frac{\rho h}{2} \int_0^a \int_0^a w_t^2 dxdy . \qquad [1.47]$$

Using the ansatz

$$w(x,y,t)=W(x,y)\exp(i\omega t),$$

we obtain from equations [1.45]–[1.47]

$$\lambda^2=\omega^2 a^4\frac{\rho h}{D}$$
$$=a^4\left[\int_0^a\int_0^a\left(W_{xx}^2+W_{yy}^2+2\nu W_{xx}^2W_{yy}^2+2(1-\nu)W_{xy}^2\right)dxdy\right]\left[\int_0^a\int_0^a W^2dxdy\right]^{-1}. \quad [1.48]$$

The expression for the eigenfunction $W(x,y)$ obtained using DEEM has the form

$$W(x,y)=W_0(x,y)+W_1(x)\sin(\beta_2 y+l_2)+W_2(y)\sin(\beta_1 x+l_1), \quad [1.49]$$

where

$$W_0(x,y)=\sin(\beta_1 x+l_1)\sin(\beta_2 y+l_2), \quad [1.50]$$

$$W_1(x)=C_{11}\exp\left[\alpha_1(x-a)\right]+C_{12}\exp(-\alpha_1 x), \quad [1.51]$$

$$W_2(y)=C_{21}\exp\left[\alpha_2(y-a)\right]+C_{22}\exp(-\alpha_2 y). \quad [1.52]$$

On satisfying the boundary conditions [1.43] and [1.44] to determine the wave numbers, we obtain a system of transcendental equations

$$\beta_i a=2l_i+m\pi,\quad i=1,2;\quad m=0,1,2,..., \quad [1.53]$$

where $l_i=\arctan\left[\frac{\beta_i}{\alpha_i}\left(\frac{\beta_i^2+(2-\nu)\beta_k^2}{\beta_i^2+\nu\beta_k^2}\right)^2\right]$, $\alpha_i=\left(\beta_i^2+2\beta_k^2\right)^{1/2}$, $i,k=1,2;\ i\neq k.$

For constants C_{ij}, we obtain

$$C_{i1}=\frac{\alpha_i^2\sin l_i}{\alpha_i^2-\nu\beta_k^2},\quad C_{i2}=\frac{\alpha_i^2\sin(\beta_i a+l_i)}{\alpha_i^2-\nu\beta_k^2},\quad i,k=1,2;\quad i\neq k. \quad [1.54]$$

Using the DEEM solution [1.49]–[1.54], we can determine the desired frequency from expression [1.48].

The square of dimensionless frequencies λ for $\nu = 0.225$ and various m, obtained by RRM (Gontkevich 1964), RRBM and DEEM are shown in Table 1.1. Wave forms along a cylindrical surface are not considered since in this case an exact solution can be obtained. The numbers corresponding to the indicated modes of vibration are omitted in Table 1.1.

m	λ, RRM (Gontkevich 1964)	λ, RRBM	Discrepancy with Gontkevich (1964), %	λ, DEEM	Discrepancy with Gontkevich (1964), %
1	14.10	14.48	2.7	12.41	13.6
3	35.96	36.68	2.0	34.60	3.9
5	65.24	66.33	1.7	63.44	2.8
6	74.45	75.28	1.1	73.59	2.5
7	109.30	109.10	0.2	106.30	2.8

Table 1.1. *Comparison of frequencies obtained using various approximation methods*

RRBM gives more accurate results than DEEM for the first natural frequencies. When m increases, both solutions asymptotically approach the exact one, namely, from above in the case of applying RRMB and from below in the case of using DEEM.

RRBM can also be used for stability problems of plates and shells with complicated boundary conditions. This method was applied to plates of complicated form (skew, circle, sector (Andrianov and Krizhevskiy 1988, 1989, 1991)) and structures (Andrianov and Krizhevskiy 1987, 1993).

An interesting modification of DEEM for determining natural frequencies and mode shapes of isotropic and orthotropic rectangular plates with various types of boundary conditions was given in Pevzner *et al.* (2000). This approach does not postulate the formula for the eigenfrequency, but rather it is based on the condition that the frequency obtained from the governing differential equations has to be equal to that given by the Rayleigh method. The paper by Pevzner *et al.* (2000) claims that this modification is more straightforward and computationally faster, and the mode shapes derived are valid on a larger part of the plate.

1.6. Quasi-separation of variables and normal modes of nonlinear oscillations of continuous systems

Bolotin did not give an exact definition of the concept of quasi-separation of variables. Intuitively, this means that the difference between solutions of boundary value problems with separated and quasi-separated variables is sufficient only near the boundaries. In other words, the energy accumulated in the EE zone is small compared to that accumulated in the inner zone. This allows us to not take into account DEE when expanding the natural mode of vibration during the calculation of forced oscillations. Bolotin's conception of quasi-separation of variables (Bolotin 1961c, 1984) can be used in the theory of normal modes of nonlinear oscillations for continuous systems.

When studying linear oscillatory systems with a finite number of DOF, normal oscillation modes play a key role. Kauderer (1958) indicated the existence of solutions in a nonlinear system, which were, in a sense, similar to the normal modes of linear systems. He called these solutions the principal ones and showed how to construct their trajectories in the configuration space. Rosenberg (1962) defined normal vibrations of nonlinear systems with a finite number of DOF, formulated the problem in the configuration space and found several classes of nonlinear systems that allowed solutions with straight-line trajectories (for details, see Mikhlin and Avramov 2011; Avramov and Mikhlin 2013). Generalizations of this concept to continuous systems are related to the exact separation of spatial and time variables (Wah 1964; Avramov and Mikhlin 2013), i.e. to the possibility of representing the sought solutions in the form

$$U(\mathbf{x},t)=X(\mathbf{x})T(t).$$

The restriction of this approach is clear since the separation of variables only works for some boundary conditions. Based on Bolotin's conception of the quasi-separation of variables, we can propose the following definition (Andrianov 2008): a function $U(\mathbf{x},t)$ is called the normal mode of nonlinear oscillations of a continuous system if

$$U(\mathbf{x},t)=X(\mathbf{x})T(t)+Y(\mathbf{x},t),$$

where $T(t)$ and $Y(\mathbf{x},t)$ are the periodic and quasi-periodic functions in time, respectively; and function $Y(\mathbf{x},t)$ is small compared to function $X(\mathbf{x})T(t)$ in some energy norm. The last condition can be verified both *a priori* and *a posteriori*.

1.7. Short-wave (high-frequency) asymptotics. Possible generalizations of DEEM

DEEM can be considered a special case of short-wave (high-frequency) asymptotics. The corresponding algorithms are known as the method of geometric optics, the ray method, the semi-classical approximation, the WKBJ (Wentzel–Kramers–Brillouin–Jeffreys) approach, the method of edge waves, the Keller–Rubinow method, etc. (Keller and Rubinow 1960; Maslov and Fedoryuk 1981; Babich *et al.* 1985; Babich and Buldyrev 1991; Chen *et al.* 1991, 1992; Chen and Zhou 1993; Bauer *et al.* 2015). They were independently developed in various fields of mathematics, mechanics and physics.

Note an interesting fact: Ufimtsev proposed the asymptotic method of edge waves (Ufimtsev 1962, 2003, 2014). According to Rich and Janos (1994) and Mitzner (2003), this theory played a critical role in the design of American stealth aircrafts F-117 and B-2. It is a fascinating example of the direct application of asymptotic formulas in engineering practice!

The key to short-wave asymptotics is the ansatz $\varphi(x)\exp\left(i\varepsilon^{-1}S(x)\right)$, in the nonlinear case – $\varphi(x)\Phi\left(i\varepsilon^{-1}S(x)\right)$, where $i=\sqrt{-1},\ 0<\varepsilon<<1$. In the DEE method, $\varepsilon=1/\lambda$, where λ is the nondimensional frequency. As a result, the construction of the asymptotics can be reduced to solving the eikonal and transport equations (Birger and Panovko 1968). However, short-wave (high-frequency) asymptotics can be treated as a multiscale approach (Maslov and Fedoryuk 1981).

Let us show the generalization of DEEM based on the WKBJ approach using the toy problem – natural oscillation of a beam of variable cross-section (Bauer *et al.* 2015):

$$\frac{d^2}{dx^2}\left[EI\varphi_1(x)\frac{d^2w}{dx^2}\right]-\rho\varphi_2(x)F\omega^2w=0, \qquad [1.55]$$

with clamped edges.

In dimensionless variables, we obtain

$$\varepsilon^4\frac{d^2}{d\xi^2}\left[\varphi_1\frac{d^2w}{d\xi^2}\right]-\varphi_2w=0, \qquad [1.56]$$

$$w=\frac{dw}{d\xi}=0 \quad \text{at } \xi=0,1, \qquad [1.57]$$

where $\varepsilon^4 = \dfrac{EI}{\omega^2 \rho F L^4}$, $\xi = \dfrac{x}{L}$.

Functions φ_1, φ_2 are supposed to be smooth enough to avoid turning points.

The solution to equation [1.56] is sought in the form:

$$w = \exp\left(\varepsilon^{-1}\int_0^{\xi}\psi(\tau)d\tau\right)\left[u_0(\xi)+\varepsilon u_1(\xi)+\varepsilon^2 u_2(\xi)+\ldots\right]. \qquad [1.58]$$

Substituting ansatz [1.58] into equation [1.56] and applying ε-splitting, we obtain a recurrent system of equations:

$$\left(\varphi_1\psi^4 - \varphi_2\right)u_0 = 0, \qquad [1.59]$$

$$4\varphi_1\psi^3 u_0' + 6\varphi_1\psi^2\psi' u_0 + 2\psi^3\varphi_1' u_0 = 0, \qquad [1.60]$$

The eikonal equation [1.59] has the following solutions:

$$\psi_{1,2} = \pm\left(\frac{\varphi_2}{\varphi_1}\right)^{1/4}, \quad \psi_{1,2} = \pm i\left(\frac{\varphi_2}{\varphi_1}\right)^{1/4}.$$

From the transport equation [1.60], we obtain

$$u_0(\xi) = \frac{1}{\psi^{3/2}\varphi_1^{1/2}}.$$

In the first approximation, the general solution of equation [1.56] can be written as follows:

$$w = C_1\sin\left(\varepsilon^{-1}\int_0^{\xi}\psi(\tau)d\tau\right)u_0(\xi) + C_2\cos\left(\varepsilon^{-1}\int_0^{\xi}\psi(\tau)d\tau\right)u_0(\xi)$$
$$+ C_3\exp\left(-\varepsilon^{-1}\psi(0)\xi\right)u_0(0) + C_4\exp\left(-\varepsilon^{-1}\psi(1)(1-\xi)\right)u_0(1). \qquad [1.61]$$

In expression [1.61], function $u_0(\xi)$ is "frozen" at either end of the interval (i.e. we can change $u_0(\xi)$ to $u_0(0)$ or $u_0(\xi)$ to $u_0(1)$) for rapidly decaying components).

Using solution [1.61] and boundary conditions [1.57], we obtain the frequency of oscillations:

$$\omega = \pi^2 (n+0.5)^2 \sqrt{\frac{EI}{\rho F L^4}} \left[\int_0^1 \left(\frac{\varphi_2(x)}{\varphi_1(x)} \right)^{1/4} dx \right]^{-2}, \quad n = 1,2,\ldots \qquad [1.62]$$

Formula [1.62] at $\varphi_1 = \varphi_2 = 1$ coincides with Bolotin's formula [1.18]. Thus, the WKBJ method generalizes the DEE method to the problems with variable coefficients.

As it is mentioned in Chen *et al.* (1991, 1992), the short-wave (high-frequency) asymptotics gives the same results as the DEE approach for domains of simple geometry. At the same time, DEE does not cover the cases of different geometries (circular, elliptical, etc.) or non-self-adjoint problems. The short-wave asymptotics in the form of the Keller–Rubinow approach (Keller and Rubinow 1960) in Chen *et al.* (1991) allows more ready extension to other geometries and is more aptly generalizable to dissipative boundary conditions. In other words, it gives the possibility to overcome degeneration of the DEE case.

1.8. Conclusion: DEEM, highly recommended

The importance and usefulness of a particular calculation method is determined by its wide application when studying practically important systems and phenomena. From this point of view, the importance of DEEM is not in doubt.

From the very beginning, DEEM was originated by Bolotin for the analysis of overhead power lines (Bolotin *et al.* 1958).

DEEM was used to obtain estimates for the density of natural frequencies of shallow shell vibrations (Gavrilov 1961b; Bolotin 1963; Stearn 1970; Zhinzher and Khromatov 1971, 1972a, 1972b; Moskalenko 1972, 1975). This is very important during the study of random vibrations of elastic structures (Bolotin 1966; Birger and Panovko 1968; Moskalenko 1968; Bolotin 1984; Elishakoff *et al.* 1994).

The influence of the magnetic field on the distribution of plate and shell vibration frequencies was studied in Bagdasaryan (1986), Koreshkova and Khromatov (2009), Golubeva *et al.* (2013) and Khromatov and Golubeva (2013).

A lot of research devoted to the problems of aeroelastic stability and supersonic flutter can be mentioned (Zhinzher 1983; Zhinzher and Kadarmetov 1984, 1986; Dubovskikh *et al.* 1996).

We also mention the optimal control problem for continuous systems (Andrianov and Iskra 1991).

DEEM and its generalizations are important particular cases of high-frequency asymptotics. The effectiveness of this method for analyzing the main types of plates and shells used in engineering practices has been proven through experience. The main advantage of DEEM consists of its simplicity and good compatibility with variational approaches.

Naturally, DEEM is not a panacea. For example, when considering a mixed boundary value problem with many points of change in the boundary conditions, the method based on the homotopy parameter (Andrianov *et al.* 2014) seems more suitable.

Nevertheless, in general, we hope that our review has convinced researchers that DEEM and its generalizations occupied an honorable place in the arsenal of analytical methods for solving the dynamics and stability problems of thin-walled structures.

1.9. Acknowledgments

Several years ago, Professor I. Elishakoff pointed out that it would be useful to prepare a new review of Bolotin's method, since his previous review on this topic was written in 1976. We are grateful to him for this idea.

CONFLICTS OF INTEREST.– The authors declare no conflict of interest.

1.10. Appendix

Professor Elishakoff enjoys historiography of science and his historical research is read with great interest. Bubnov or Galerkin? Timoshenko or Ehrenfest? The chicken or the egg?

We also provided a little historical research. Bolotin wrote about the possibility of extending the range of applicability of DEEM to PDEs with variable coefficients (Bolotin *et al.* 1961, Chapter II): "If the coefficients of the equation change slowly, then it is advisable to combine this method with the Wentzel–Brillouin–Kramers method or its related Blumenthal-Shtaerman approach" (*translated by us*). After reading the relevant papers, we were convinced that Blumenthal used the "WKB method", created to solve problems of quantum mechanics in 1926, already in 1912 (Blumenthal 1912, 1914). He created this asymptotic method for solving problems of the shell theory. H. Reissner used Blumenthal's approach in 1912 (Reissner 1912), as did Shtaerman in 1924 (Shtaerman 1924).

With these remarks, we are certainly not going to interfere with the complex priority history of the WKB approach (Wikipedia 2020). We recall Nayfeh's remark concerning one well-known asymptotic method (Nayfeh 2000, p. 232): "The method of multiple scales is so popular that it is being rediscovered just about every 6 months". A lot of phenomena in completely different fields of science are described using similar or directly identical equations. Researchers, as a rule, do not search for methods of their solution in areas far from them, but simply rediscover them. The corresponding methods are naturally given different names in different fields of science. Surprisingly, this does not lead to the "Tower of Babel effect".

1.11. References

Abramowitz, M. and Stegun, I.A. (1965). *Handbook of Mathematical Functions, with Formulas, Graphs, and Mathematical Tables*. Dover Publications, New York.

Andrianov, I.V. (2008). Asymptotic construction of nonlinear normal modes for continuous systems. *Nonl. Dyn.*, 51(1–2), 99–109.

Andrianov, I.V. and Iskra, V.S. (1991). Use of Bolotin's asymptotic method in the optimal control problem. *Probl. Mashinostr.*, 36, 79–82.

Andrianov, I.V. and Kholod, E.G. (1985). Natural nonlinear oscillations of shallow shells. *Struct. Mech. Theory Struct.*, 4, 51–54.

Andrianov, I.V. and Kholod, E.G. (1993a). Intermediate asymptotical forms in nonlinear dynamics of shells. *Mech. Solids*, 28(2), 160–165.

Andrianov, I.V. and Kholod, E.G. (1993b). Non-linear free vibration of shallow cylindrical shell by Bolotin's asymptotic method. *J. Sound Vib.*, 165(1), 9–17.

Andrianov, I.V. and Kholod, E.G. (1995). Bolotin's asymptotic method for nonlinear free vibration of shells. *SAMS*, 18–19, 211–213.

Andrianov, I.V. and Krizhevskiy, G.A. (1987). Modified asymptotic method for the problems of stiffened constructions dynamics. *Struct. Mech. Theory Struct.*, 2, 66–68.

Andrianov, I.V. and Krizhevskiy, G.A. (1988). Calculation of skew plate natural oscillation by approximate method. *Izv. VUZov. Civil Eng. Archit.*, 12, 46–49.

Andrianov, I.V. and Krizhevskiy, G.A. (1989). Analytical investigation of geometrically nonlinear oscillation of sector plates, reinforced by radial ribs. *Dokl. AN Ukr. SSR, ser. A*, 11, 30–33.

Andrianov, I.V. and Krizhevskiy, G.A. (1991). Investigation of natural oscillation of circle and sector plates with consideration of geometrical nonlinearity. *Mech. Solids*, 26(2), 143–148.

Andrianov, I.V. and Krizhevsky, G.A. (1993). Free vibration analysis of rectangular plates with structural inhomogeneity. *J. Sound Vib.*, 162(2), 231–241.

Andrianov, I.V., Manevitch, L.I., Kholod, E.G. (1979). On the nonlinear oscillation of rectangular plates. *Struct. Mech. Theory Struct.*, 5, 48–51.

Andrianov, I.V., Awrejcewicz, J., Manevitch, L.I. (2004). *Asymptotical Mechanics of Thin-Walled Structures: A Handbook*. Springer-Verlag, Heidelberg, Berlin.

Andrianov, I.V., Awrejcewicz, J., Danishevs'kyy, V.V., Ivankov, A.O. (2014). *Asymptotic Methods in the Theory of Plates with Mixed Boundary Conditions*. John Wiley & Sons, Chichester.

Avramov, K.V. and Mikhlin, Y.V. (2013). Review of applications of nonlinear normal modes for vibrating mechanical systems. *Appl. Mech. Rev.*, 65(2), 020801-20.

Awrejcewicz, J., Andrianov, I.V., Manevitch, L.I. (1998). *Asymptotic Approaches in Nonlinear Dynamics: New Trends and Applications*. Springer-Verlag, Heidelberg, Berlin, New York.

Babich, V.M. and Buldyrev, V.S. (1991). *Asymptotic Methods in Short-Wavelength Diffraction Theory*. Springer, Berlin.

Babich, V.M., Buldyrev, V.S., Molotkov, I.A. (1985). *A Space-Time Ray Method*. Leningrad University, Leningrad.

Bagdasaryan, G.E. (1986). Application of V.V. Bolotin's asymptotic methods for investigation of magnetoelastic vibration of rectangular plates. *Probl. Mashinostr.*, 25, 63–68.

Bauer, S.M., Filippov, S.B., Smirnov, A.L., Tovstik, P.E., Vaillancourt, R. (2015). *Asymptotic Methods in Mechanics of Solids*. Birkhäuser, Basel.

Birger, I.A. and Panovko, Y.G. (1968). *Prochnost. Ustoichivost. Kolebaniya (Strength. Stability. Oscillations Handbook) 3*. Mashinostroyenie, Moscow.

Blumenthal, O. (1912). Über asymptotische Integration von Differentialgleichungen mit Anwendung auf eine asymptotische Theorie der Kugelfunctionen. *Archiv Math. Physik, ser. 3*, 19, 136–174.

Blumenthal, O. (1914). Über asymptotische Integration von Differentialgleichungen mit Anwendung auf die Berechnung von Spannungen in Kugelschalen. *Z. Math. Physik*, 62, 343–358. Extract previously appeared in *Proc. Fifth Intern. Cong. Math.*, Cambridge (1913), II, 319–327.

Bolotin, V.V. (1960a). Dynamic edge effect in the elastic vibrations of plates. *Inzh. Sb.*, 31, 3–14.

Bolotin, V.V. (1960b). The edge effect in the oscillations of elastic shells. *J. Appl. Math. Mech.*, 24(5), 1257–1272.

Bolotin, V.V. (1961a). A generalization of the asymptotic method of the eigenvalue problems for rectangular regions. *Inzh. Zh.*, 1(3), 86–92.

Bolotin, V.V. (1961b). An asymptotic method for the study of the problem of eigenvalues of rectangular regions. *Problems of Continuum Mechanics*, SIAM, 56–68.

Bolotin, V.V. (1961c). Asymptotic method in the theory of oscillations of elastic plates and shells. *Tr. Konf. po Teorii Plastin i Obolochek*, Kazan State University, 21–26.

Bolotin, V.V. (1961d). The natural oscillations of a rectangular elastic parallelepiped. *J. Appl. Math. Mech.*, 25(1), 220–227.

Bolotin, V.V. (1963). On the density of the distribution of natural frequencies of thin elastic shells. *J. Appl. Math. Mech.*, 27(2), 538–543.

Bolotin, V.V. (1966). Broadband random vibrations of elastic systems. *Int. J. Solids Struct.*, 2(1), 105–124.

Bolotin, V.V. (1970). Application of edge effect theory to forced vibration analysis of elastic systems. *Trudy Moscow Energet. Inst. Dyn. Soprot. Mater.*, 74, 180–192.

Bolotin, V.V. (1984). *Random Vibrations of Elastic Systems.* Springer, Dordrecht.

Bolotin, V.V. (2006). 80th birthday tribute. *J. Appl. Math. Mech.*, 70(2), 161–175.

Bolotin, V.V., Marein N.S., Vinokurov A.I., Poznyak E.L., Ivovich V.A. (1958). Vibration and vibrational strength of overhead power lines. *Nauch. Dokl. Vish. Shkoly. Energetika*, 2, 55–62.

Bolotin, V.V., Makarov, V.P., Mishenkov, G.V., Shveiko, Yu.Yu. (1960). Asymptotic method of investigating the eigenfrequency spectrum of elastic plates. *Rasch. Prochn.*, 6, 231–253.

Bolotin, V.V., Gol'denblat, I.I., Smirnov, A.F. (1961). *Modern Problems of Structural Mechanics*. Stroyizdat, Moscow.

Chen, G. and Zhou, J. (1993). *Vibration and Damping in Distributed Systems Vol. II: WKB and Wave Methods, Visualization and Experimentation.* CRC Press, Boca Raton.

Chen, G., Coleman, M.P., Zhou, J. (1991). Analysis of vibration eigenfrequencies of a thin plate by the Keller-Rubinow wave method I: Clamped boundary conditions with rectangular or circular geometry. *SIAM J. Appl. Math.*, 51(4), 967–983.

Chen, G., Coleman, M.P., Zhou, J. (1992). The equivalence between the wave propagation method and Bolotin's method in the asymptotic estimation of eigenfrequencies of a rectangular plate. *Wave Motion*, 16(3), 285–297.

Crighton, D.G. (1994). Asymptotics – An indispensable complement to thought, computation and experiment in applied mathematical modelling. In *Seventh Europ. Conf. Math. Ind.*, Fasano, A., Primicerio, M.B., Teubner, G. (eds). B.G. Teubner, Stuttgart.

Dickinson, S.M. (1971). The flexural vibration of rectangular orthotropic plates subject to in-plane forces. *J. Appl. Mech.*, 38(3), 699–700.

Dickinson, S.M. (1975a). Bolotin's method applied to the buckling and lateral vibration of stressed plates. *AIAA J.*, 13(1), 109–110.

Dickinson, S.M. (1975b). Modified Bolotin's method applied to buckling and vibration of stressed plates. *AIAA J.*, 13(12), 1672–1673.

Dickinson, S.M. and Warburton, G.B. (1967). Natural frequencies of plate systems using the edge effect method. *J. Mech. Eng. Sci.*, 9(4), 318–324.

Dubovskikh, Y.A., Khromatov, V.E., Chirkov, V.E. (1996). Asymptotic analysis of stability and postcritical behavior of elastic panels in a supersonic flow. *Mech. Solids*, 31(3), 65–75.

Elishakoff, I. (1974). Vibration analysis of clamped square orthotropic plate. *AIAA J.*, 12, 921–924.

Elishakoff, I. (1976). Bolotin's dynamic edge-effect method. *Shock Vibr. Digest*, 8(1), 95–104.

Elishakoff, I. and Steinberg, A. (1979). Eigenfrequencies of continuous plates with arbitrary number of equal spans. *J. Appl. Mech.*, 46, 656–662.

Elishakoff, I. and Wiener, F. (1976). Vibration of an open shallow cylindrical shell. *J. Sound Vibr.*, 44, 379–392.

Elishakoff, I., Steinberg, A., van Baten, T. (1993). Vibration of multispan stiffened plates via modified dynamic edge effect method. *Comp. Meth. Appl. Mech. Eng.*, 105, 211–223.

Elishakoff, I., Lin, Y.K., Zhu, L.P. (1994). *Probabilistic and Convex Modelling of Acoustically Excited Structures*. Elsevier, Amsterdam.

Emmerling, F.A. (1979). Ermittlung von Eigenkreisfrequenzen schwingender Rechteckplatten mit Hilfe der asymptotishen Methode von Bolotin. *Stahlbau*, 49(11), 327–334.

Gavrilov, Y.V. (1961a). Determination of natural vibration frequencies of elastic circular cylindrical shells. *Izv. AN SSSR OTN Mech. Mashin.*, 1, 161–163.

Gavrilov, Y.V. (1961b). Investigation of the spectrum of natural oscillations of elastic cylindrical shells. *Tr. Konf. po Teorii Plastin i Obolochek*, Kazan State University, 72–76.

Gibigaye, M., Yabi, C.P., Alloba, I.E. (2016). Dynamic response of a rigid pavement plate based on an inertial soil. *Int. Schol. Res. Not.*, 1–9.

Golubeva, T.N., Korobkov, Y.S., Khromatov, V.E. (2013). The influence of a longitudinal magnetic field on the frequency spectra of oscillations of ferromagnetic plates. *Electrotechnika*, 3, 44–48.

Gontkevich, V.S. (1964). *Natural Oscillations of Plates and Shells*. Naukova Dumka, Kiev.

Kantorovich, L.V. and Krylov, V.I. (1958). *Approximate Methods of Higher Analysis.* Noordhoff, Groningen.

Kauderer, H. (1958). *Nichtlineare Mechanik.* Springer, Berlin, Göttingen, Heidelberg.

Kaza, V. and Ramaiah, G.K. (1978). Use of asymptotic solutions from a modified Bolotin method for obtaining natural frequencies of clamped rectangular orthotropic plates. *J. Sound Vib.*, 59(3), 335–347.

Keller, J.B. and Rubinow, S.I. (1960). Asymptotic solution of eigenvalue problems. *Ann. Phys.*, 9(1), 24–75. Errata, *Ann. Phys.*, 9(2).

Khromatov, V.E. (1972a). Properties of spectra of thin circular cylindrical shells oscillating near momentless stress state. *Mech. Solids*, 7(2), 103–108.

Khromatov, V.E. (1972b). Density of frequencies of natural oscillations of thin spherical shells in momentless stress state. *Trudy Moscow Energet. Inst.*, 101, 148–153.

Khromatov, V.E. and Golubeva, T.N. (2013). Oscillations and stability of a ferromagnetic cylindrical shell in a magnetic field. *Vestnik Moscow Avia. Inst.*, 20(3), 212–219.

King, W.W. and Lin, C.-C. (1974). Application of Bolotin's method to vibrations of plates. *AIAA J.*, 12(3), 399–401.

Kline, S.J. (1965). *Similitude and Approximation Theory.* McGraw-Hill, New York.

Koreshkova, N.S. and Khromatov, V.E. (2009). On the influence of a transverse magnetic field on the vibration spectra of shallow shells. *Mech. Solids*, 44, 632–638.

Krizhevskii, G.A. (1988). Combination of Rayleigh and dynamic edge effect methods in studying vibrations of rectangular plates. *J. Appl. Mech. Techn. Phys.*, 29(6), 919–921.

Krizhevskii, G.A. (1989). Vibration and stability of orthotropic rectangular plates. *Sov. Appl. Mech.*, 25(8), 822–825.

Kudryavtsev, E.P. (1960). Influence of shear deformation and rotary inertia on flexural vibration of an elastic beam. *Izv. AN SSSR OTN Mech. Mashin.*, 5, 156–159.

Kudryavtsev, E.P. (1964). Application of asymptotic method for investigating the eigenfrequencies of elastic rectangular plates. *Rasch. Prochn.*, 10, 352–362.

Lin, C.C. and King, W.W. (1974). Free transverse vibrations of rectangular unsymmetrically laminated plates. *J. Sound Vib.*, 36(1), 91–103.

Maslov, V.P. and Fedoryuk, M.V. (1981). *Semi-classical Approximation in Quantum Mechanics.* Kluwer, Dordrecht.

Meilani, M. (2012). Modified Bolotin method to obtain the natural frequency of stiffened plate with semirigid support. *Procedia Eng.*, 50, 110–121.

Meilani, M. (2015). Obtaining the natural frequency of stiffened plate with modified Bolotin method. *Int. J. Appl. Eng. Res.*, 9(23), 21501–21512.

Mikhlin, Y.V. and Avramov, K.V. (2011). Nonlinear normal modes for vibrating mechanical systems. Review of theoretical developments. *Appl. Mech. Rev.*, 63(6), 060802–21.

Mitzner, K.M. (2003). Foreword. In *Theory of Edge Diffraction in Electromagnetics*, Ufimtsev, P.Y. (ed.). Tech Science Press, Encino, California.

Moskalenko, V.N. (1961). On the application of refined theories of bending of plates in free vibration problems. *Inzh. Zh.*, 1(3), 93–101.

Moskalenko, V.N. (1968). Random vibrations of multi-span plates. *Mech. Solids*, 3(4), 79–84.

Moskalenko, V.N. (1969). On the vibrations of multispan plates. *Rasch. Prochn.*, 14, 360–367.

Moskalenko, V.N. (1972). On the frequency spectra of natural vibrations of shells of revolution. *J. Appl. Math. Mech.*, 36(2), 279–283.

Moskalenko, V.N. (1975). Frequency spectra and modes of free vibrations of doubly periodic systems. *J. Appl. Math. Mech.*, 39, 503–510.

Moskalenko, V.N. and Chen, D.L. (1965). On natural vibrations of multispan uncut plates. *Prikl. Mekh. (Appl. Mech.)*, 1(3), 59–66.

Nayfeh, A.H. (2000). *Perturbation Methods*. Wiley, New York.

Nelson, H.M. (1978). High frequency flexural vibration of thick rectangular bars and plates. *J. Sound Vib.*, 60, 101–118.

Pevzner, P., Berkovits, A., Weller, T. (2000). Further modification of Bolotin method in vibration analysis of rectangular plates. *AIAA J.*, 38(9), 1725–1729.

Reissner, H.J. (1912). Spannungen in Kugelschalen (Kuppeln). *Festschrift Heinrich Müller-Breslau gewidmet nach Vollendung seines sechzigsten Lebensjahres*. Alfred-Kröner Verlag, Leipzig, 181–193.

Rich, B. and Janos, L. (1994). *Skunk Works: A Personal Memoir of My Years at Lockheed*. Little Brown, Boston.

Rosenberg, R.M. (1962). The normal modes of nonlinear n-degree-of-freedom systems. *J. Appl. Mech.*, 29, 7–14.

Shtaerman, I.Y. (1924). On the application of the method of asymptotic integration to the calculation of elastic shells. *Izv. Kievskogo Polit. & S.-H. Inst.*, 1(2), 75–99.

Stearn, S.M. (1970). Spatial variation of stress strain and acceleration in structures subject to broad frequency band excitation. *J. Sound Vib.*, 12, 85–97.

Ueng, C.E.S. and Nickels Jr., R.C. (1978). Dynamic response of structural panel by Bolotin's method. *Int. J. Solids Struct.*, 14(7), 571–578.

Ufimtsev, P.Y. (1962). *Method of Edge Waves in the Physical Theory of Diffraction*, translated by Foreign Technology Division Wright-Patterson AFB. Def. Techn. Inf. Center, Cameron Station, Alexandria.

Ufimtsev, P.Y. (2003). *Theory of Edge Diffraction in Electromagnetics*. Tech Science Press, Encino, California.

Ufimtsev, P.Y. (2014). *Fundamentals of the Physical Theory of Diffraction*. John Wiley & Sons, Hoboken, New Jersey.

Vakhromeev, Y.M. and Kornev, V.M. (1972). Dynamic edge effect in beams. Formulation of truncated problems. *Mech. Solids*, 7(4), 95–103.

Vijaykumar, K. and Ramaiah, G.K. (1978a). Analysis of vibration of clamped square plates by the Rayleigh-Ritz method with asymptotic solutions from a modified Bolotin method. *J. Sound Vib.*, 56(1), 127–135.

Vijaykumar, K. and Ramaiah, G.K. (1978b). Use of asymptotic solutions from a modified Bolotin method for obtaining natural frequencies of clamped rectangular orthotropic plates. *J. Sound Vib.*, 59(3), 335–347.

Wah, T. (1964). The normal modes of vibration of certain nonlinear continuous systems. *J. Appl. Mech.*, 31(1), 139–140.

Weaver Jr., W., Timoshenko, S.P., Young, D.H. (1990). *Vibration Problems in Engineering*, 5th edition. John Wiley & Sons, New York.

Wikipedia (2020). WKB approximation [Online]. Available at: https://en.wikipedia.org/wiki/WKB_approximation [accessed July 2020].

Zhinzher, N.I. (1975). Dynamic edge effects in orthotropic elastic shells. *J. Appl. Math. Mech.*, 39(4), 723–726.

Zhinzher, N.I. (1983). Asymptotic method in problems of aeroelastic stability. *Probl. Ust. Predel. Nesushch. Sposobnosti Konstr.* Leningrad, 44–53.

Zhinzher, N.I. and Denisov, V.N. (1983). Asymptotic method in a problem of nonlinear shell vibrations. *Strength Mater.*, 15(9), 1219–1223.

Zhinzher, N.I. and Denisov, V.N. (1985). Asymptotic method in the problem of nonlinear oscillations of isotropic rectangular plates. *Mech. Solids*, 20(1), 152–158.

Zhinzher, N.I. and Kadarmetov, I.M. (1984). Application of the asymptotic method to the problem of supersonic flutter of a cylindrical shell. *Koleb. Uprug. Konstr. s Zhidkost.* Moscow, 114–118.

Zhinzher, N.I. and Kadarmetov, I.M. (1986). Application of the asymptotic method to the problem of the flutter of an orthotropic cylindrical shell. *Izv. AN ArmSSR Mech.*, 39(2), 31–39.

Zhinzher, N.I. and Khromatov, V.E. (1971). Application of the asymptotic method to the study of vibration spectra of orthotropic circular cylindrical shells. *Mech. Solids*, 6(6), 72–82.

Zhinzher, N.I. and Khromatov, V.E. (1984). Asymptotic method in problems of nonlinear vibration of rectangular slightly orthotropic plates. *Sov. Appl. Mech.*, 20(11), 742–746.

Zhinzher, N.I. and Khromatov, V.E. (1990). Oscillation of shallow shells with finite amplitudes. *Sov. Appl. Mech.*, 26(11), 1100–1104.

2

On the Principles to Derive Plate Theories

2.1. Introduction

Plates are defined as plane structural elements with a small thickness compared to the planar dimensions (Timoshenko and Woinowsky-Krieger 1959; Altenbach *et al.* 2016). The plate can be presented by the plane surface. The position of this surface is usually in the mid of the thickness, which can be constant or slightly changing. In contrast to the plane structures, thin structures with a curved midsurface can be applied as a structural element. In this chapter, we limit ourselves to plane structures.

When halving the thickness of a thin-walled structure at every point of the planar dimensions, all points lie on the bisecting surface. Structural elements can thus be described geometrically by the geometry of the middle surface and the specification of the element thickness. If the middle surface is a plane, the element is called a disk or a plate. In the case of a disk, all external stresses only act in the middle area. The middle surface is distorted, and there are no bending effects. In the case of a plate, bending deformations must be taken into account. The external forces act primarily orthogonal to the middle surface.

In the simplest case of a first-order structural theory, the disk and plate effects are calculated independently of each other. All governing equations are formulated for the reference configuration. However, there are also numerous applications for which such decoupling is not possible and we have to apply a second- or third-order structural theory. A typical second-order structural theory is the stability problem for the plate or the so-called von Kármán theory taking into account large deflections. In the latter, the equilibrium equations are presented in the current configuration.

The aim of this chapter is to briefly discuss the formulation approaches for plate theories. From the above-mentioned statement, it is obvious that if a plate-like body

---

Chapter written by Marcus AẞMUS and Holm ALTENBACH.

geometry can be reduced to the geometry of a plane, the mechanical behavior can be presented by a two-dimensional theory instead of the three-dimensional approach. The reduction of the set of three-dimensional coupled partial differential equations to a two-dimensional case is not trivial. Some reduction methods will be presented and discussed below.

2.2. Some historical remarks

The starting point for the development of any structural mechanics theory based on dimensionally reduced equations (one-dimensional theories for rods, beams, arcs, etc., or two-dimensional theories for plates, shells, folded structures, etc.) are the ideas of two mathematicians: Jacob Bernoulli and Leonhard Euler. They presented a first proper beam theory containing all necessary elements: kinematics, reaction to external loadings and equilibrium statements based on the balances of momentum and of moment of momentum. They introduced several crucial aspects, including:

- kinematic hypotheses;
- forces and moments, which are (as shown later) stress resultants.

In addition, they understood intuitively that forces and moments can be independent (in contrast to the point of view that a moment is, in the simplest case, force multiplied by length). The Bernoulli–Euler approach results in a theory fitting the needs of applications. However, their beam theory is an approximate theory and the application bounds are not clear. But the resulting set of coupled beam equations allows solutions in a closed form.

One possible way to prove the validity of a theory is related to the performance of experiments. In the case of plates, experiments were used not only to verify theoretical results. The studies of Ernst Florens Friedrich Chladni on patterns of a guitar backplate (see Chladni (1787)) initiated the development of theories allowing the mathematically correct description of these patterns. With respect to the fact that the sound of music results in vibrations, first suggestions for vibration equations were made. Sophie Germain had presented the first closed theory in a submission to the Paris Academy of Sciences. She formulated the first vibration equation for a plane lamina (Germain 1821)

$$N^2\left(\frac{\partial^4 z}{\partial x^4}+2\frac{\partial^4 z}{\partial x^2\partial y^2}+\frac{\partial^4 z}{\partial y^4}\right)+\frac{\partial z^2}{\partial t^2}=0, \qquad [2.1]$$

where N^2 is a constant[1].

1. It should be mentioned that equation [2.1] is due to Lagrange who corrected the wrong plate equation of Germain (Bucciarelli and Dworsky 1980; Altenbach 2020), and sometimes the equation is named the Germain–Lagrange equation instead of the Germain equation.

Possibly the next important step in the development of the theory of plates was made by Poisson (1829). He showed that the normal deflection of the plate satisfies the non-homogeneous biharmonic equation of Germain–Lagrange. On the other hand, he formulated three boundary conditions. As was mentioned, for example, by Zhilin (1992), this is strange: the excellent mathematician and specialist in the theory of partial differential equations allows such a discrepancy between the order of the differential equation and the number of boundary conditions.

Several researchers discussed this problem later and tried to develop better theories. Here, we mention only the contribution of Kirchhoff (1850), Reissner (1945, 1947) and Mindlin (1951). It is not possible to present all variants of the plate theory. Here, we will focus on general formulation principles.

2.3. Possibilities to formulate plate theories

Any plate-like body can be analyzed in the simplest way by a three-dimensional theory of linear isotropic elasticity. It is obvious that in this case numerical methods must be used to solve problems. There are several possibilities, among which the finite element method is the most popular, which can provide solutions with the desired accuracy. However, one- and two-dimensional theories are studied worldwide since the lower-dimensional set of equations has two advantages (Schneider and Kienzler 2017):

– these theories may have closed-form solutions;

– they provide the basis for lower-dimensional finite elements saving a great deal of computational time.

The disadvantage is that they are approximate and have a systematic error.

Let us limit ourselves to the following restrictions:

– the thinness hypothesis, which allows us to present the approximate theory using only two valid coordinates;

– isotropic linear elastic material behavior.

The second limitation is only for making the presentation straightforward.

The following formulation concepts in the theory of plates should be mentioned:

– the engineering approach based on the introduction of hypotheses;

– mathematical techniques;

– the direct approach;

– the consistent approach.

The first three approaches were widely discussed in the literature, with the last one mainly developed in the last two decades. Review papers, with respect to the first three approaches, were published by Grigolyuk and Kogan (1972), Naghdi (1972), Grigolyuk and Seleznev (1973), Rothert (1973), Wunderlich (1973), Reissner (1985), Altenbach (1998), Altenbach *et al.* (2010), Altenbach and Eremeyev (2017). The last approach was introduced only some years ago by Kienzler (2002), Schneider and Kienzler (2011), Schneider *et al.* (2014), Kienzler and Schneider (2016). Some details of the various approaches are presented below.

2.3.1. *Theories based on hypotheses*

Reduction of the set of three-dimensional partial differential equations can be performed using hypotheses related to the stress state, strain state and/or kinematics. In the case of Kirchhoff's plate theory, two assumptions were introduced:

– straight orthogonal lines to the midsurface in the undeformed state remain straight and orthogonal to the midsurface in the deformed state;

– the midsurface is dilatation-free.

With the deflection w, which is constant over the thickness, the Kirchhoff plate equation can be expressed as (Altenbach *et al.* 2016)

$$\Delta\Delta w + \frac{\rho}{D}\frac{\partial^2 w}{\partial t^2} = \frac{q}{D} \qquad [2.2]$$

The deflection, density ρ and orthogonal force surface load q are functions of the coordinates of the midplane and time t, and D is the Kirchhoff plate (bending) stiffness. To solve the conflict of the dynamic boundary conditions (three stress resultants are prescribed), Kirchhoff suggested two boundary conditions: a bending moment and the sum of transverse force and the edgewise rate of change of a twisting couple.

The Kirchhoff hypotheses imply the following displacement field approximation:

$$\boldsymbol{u}(\boldsymbol{x}) = x_3\tilde{\boldsymbol{\nabla}} w + w\boldsymbol{n} \qquad [2.3]$$

where $\boldsymbol{u}(\boldsymbol{x})$ is the three-dimensional displacement vector, which depends on the three-dimensional radius vector, x_3 is the coordinate in the thickness direction, $\tilde{\boldsymbol{\nabla}}$ is the two-dimensional nabla operator and $\boldsymbol{n}$ is the normal to the midsurface. It is easy to see that the strains can be computed in the standard manner

$$\boldsymbol{\varepsilon} = \frac{1}{2}\left[\boldsymbol{\nabla}\boldsymbol{u} + (\boldsymbol{\nabla}\boldsymbol{u})^{\mathrm{T}}\right] \qquad [2.4]$$

With these strains from Hooke's law, the stresses can be computed.

REMARK 2.1.– *It should be noted that there is no constitutive equation for the transverse shear force vector. The forces can be calculated only from the equilibrium equations.*

REMARK 2.2.– *It is easy to show that the von Kármán theory can be obtained using a nonlinear strain expression instead of* [2.5]. *However, in this case, the equilibrium equations should be established in the deformed state.*

The next step in the establishment of plate theories based on hypotheses was developed by Reissner and Mindlin. In the literature, often the Mindlin–Reissner theory is mentioned, but the foundations of both theories are different (Reissner 1944, 1945; Mindlin 1951). However, both theories are extensions of the Kirchhoff plate theory since they take into account transverse shear strains. In the case of Mindlin's approach, the first Kirchhoff assumption is less restrictive:

– straight orthogonal lines to the midsurface in the undeformed state remain straight, but no more orthogonal to the midsurface in the deformed state.

Note that Reissner's theory leads to approximate values not for the deflection of the midsurface of the plate but for a weighted average across the thickness of the deflections of all points of the plate that lie on a normal to the midsurface (Reissner 1947).

As mentioned by Elishakoff (2020), a similar theory to Mindlin's was proposed by Uflyand (1948) using an extended kinematics in comparison with Kirchhoff. Both the Reissner and the Uflyand–Mindlin theories are intended for thick and transverse shear-deformable plates. In addition to accounting for the transverse shear strain, these theories allow us to estimate the transverse shear forces by constitutive laws, and three boundary conditions can be fulfilled in the general case. In the literature, these theories belong to the so-called first-order shear deformation theories (FOSDT), among which we should mention the theories presented in Hencky (1947), Bollé (1947a, b), Kromm (1953), Vlasov (1957) and Ambarcumyan (1958). Since a first-order shear deformation theory implies a linear displacement variation through the thickness, it is incompatible with Reissner's static plate theory.

The continuation of Mindlin's approach are the so-called higher-order shear deformation theories. In the 1960s and 1970s, several suggestions were made. In Lo *et al.* (1977), this approach was finalized. However, only the third-order shear deformation theories (see, for example, Reddy (1984)) are used in many applications. The reason is simple – this theory has the advantage that no shear correction must be introduced. On the other hand, higher-order theories include higher-order stress resultants. Their physical interpretation is connected with difficulties.

2.3.2. *Reduction of the governing equations by mathematical techniques*

This direction is connected with several ideas, including:

– power series representations;

– trigonometric function representations;

– special function representations;

– asymptotic integration techniques.

The first approach is presented, for example, in Preußer (1984). Like other theories based on hypotheses, power series theories, with respect to a small parameter, allow the presentation of all governing equations with a given accuracy. The trigonometric function approach was suggested, for example, in Touratier (1988, 1991), using sin or cos functions over the thickness. In Vekua (1985), special functions (here Legendre's polynomials) were applied. In addition, a hierarchical structure of the governing equations was established. Asymptotic integration techniques of the three-dimensional elasticity equations were used, for example, in Goldenveizer (1962), Aghalovyan (2015).

The reduction of the governing equations by mathematical techniques is now well discussed in the literature. Practical applications of such theories are somewhat limited because the use of these approaches needs, in part, deep knowledge of the mathematical background. In addition, sometimes, limitations are obvious since not all boundary conditions, etc. can be included. In any case, these approaches allow estimation of the accuracy of other theories and numerical solution techniques.

2.3.3. *Direct approach*

The direct approach is based on the idea that the mechanical behavior of a plate-like body can be presented by a deformable plane surface. In this case, the stress resultants are introduced *a priori* and the kinematics in each point of the surface is based on the introduction of displacements and independent rotations. The two-dimensional plate theory can be established so strong like three-dimensional continuum mechanics introducing the balance equations for the two-dimensional continuum. The specific constitutive equations are introduced in the conventional way assuming, for example, a two-dimensional strain energy. The disadvantage of such a theory is that the parameters or parameter functions in the constitutive equations must be identified from the properties of the *real* three-dimensional world, which is not trivial.

The ideas of the direct approach are connected with the continuum model of the Cosserat brothers (Cosserat and Cosserat 1909), which was a rebirth of the Bernoulli–Euler approach in mechanics. Possibly the first paper in this direction was by Ericksen and Truesdell (1958). Following Green and Naghdi (1967) and others, a direct theory of shells was suggested by Zhilin (1976). In the case of elastic isotropic plates, we can assume that the interaction of different parts of the plate can be presented by a force vector and a moment tensor and the kinematics is based on the deflections and a special two-dimensional rotation vector (the rotation about the normal to the surface is ignored). Instead of equation [2.1] or [2.2], the equation of motion can be written as (Altenbach 1984)

$$\left(\Delta - \frac{\rho}{\Gamma}\frac{\partial^2}{\partial t^2}\right)\left(D\Delta - < \rho_* z^2 > \frac{\partial^2}{\partial t^2}\right) w + \rho\frac{\partial^2 w}{\partial t^2} = \left(1 - \frac{D}{\Gamma}\Delta + \frac{< \rho * z^2 >}{\Gamma}\frac{\partial^2}{\partial t^2}\right) q + M \qquad [2.5]$$

where t is the time, $\Delta = \nabla \cdot \nabla$ is the Laplace operator, $< \rho_* z^2 >$ is the rotatory inertia (ρ_* is the density of the three-dimensional continuum, z is the thickness coordinate and $< \ldots >$ denotes the averaging over the thickness), ρ is the two-dimensional density ($< \rho_* >= \rho$), Γ is the transverse stiffness and M is an auxiliary function that reflects the action of the surface moments $\boldsymbol{m}$ - $M = \nabla \cdot (\boldsymbol{m} \times \boldsymbol{n})$. Equation [2.5] has five important special cases:

– rotatory inertia is zero ($< \rho_* z^2 >= 0$)

$$\left(\Delta - \frac{\rho}{\Gamma}\frac{\partial^2}{\partial t^2}\right) D\Delta w + \rho\frac{\partial^2 w}{\partial t^2} = \left(1 - \frac{D}{\Gamma}\Delta\right) q + M \qquad [2.6]$$

– transverse shear is ignored ($\Gamma \to \infty$)

$$\Delta\left(D\Delta - < \rho_* z^2 > \frac{\partial^2}{\partial t^2}\right) w + \rho\frac{\partial^2 w}{\partial t^2} = q + M \qquad [2.7]$$

– rotatory inertia is zero and transverse shear is ignored (Kirchhoff-type theory)

$$\Delta\left(D\Delta - < \rho_* z^2 > \frac{\partial^2}{\partial t^2}\right) w + \rho\frac{\partial^2 w}{\partial t^2} = q + M \qquad [2.8]$$

– eigenvibration equation ($q = 0, M = 0$)

$$\left(\Delta - \frac{\rho}{\Gamma}\frac{\partial^2}{\partial t^2}\right)\left(D\Delta - < \rho_* z^2 > \frac{\partial^2}{\partial t^2}\right) w + \rho\frac{\partial^2 w}{\partial t^2} = 0 \qquad [2.9]$$

– static bending equation (Reissner-type theory)

$$D\Delta\Delta w = \left(1 - \frac{D}{\Gamma}\Delta\right) q + M \qquad [2.10]$$

The full set of equations is presented, for example, in Palmow and Altenbach (1982), while the subclasses with relations to classical theories are given in Aßmus *et al.* (2019).

Let us briefly discuss the constitutive equations in this class of theories. In contrast to theories based on hypotheses or mathematical techniques, the constitutive parameters cannot be derived directly. The constitutive equations can be established in the traditional manner, making constitutive assumptions for the behavior of the two-dimensional deformable surface. However, the identification of the constitutive parameters (in the case discussed above, we have two stiffness parameters D and Γ) is not trivial. It was shown by Zhilin (1976) that the comparison of the eigenfrequencies of the three-dimensional plate-like body and the two-dimensional deformable surface allows the estimation of the stiffness parameters. In Altenbach (1987), the parameters were estimated for inhomogeneous in thickness direction plates based on the solution of three- and two-dimensional boundary value problems. The in-plane, out-of-plane and mixed stiffness parameters can be expressed as weighted sums of the layer parameters (for laminates) or as integrals over the thickness. For the transverse shear stiffness, this is not valid.

The direct approach is useful in discussions of the accuracy of other theories and numerical solution techniques. Limitations are related to the constitutive parameters estimations and the representation of the kinematic variables by the three-dimensional displacement vector. In contrast to the stress resultant–stress tensor relations (they are integrals over the thickness of the stress tensor or the stress tensor multiplied by the thickness coordinates), the two-dimensional displacements and rotations cannot be established by averaging the three-dimensional displacement vector.

2.3.4. *Consistent approach*

Kienzler (2002) postulated the so-called uniform-approximation approach. Later, he got carried away and declared this approach consistent, which, as already known, is a problematic term. Displacements, strains and stresses are developed in the thickness direction in the Taylor series. Introduction of dimensionless measures and integration across the thickness result in a natural parameter for the thickness-to-length ratio of the structure. The elastic potential appears as a power series in this parameter. Recently, attempts have been made to evaluate the approximation quality of various plate theories using this approach (Schneider and Kienzler 2020).

Since this approach is also not possible without assumptions, it is referred to here as pseudo-consistent. Its procedure is as inconsistent as all other approaches described in this chapter. The procedure furthermore lacks generality, as coupling of the three "classical" deformation states (in-plane state, bending state, transverse shear state → shear coupling, bending–drill coupling, bending–expansion coupling), which may be initiated by geometrical asymmetry, does not seem fully achievable. Thus, this approach is another special case in the inscrutable thicket of derivation procedures.

2.4. Shear correction

Shear correction factors are tools for corrections to the share of the transverse shear energy contribution. They are introduced in the constitutive equation for the transverse shear force vector (see, for example, Palmow and Altenbach 1982)

$$\boldsymbol{F} = \Gamma(\tilde{\boldsymbol{\nabla}} w - \boldsymbol{\phi}), \qquad [2.11]$$

where $\boldsymbol{\phi}$ is the vector of independent rotations. Transverse shear stiffness Γ can include the shear correction factor. The introduction of this factor becomes necessary since the transverse shear stress curve takes, for example, a parabolic course in the thickness direction. This observation becomes necessary when the shear stiffness of the investigated structure does not tend towards infinity, i.e. when materials with transverse shear sensitivity are considered. It is obvious that in the case of transverse shear-rigid theories, $\Gamma \to \infty$ holds and the rotations are the first derivatives of the deflection since the term in brackets in equation [2.11] must be zero. On the other hand, the transverse shear force vector must be finite and, if the theory does not contain any constitutive equation for this vector, the transverse shear forces follow from the equilibrium equations.

Since the transverse shear stiffness in a shear deformation plate theory only takes into account a kind of mean value, as it is related to the reference surface of the structure solely, only a constant course is considered. Therefore, a correction to correlate both courses, i.e. of reality and theory, is needed. The idea originally came from Timoshenko, who was faced with similar problems in beam theory as early as 1921 (see Timoshenko (1921, 1922)).

However, if Γ has a finite value, several suggestions are made. It follows from the dimensional analysis (Niordson 1971) that the transverse shear stiffness is related to the shear modulus G and the plate thickness h. The associated stiffness Gh needs to be corrected, i.e. when κ is a correction factor, $\Gamma = \kappa G h$ holds.

However, there are three classical procedures to determine shear correction factors. These are:

1) equating the shear strain energies (Reissner 1945);

2) calibration with natural frequency investigations (Mindlin 1951; Zhilin 1976);

3) comparing similar three- and two-dimensional boundary value problems (Altenbach and Zhilin 1988).

However, Reissner (1947) obtained $5/6$ for the shear correction factor from a static formulation of a first-order shear deformation theory and Mindlin (1951) calculated the correction $\pi^2/12$ based on a dynamic problem. Both values are very close. The challenge is to find a principle for the correction since the simplest approach using a parabolic distribution law for the transverse shear stresses is only a rough estimate of the shear correction.

The common procedure seems to be based on the first item of the preceding enumeration. The determination of the shear correction factor is then computed by the subsequent three steps:

– calculation of the complementary strain energy with the parabolic distribution law over the thickness;

– calculation of the complementary strain energy with the constant distribution law over the thickness and the shear correction;

– comparison of both expressions.

Usually these correction factors have only a limited range of validity and are linked to conditions for the load spectrum and the supports. Thus, shear correction factors are tuning parameters.

Some authors limit these factors to a range between 0 and 1. In other publications, mainly in the field of multi-layered plates, shear correction factors larger than 1 are reported which seems questionable at least from a physical point of view. In the meantime, however, the identification of such factors is the subject of specific research efforts (see Vlachoutsis (1992), Klarmann and Schweizerhof (1993), Altenbach (2000), Birman and Bert (2002), Gruttmann and Wagner (2016)).

REMARK 2.3.– *In the case of plates, the Poisson effect, which is ignored for beams, must be taken into account. This means the shear correction factor should also take into account the Poisson ratio. In Zhilin (1976), this was not considered, but in Zhilin (2006), an improvement was suggested as* $5Gh/(6-\nu)$.

REMARK 2.4.– *In Zhilin (1976), an alternative approach was discussed. The transverse shear stiffness is not corrected but calculated from the comparison of the eigenfrequencies of an elastic plate-like body and a deformable surface. In this case, the transverse shear stiffness is equal to* $\Gamma = \pi^2 Gh/12$. *This value is similar to Mindlin's expression.*

REMARK 2.5.– *The rst-order shear deformation concept can be extended to the case of inelastic materials. As an example, the creep behavior is discussed in Altenbach and Naumenko (2002).*

2.5. Conclusion

In this chapter, the four most important approaches to derive plate theories are presented. All approaches have advantages and disadvantages. All theories are approximative. The choice of any theory is ruled by the application cases, and this is the reason that theories based on hypotheses are preferred by engineers.

The principles listed here are accompanied by a variety of further specializations. This is associated with the development of special theories for plates. With the abundance of relevant publications, it is nearly possible to get an overview of all developments. This is due, not least, to the different starting points and viewpoints of the disciplines involved, i.e. mathematics, engineering and physics. Obviously, none of these publications have yet addressed the highest scientific goal, a systematization of approaches to a generalized approach, etc.

Furthermore, as is the standpoint of the authors, consistency is obtained while changing the starting point of the derivations. This is classically given as a three-dimensional Cauchy continuum. As shown by Neff *et al.* (2010), consistency is gained when starting from a three-dimensional Cosserat continuum.

2.6. References

Aghalovyan, L.A. (2015). *Asymptotic Theory of Anisotropic Plates and Shells*. World Scientific, Singapore.

Altenbach, H. (1984). Die Grundgleichungen einer linearen Theorie für dünne, elastische Platten und Scheiben mit inhomogenen Materialeigenschaften in Dickenrichtung. *Technische Mechanik*, 5(2), 51–58.

Altenbach, H. (1987). Definition of elastic moduli for plates made from thickness-uneven anisotropic material. *Mechanics of Solids*, 22(1), 135–141.

Altenbach, H. (1998). Theories for laminated and sandwich plates. *Mechanics of Composite Materials*, 34(3), 243–252.

Altenbach, H. (2000). An alternative determination of transverse shear stiffnesses for sandwich and laminated plates. *Int. J. Solids Struct*, 37(25), 3503–3520.

Altenbach, H. (2020). Germain, Marie-Sophie. In *Encyclopedia of Continuum Mechanics*, Altenbach, H. and Öchsner, A. (eds). Springer, Berlin, Heidelberg.

Altenbach, H. and Eremeyev, V. (2017). Thin-walled structural elements: Classification, classical and advanced theories, new applications. In *Shell-like Structures Advanced Theories and Applications*, Altenbach, H. and Eremeyev, V. (eds). Springer, Vienna.

Altenbach, H. and Naumenko, K. (2002). Shear correction factors in creep-damage analysis of beams, plates and shells. *JSME International Journal Series A Solid Mechanics and Material Engineering*, 45(1), 77–83.

Altenbach, H. and Zhilin, P.A. (1988). A general theory of elastic simple shells [in Russian]. *Advances in Mechanics*, 11(4), 107–148.

Altenbach, J., Altenbach, H., Eremeyev, V.A. (2010). On generalized Cosserat-type theories of plates and shells: A short review and bibliography. *Archive of Applied Mechanics*, 80(1), 73–92.

Altenbach, H., Altenbach, J., Naumenko, K. (2016). *Ebene Flächentragwerke 2*. Springer, Berlin.

Ambarcumyan, S.A. (1958). On the theory of bending of anisotropic plates and shallow shells [in Russian]. *Izv. AN SSSR. Otd. Tekhn. Nauk*, (5), 69–77.

Aßmus, M., Naumenko, K., Altenbach, H. (2019). Subclasses of mechanical problems arising from the direct approach for homogeneous plates. In *Recent Developments in the Theory of Shells*, Altenbach, H., Chróścielewski, J., Eremeyev, V.A., Wiśniewski, K. (eds). Springer, Singapore.

Birman, V. and Bert, C.W. (2002). On the choice of shear correction factor in sandwich structures. *Journal of Sandwich Structures and Materials*, 4(1), 83–95.

Bollé, L. (1947a). Contribution au problème linéaire de flexin d'une plaque élastique. *Bull. Techn. Suisse Romande*, 73(21), 281–285.

Bollé, L. (1947b). Contribution au problème linéaire de flexin d'une plaque élastique. *Bull. Techn. Suisse Romande*, 73(22), 293–298.

Bucciarelli, L.L. and Dworsky, N. (1980). *Sophie Germain – An Essay in the History of the Theory of Elasticity*. D. Reidel Publishing Company, Dordrect.

Chladni, E.F.F. (1787). *Entdeckungen über die Theorie des Klanges*. Weidmanns Erben und Reich, Leipzig.

Cosserat, E. and Cosserat, F. (1909). *Théorie des corps déformables*. Herman et Fils, Paris.

Elishakoff, I. (2020). *Handbook on Timoshenko-Ehrenfest Beam and U yand-Mindlin Plate Theories*. World Scientific, New Jersey.

Ericksen, J.L. and Truesdell, C. (1958). Exact theory of stress and strain in rods and shells. *Archive for Rational Mechanics and Analysis*, 1(1), 195–323.

Germain, S. (1821). *Recherches sur la théorie des surfaces élastique*. Courcier Libraire pour les Sciences, Paris.

Goldenveizer, A.L. (1962). Formulation of approximative theory of shells with the help of the asymptotic integration of the equations of the theory of elasticity [in Russian]. *Prikl. Mat. i Mekh.*, 26(4), 668–686.

Green, A.E. and Naghdi, P.M. (1967). Linear theory of an elastic Cosserat plate. *Proc. Camb. Philo. Soc.*, 63(2), 537–550.

Grigolyuk, E.I. and Kogan, F.A. (1972). State of the art of the theory of multilayer shells. *Sov. Appl. Mech.*, 8(6), 583–595.

Grigolyuk, E.I. and Seleznev, I.T. (1973). *Nonclassical Theories of Oscillations of Rods, Plates, and Shells* [in Russian]. VINITI, Moscow.

Gruttmann, F. and Wagner, W. (2016). Shear correction factors for layered plates and shells. *Comput. Mech.*, 59, 129–146.

Hencky, H. (1947). Über die Berücksichtigung der Schubverzerrung in ebenen Platten. *Ingenieur-Archiv*, XVI, 72–76.

Kienzler, R. (2002). On consistent plate theories. *Arch. Appl. Mech.*, 72, 229–247.

Kienzler, R. and Schneider, P. (2016). Direct approach versus consistent theory, In *Advanced Methods of Continuum Mechanics for Materials and Structures*, Naumenko, K. and Aßmus, M. (eds). Springer, Singapore.

Kirchhoff, G.R. (1850). Über das Gleichgewicht und die Bewegung einer elastischen Scheibe. *Crelle's Journal für die reine und angewandte Mathematik*, 40(1), 51–88.

Klarmann, R. and Schweizerhof, K. (1993). *A priori* improvement of shear correction factors for the analysis of layered anisotropic shell structures. *Arch. Appl. Mech.*, 63, 73–85.

Kromm, A. (1953). Verallgemeinerte Theorie der Plattenstatik. *Ingenieur-Archiv*, XXI, 266–286.

Lo, K.H., Christensen, R.M., Wu, E.M. (1977). A high-order theory of plate deformation. Part I: Homogeneous plates. *Trans. ASME. J. Appl. Mech.*, 44(4), 663–668.

Mindlin, R.D. (1951). Influence of rotatory inertia and shear on flexural motions of isotropic, elastic plates. *ASME J. Appl. Mech.*, 18(2), 31–38.

Naghdi, P.M. (1972). The theory of shells and plates. In *Handbuch der Physik*, Flügge, S. (ed.). Springer, New York.

Neff, P., Hong, K.-I., Jeong, J. (2010). The Reissner–Mindlin plate is the Γ-limit of Cosserat elasticity. *Math. Models. Methods. Appl. Sci.*, 20(09), 1553–1590.

Niordson, F.I. (1971). A note on the strain energy of elastic shells. *Int. J. Solids Struct.*, 7(11), 1573–1579.

Palmow, W.A. and Altenbach, H. (1982). Über eine Cosseratsche Theorie für elastische Platten. *Technische Mechanik*, 3(3), 5–9.

Poisson, S.D. (1829). Mémoire sur l'équilibre et le mouvement des corps élastiques. *Mémoires de l'Académie Royal des Sciences de l'Institut de France*, 8, 357–570.

Preußer, G. (1984). Eine systematische Herleitung verbesserter Plattentheorien. *Ingenieur-Archiv*, 54, 51–61.

Reddy, J.N. (1984). A simple higher-order theory for laminated composite plates. *Trans. ASME J. Appl. Mech.*, 51, 745–752.

Reissner, E. (1944). On the theory of bending of elastic plates. *J. Math. Phys.*, 23(1–4), 184–191.

Reissner, E. (1945). The effect of transverse shear deformation on the bending of elastic plates. *ASME J. Appl. Mech.*, 12, A68–A77.

Reissner, E. (1947). On bending of elastic plates. *Q. Appl. Math.*, 5(1), 55–68.

Reissner, E. (1985). Reflection on the theory of elastic plates. *Appl. Mech. Rev.*, 38(11), 1453–1464.

Rothert, H. (1973). Direkte Theorie von Linien- und Flächentragwerken bei viskoelastischen Werkstoffverhalten. In *Techn.-Wiss. Mitteilungen des Instituts für Konstruktiven Ingenieurbaus*. Ruhr-Universität, Bochum.

Schneider, P. and Kienzler, R. (2011). An algorithm for the automatisation of pseudo reductions of PDE systems arising from the uniform-approximation technique. In *Shell-like Structures: Non-classical Theories and Applications*, Altenbach, H. and Eremeyev, V.A. (eds). Springer, Berlin, Heidelberg.

Schneider, P. and Kienzler, R. (2017). A Reissner-type plate theory for monoclinic material derived by extending the uniform-approximation technique by orthogonal tensor decompositions of nth-order gradients. *Meccanica*, 52, 2143–2167.

Schneider, P. and Kienzler, R. (2020). *A priori* estimation of the systematic error of consistently derived theories for thin structures. *Int. J. Solids Struct.*, 190, 1–21.

Schneider, P., Kienzler, R., Böhm, M. (2014). Modeling of consistent second-order plate theories for anisotropic materials. *ZAMM – Journal of Applied Mathematics and Mechanics/Zeitschrift für Angewandte Mathematik und Mechanik*, 94(1–2), 21–42.

Timoshenko, S.P. (1921). On the correction for shear of the differential equation for transverse vibrations of prismatic bars. *Philos. Mag.*, 41(245), 744–746.

Timoshenko, S. P. (1922). On the transverse vibrations of bars of uniform cross-section. *Philos. Mag.*, 43(253), 125–131.

Timoshenko, S.P. and Woinowsky-Krieger, S. (1959). *Theory of Plates and Shells*. McGraw–Hill, New York.

Touratier, M. (1988). A refined theory of thick composite plate. *Mech. Res. Commun.*, 15(4), 229–236.

Touratier, M. (1991). An efficient standard plate theory. *Int. J. Eng. Sci.*, 29(8), 901–916.

Uflyand, Y.S. (1948). Wave propagation by transverse vibrations of beams and plates [in Russian]. *Prikladnaya Matematika i Mekhanika*, 12, 287–300.

Vekua, I.N. (1985). *Shell Theory: General Methods of Construction*. Pitman, Boston.

Vlachoutsis, S. (1992). Shear correction factors for plates and shells.*Int. J. Num. Methods Eng.*, 33(7), 1537–1552.

Vlasov, B.F. (1957). On the equations in the plate bending theory [in Russian].*Izv. AN SSSR, Otd. Tekhn. Nauk*, 12, 57–60.

Wunderlich, W. (1973). Vergleich verschiedener Approximationen der Theorie dünner Schalen (mit numerischen Beispielen). In *Techn.-Wiss. Mitteilungen des Instituts für Konstruktiven Ingenieurbaus*. Ruhr-Universität, Bochum.

Zhilin, P.A. (1976). Mechanics of deformable directed surfaces. *Int. J. Solids Struct.*, 12, 635–648.

Zhilin, P.A. (1992). On theories of Poisson and Kirchhoff from the point of modern theory of plates [in Russian]. *Mech. Solids*, 27(3), 48–64.

Zhilin, P.A. (2006). *Applied Mechanics – Foundations of the Shell Theory*. State Polytechnical University Publisher, St. Petersburg.

3

A Softening–Hardening Nanomechanics Theory for the Static and Dynamic Analyses of Nanorods and Nanobeams: Doublet Mechanics

3.1. Introduction

Nowadays, nanostructures are used as non-traditional materials in engineering, electronics and medical applications. The classical theory of elasticity does not describe the size effects of materials when their characteristic sizes scale down to the order of microns (Fleck *et al.* 1994; Lam *et al.* 2003). In order to model the mechanical behavior of these nanostructures such as beams, plates and shells, different size-dependent continuum theories have been used in the literature. In such microscale structures, the characteristic length scales become significant and the mechanical behavior of microstructures becomes non-homogeneous. Accordingly, some size-dependent continuum theories (Cosserat 1909; Mindlin and Triersten 1964; Eringen 1966; Mindlin and Eshel 1968) have been developed. A common property of the aforementioned higher-order theories is their complication in the engineering applications. Mostly, two main size-dependent continuum theories have been applied in the mechanical analysis of nanoscale structures. First, strain gradient elasticity theory has been proposed by Aifantis (1992) and Altan and Aifantis (1997), in order to introduce the length scale parameter that represents the microstructural properties and non-homogeneous phenomena of the material. However, length scale parameter is not derived directly from the microstructures. Thus, this continuum model can be accepted as phenomenological. In strain gradient theory, stress in a point is related to strain in the same and adjacent points, in contrast to classical

Chapter written by Ufuk GUL and Metin AYDOGDU.

elasticity theory, which only considers the strains and stresses in the same points while defining the constitutive equation. Some anomalies occur depending on the sign of the gradient term in the strain gradient theory. For instance, a strain gradient with a negative sign guarantees the stability and uniqueness for static analysis; however, it gives imaginary frequencies and infinite group velocities in the dynamic problems (Askes *et al.* 2002). Moreover, a strain gradient with a positive sign length scale parameter is obtained from a discrete system. However, this model has some anomalies about singularities or discontinuities compared to strain gradients with negative signs. Thus, some confusion still exists about the use of the strain gradient theory in static and dynamic problems.

Second, nonlocal stress gradient theory has been introduced by Eringen (1972a, 1972b, 1983) in order to model periodic structures (polymers, granular materials and crystal lattices). In Eringen's nonlocal theory, stress is not only a function of the strain at a related point, but also depends on the strain in a closer volume, near to that point. Eringen's integral-type nonlocal model is complicated for one-dimensional and two-dimensional problems. Therefore, a differential type equation was proposed by Eringen (1972b) for the simplicity of a one-dimensional wave propagation in an unbounded medium. Then, this model was used on many nonlocal problems, which can be considered as the size effects in nanostructures. The differential form of Eringen's nonlocal theory has been applied by Peddieson *et al.* (2003) and Sudak (2003) to investigate the size effects in finite length nanobeams. However, it has recently been shown that the differential form of a nonlocal model does not consider the size effect in nanomaterials under the point loads and stiffening behavior predicted for clamped-free boundary conditions, in contrast to other boundary conditions, with softening behavior obtained at the nanoscale. To overcome these ill-posedness problems, some recent modified nonlocal elasticity models have been proposed (Challamel and Wang 2008; Li *et al.* 2015; Barretta *et al.* 2016; Fernandez-Savez *et al.* 2016; Romano and Barretta 2016, 2017; Apuzzo *et al.* 2017; Romano *et al.* 2017). Near the boundary layers of non-locally thin structures, the effect of boundary layers becomes essential. Chebakov *et al.* (2016) deduced a first-order nonlocal correction to the boundary conditions on a free surface. In another study, Chebakov *et al.* (2017) studied the variation of nonlocal properties across the plate thickness using Eringen's integral-type nonlocal model. Moreover, some studies related to the asymptotic approach, using direct integration through the thickness, are adopted (Goldenveizer *et al.* 1993; Kaplunov *et al.* 2000, 2006).

Furthermore, some other size-dependent continuum models have been used for the design of the microstructures, such as modified couple stress theory (Cosserat 1909) and peridynamics (Silling 2000).

Another popular scale-dependent continuum theory, known as doublet mechanics (DM), was first introduced by Granik (1978). DM is not a phenomenological theory, in contrast to the other size-dependent models aforementioned above. The length scale parameter in DM is determined by considering the microstructure of the solid. The bonding length of atoms in the solid becomes a length scale parameter for each considered material in DM theory. In DM theory, microstrains and microstresses are defined as vectors, and then they are transformed into micro–macro stress relations. DM theory plays a significant role between the lattice dynamics and continuum mechanics at the two limits. Microstrains and microstresses are obtained on the basis of the Taylor series expansion in DM. Essentially, two important points affect the material behavior in this size-dependent model. First, the number of terms considered in the Taylor series expansion is chosen as an appropriate atomic structure of the material, and second, the composition of the lattice structure of atoms in the material affects the characteristic behavior of the solid. In light of this information, the discreteness of the solid can be considered in DM. Another remarkable observation in DM is that hardening or softening material behaviors can be predicted similarly to experimental studies. This important behavior will be shown for the static and dynamic analyses of periodic nanostructures in this study.

DM theory was first used to derive microequations for granular materials by Granik and Ferrari (1993). Then, kinematic relations of DM theory have been presented in detail for granular materials by Ferrari *et al.* (1997). A finite element formulation has been developed for two-dimensional microstructures by using the DM theory (Kojic *et al.* 2011). The obtained finite element formulation basis of DM was validated in that study. This is significant to provide the solution of complicated microstructural problems. Lin and Shen (2005) investigated the microstress fields of a half-plane subjected to moving loads, considering DM theory. They showed that the micro- and macro-stress field pattern in the half-plane is different from the classical theory of elasticity.

Recently, DM has been applied to nanomaterials (nanorods, nanobeams, nanoplates) to investigate their mechanical characterization. Vajari and Imam (2016a, 2016b) studied the axial and torsional vibration of single-walled carbon nanotubes and they compared the DM results with the classical elasticity solution. Gul *et al.* (2017) investigated the longitudinal vibration of nanorods, by taking the elastic medium into account. They showed that DM theory agrees well with the lattice dynamics results. In another study, wave dispersion relations for double-walled carbon nanotubes have been examined using DM by Gul and Aydogdu (2017). The predicted DM results were compared to experiment dynamics, and good agreement was obtained in that study. Gul and Aydogdu (2018a) studied the static and dynamic analyses of nanorods and nanobeams based on DM theory. It was concluded that DM theory gives excellent results with experiment and lattice dynamics in wave propagation analysis opposed to other scale-dependent continuum theories. This approves the validation of the physical fundamentals of the DM model. Buckling and vibration characteristics have been investigated in single and double nanofibers embedded in an elastic medium via DM theory (Aydogdu and Gul 2018a; Gul *et al.* 2018). Another remarkable study related to the vibration and buckling of double-walled carbon nanotubes has been examined using DM theory (Gul and Aydogdu 2018b) and agreeable DM results obtained with other scale-dependent theories. Aydogdu and Gul (2018b) investigated the axial wave propagation in stepped nanorods via DM theory. Transmission and reflection of axial waves have been presented and their matrices have been obtained for stepped nanorods in that study. More recently, the axial vibration of nanorods has been examined considering Love's assumption based on the DM by Gul and Aydogdu (2019). By using DM, the exact frequencies of clamped–clamped and clamped-free nanorods have been acquired and the obtained vibration frequencies were validated with molecular dynamic simulation. Zigzag and armchair nanotube models have been considered in the analysis. In addition to these studies, there are some studies related to the structural modeling of nanomaterials via DM theory (Vajari and Imam 2016c; Vajari 2018).

Static and dynamic analyses of periodic nanostructures based on DM have not yet been investigated analytically, and this chapter intends to consider such analyses. In this chapter, static deformation, buckling, vibration and wave propagation of periodic nanostructures modeled as nanorods and nanobeams have been investigated by using DM theory. A periodic simple square microstructured solid has been modeled as a two-dimensional periodic structure in the present study. Governing equations are derived for this lattice structure, and the influences of the geometry of nanostructures, length scale parameter and chiral angle of the periodic structures are examined. The results obtained by DM are compared to lattice dynamics and classical elasticity theory. After a brief review of the DM theory in section 3.2, corresponding equations have been derived for static and dynamic analyses of periodic nanostructures in section 3.3. Section 3.4 finalizes the solution of the corresponding equations on the basis of DM theory in nanorods and nanobeams, respectively. Then, numerical results are presented and discussed in section 3.5. Finally, the main conclusions are summarized for this investigation in section 3.6.

3.2. Doublet mechanics formulation

DM is a scale-dependent micromechanics theory. DM theory bridges the lattice dynamics at the nanolength scale and the continuum mechanics at the macroscale limit. Any two adjacent nodes in a solid are called doublets, and they are separated from each other with doublet separation distance $\eta_\alpha = |\xi_\alpha|$, which also shows the bonding length of atoms in the considered material. Each doublet is subjected to elongation ϵ_α and/or rotation during the deformation in this theory. As a result of microdeformations, three microstresses emerge as elongation microstress $\boldsymbol{p}_\alpha$, shear microstress $\boldsymbol{t}_\alpha$ and torsional microstress $\boldsymbol{m}_\alpha$. These microstresses are vector quantities. For simplicity, shear microstress and torsional microstress $(\boldsymbol{t}_\alpha, \boldsymbol{m}_\alpha)$ are ignored and only elongation microstress/strains $(\boldsymbol{p}_\alpha, \epsilon_\alpha)$ are considered in the present study. Considering all of these microstresses and microstrains observed during the deformation makes the analysis very complicated, and such microeffects $(\boldsymbol{t}_\alpha, \boldsymbol{m}_\alpha)$ may be considered for future DM studies. For each doublet, the increment of displacement in scalar form can be defined as follows (Ferrari *et al.* 1997):

$$\Delta u_\alpha = \sum_{\chi=1}^{M} \frac{(\eta_\alpha)^\chi}{\chi!} \tau_{\alpha k_1}^o \dots \tau_{\alpha k_\chi}^o \frac{\partial^\chi u_i}{\partial x_{k_1} \dots \partial x_{k_\chi}} \quad [3.1]$$

where each of the subscripts $k_1, \dots k_\chi$ are integers that take the values of 1, 2 and 3 with respect to three-dimensional Cartesian coordinates, M is the number of terms in the Taylor series expansion which shows the discreteness of the solid, and τ_α^o is the unit vector in α-direction. It is notable that the increment of displacement Δu_α is smaller than the doublet separation distance η_α; therefore, it can be assumed that the results combine for $\tau_\alpha = \tau_\alpha^o$. For this study, the first three terms (M=3) in the Taylor series expansion are considered, while defining the microstrains and microstresses in the solid. M denotes the discreteness of the material and is not related to accuracy of the approach. The three terms (M=3) in the Taylor series expansion provide good results in DM theory (Vajari and Imam 2016a; Gul *et al.* 2017; Aydogdu and Gul 2018a; Gul and Aydogdu 2018a; Gul *et al.* 2018). It is noted that the classical elasticity theory is obtained by taking M=1 in the Taylor series expansion. By taking M>1, a multiscale theory has been obtained. The elongation microstrain ϵ_α in the α-doublet is defined as follows (Ferrari *et al.* 1997):

$$\epsilon_\alpha = \tau_{\alpha_i}^o \sum_{\chi=1}^{M} \frac{(\eta_\alpha^o)^{\chi-1}}{\chi!} \tau_{\alpha k_1}^o \dots \tau_{\alpha k_\chi}^o \frac{\partial^\chi u_i}{\partial x_{k_1} \dots \partial x_{k_\chi}} \quad [3.2]$$

where the superscript ^{o} represents the initial configuration of the system. For M=3, the microstrain is

$$\epsilon_\alpha = \tau_{\alpha i}^o \tau_{\alpha j}^o \frac{\partial u_i}{\partial x_j} + \tau_{\alpha i}^o \frac{\eta_\alpha}{2} \tau_{\alpha j}^o \tau_{\alpha k}^o \frac{\partial^2 u_i}{\partial x_j \partial x_k} + \tau_{\alpha i}^o \frac{{\eta_\alpha}^2}{6} \tau_{\alpha j}^o \tau_{\alpha k}^o \tau_{\alpha l}^o \frac{\partial^3 u_i}{\partial x_j \partial x_k \partial x_l} \quad [3.3]$$

The microstress p_α and the microstrain ϵ_α relations acquire the form:

$$p_\alpha = \sum_\beta B_{\alpha\beta} \, \epsilon_\beta \quad [3.4]$$

where $B_{\alpha\beta}$ represents the matrix of the micromodulus of the doublet between the nodes α and β, and ϵ_β represents the axial microstrain associated with the β doublet. The matrix of microelastic modulus $B_{\alpha\beta}$ is obtained in terms of Lamé's constants in previous studies (Gul and Aydogdu 2018a, 2018b). The relationship between macrostress and microstress can be defined as follows (Ferrari *et al.* 1997):

$$\sigma_{k_1 i}{}^{(M)} = \sum_{\alpha=1}^{m} \tau_{\alpha k_1}^{o} \sum_{\chi=1}^{M} (-1)^{\chi+1} \times \left[\frac{(\eta_\alpha)^{\chi-1}}{\chi!} \tau_{\alpha k_2}^{o} \dots \tau_{\alpha k_\chi}^{o} \frac{\partial^{\chi-1} p_{\alpha i}}{\partial \chi_{k_2} \dots \partial \chi_{k_\chi}} \right] \qquad [3.5]$$

Using $p_{\alpha i} = p_\alpha \tau_{\alpha i}^{o}$, equation [3.5] becomes the following form for M=1:

$$\sigma_{ij} = \sum_{\alpha=1}^{m} \tau_{\alpha i}^{o} \tau_{\alpha j}^{o} p_\alpha \qquad [3.6]$$

where m is the number of the neighboring nodes in the doublet. By taking M=3, the relation between the macrostress and microstress becomes

$$\sigma_{ij}^{(M)} = \sum_{\alpha=1}^{n} \tau_{\alpha i}^{0} \tau_{\alpha j}^{0} \left\{ \left(p_\alpha - \frac{1}{2} \eta_\alpha \vec{\tau}_{\alpha}^{0} . (\nabla p_\alpha) + \frac{1}{6} \eta_\alpha^2 [(\vec{\tau}_{\alpha}^{0} . \nabla)(\vec{\tau}_{\alpha}^{0} . \nabla p_\alpha)] \right) \right\} \qquad [3.7]$$

It is assumed that $-\frac{\varepsilon_{33}}{\varepsilon_{11}} = -\frac{\varepsilon_{22}}{\varepsilon_{11}} = \nu = \frac{1}{3}$ and by only considering the axial strain, the stress expression of DM theory is obtained as

$$\sigma_{xx} = f_1(\theta) E \varepsilon_{11} + f_2(\theta) E \frac{\eta^2}{12} \frac{\partial^2 \varepsilon_{11}}{\partial x^2} \qquad [3.8]$$

where E is the elasticity modulus of the considered material, ε_{11} is the axial macrostrain, η is the doublet separation distance (or internal length scale parameter) and $f_1(\theta)$ and $f_2(\theta)$ represent the directional cosines that depend on the chiral angle (θ) of the atomic structure in the material. The value of the chiral angle (θ) of the atoms in a solid changes between 0° and 30°. In the present study, the simple square periodic lattice structure of the periodic material is considered (Figure 3.1), and for this lattice type, $f_1(\theta)$ and $f_2(\theta)$ become:

$$f_1(\theta) = (\cos(\theta))^4 + \left(\cos\left(\frac{\pi}{2} + \theta\right)\right)^4 + (\cos(\pi + \theta))^4 + \left(\cos\left(\frac{3\pi}{2} + \theta\right)\right)^4 - \frac{1}{3}\left[(\cos(\theta))^2 \left(\cos\left(\frac{\pi}{2} - \theta\right)\right)^2 + \left(\cos\left(\frac{\pi}{2} + \theta\right)\right)^2 (\cos(\theta))^2 + (\cos(\pi + \theta))^2 \left(\cos\left(\frac{\pi}{2} + \theta\right)\right)^2 + \left(\cos\left(\frac{3\pi}{2} + \theta\right)\right)^2 (\cos(\theta))^2\right] \qquad [3.9]$$

$$f_2(\theta) = (\cos(\theta))^6 + \left(\cos\left(\frac{\pi}{2} + \theta\right)\right)^6 + (\cos(\pi + \theta))^6 + \left(\cos\left(\frac{3\pi}{2} + \theta\right)\right)^6 - \frac{1}{3}\left[(\cos(\theta))^4 \left(\cos\left(\frac{\pi}{2} - \theta\right)\right)^2 + \left(\cos\left(\frac{\pi}{2} + \theta\right)\right)^4 (\cos(\theta))^2 + (\cos(\pi + \theta))^4 \left(\cos\left(\frac{\pi}{2} + \theta\right)\right)^2 + \left(\cos\left(\frac{3\pi}{2} + \theta\right)\right)^4 (\cos(\theta))^2\right] \qquad [3.10]$$

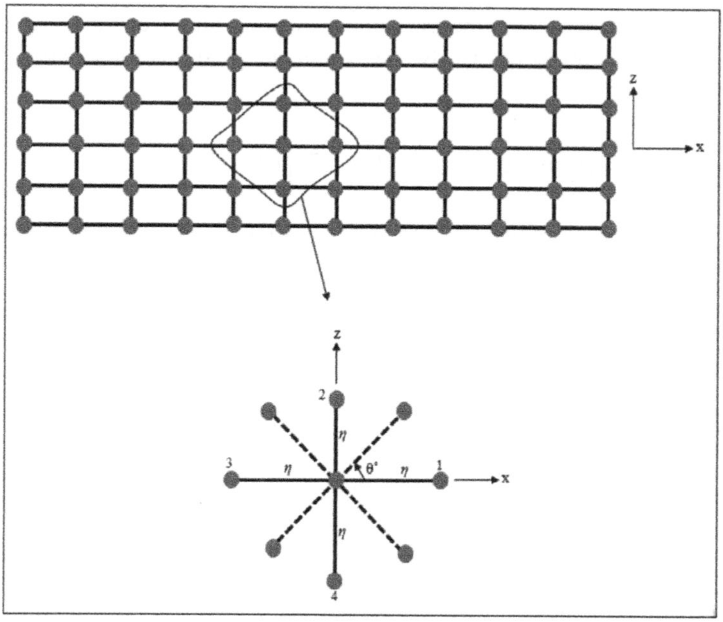

Figure 3.1. *Arrangement of atoms in a periodic nanostructure*

The first term of equation [3.8] is known as a classical Cauchy stress and the second term of equation [3.8] can be defined as a doublet stress for the present DM theory. It is important to emphasize that the coefficient of Cauchy stress $f_1(\theta)$ only changes with the chiral angle (θ), whereas the coefficient of doublet stress $f_2(\theta)$ changes both with the chiral angle (θ) and the value of the length scale parameter η. This property directly affects the mechanical behavior of the solid in the DM theory.

3.3. Governing equations

3.3.1. *Static equilibrium equations of a nanorod with periodic micro- and nanostructures*

The static governing equation and boundary conditions are obtained in a nanorod with periodic nanostructures using the minimum potential energy principle:

$$\delta(U - W) = 0 \qquad [3.11]$$

where δ, U and W represent the variational symbol, strain energy and total work done by external forces, respectively. The strain energy of the rod can be determined as

$$U = \frac{1}{2}\int_0^L \left[Ef_1(\theta)u'.\varepsilon - \frac{\eta^2}{12}Ef_2(\theta)u''.\nabla\varepsilon\right]dx \qquad [3.12]$$

where L denotes length of the nanorod, u denotes the displacement component along the axial coordinate direction, a prime denotes the partial differentiation with respect to the x coordinate parameter, and $\nabla\varepsilon = d\varepsilon/dx$.

In this way, strain energy U can be defined as

$$U = \frac{1}{2}\int_0^L AE\left[f_1(\theta)(u')^2 - \frac{\eta^2}{12}f_2(\theta)(u'')^2\right]dx \qquad [3.13]$$

The first variation of the strain energy has the form

$$\delta U = -\int_0^L AE\left(f_1(\theta)u'' + \frac{\eta^2}{12}f_2(\theta)u^{IV}\right)\delta u dx + \left[AE\left(f_1(\theta)u' + \frac{\eta^2}{12}f_2(\theta)u'''\right)\delta u\right]_0^L + \left[AE\frac{\eta^2}{12}f_2(\theta)u''\delta u'\right]_0^L \qquad [3.14]$$

Moreover, the first variation of the total work can be defined as

$$\delta W = \int_0^L f\delta u dx + [P\delta u]_0^L + [R\delta u']_0^L \qquad [3.15]$$

where P denotes the classical force or concentrated force, f denotes the distributed force, and R denotes the force couple (double force). Substituting equations [3.14] and [3.15] into equation [3.11] gives:

$$\delta(U-W) = \int_0^L \left[AE\left(f_1(\theta)u'' + \frac{\eta^2}{12}f_2(\theta)u^{IV}\right) + f\right]\delta u dx + \left[\left\{P - AE\left(f_1(\theta)u' + \frac{\eta^2}{12}f_2(\theta)u'''\right)\right\}\delta u\right]_0^L + \left[\left\{R + AE\frac{\eta^2}{12}f_2(\theta)u''\right\}\delta u'\right]_0^L \qquad [3.16]$$

In this context, each term in square brackets in equation [3.16] must be zero. The first bracket of equation [3.16] gives the corresponding equation of a nanorod in the following form:

$$AE\left[f_1(\theta)u''(x)+\frac{\eta^2}{12}f_2(\theta)u^{IV}(x)\right]+f=0 \qquad [3.17]$$

and the other brackets of equation [3.16] lead to following set of boundary conditions:

$$\left[P(L)-AE\left[f_1(\theta)u'(L)+\frac{\eta^2}{12}f_2(\theta)u'''(L)\right]\right]\delta u(L) \\ -\left[P(0)-AE\left[f_1(\theta)u'(0)+\frac{\eta^2}{12}f_2(\theta)u'''(0)\right]\right]\delta u(0) \\ =0,$$

$$\left[R(L)-AE\left[-\frac{\eta^2}{12}f_2(\theta)u''(L)\right]\right]\delta u'(L)- \\ \left[R(0)-AE\left[-\frac{\eta^2}{12}f_2(\theta)u''(0)\right]\right]\delta u'(0)=0 \qquad [3.18]$$

where the displacement u and the axial force $AE(f_1(\theta)u'+\frac{\eta^2}{12}f_2(\theta)u''')$ represent the cases of modified classical boundary conditions, and the boundary strain u' and the boundary double force $AE\frac{\eta^2}{12}f_2(\theta)u''$ represent the cases of non-classical boundary conditions in the DM theory. It is seen that the absence of length scale parameter (η) in equations [3.17] and [3.18] gives the classical elasticity equation and classical boundary conditions.

3.3.2. *Equations of motion of a nanorod with periodic micro- and nanostructures*

By using the Hamilton principle, the equation of motion and the boundary conditions of the rod can be defined as

$$\int_{t_0}^{t_1}\delta(U-T)dt-\int_{t_0}^{t_1}\delta W dt=0 \qquad [3.19]$$

where U represents the strain energy defined in equation [3.12], W represents the total work done by external forces, and T represents the kinetic energy of the rod, which is defined as follows:

$$T=\frac{1}{2}\int_0^L \rho A\dot{u}^2\,dx \qquad [3.20]$$

where ρ is the density and a dot is the partial differentiation with respect to time (t). The appropriate use of equations [3.14], [3.15], [3.19] and [3.20], leads to the variational equation (equation [3.19]) as follows:

$$\int_{t_0}^{t_1}\int_{0}^{L}\left[AE\left(f_1(\theta)u''+\frac{\eta^2}{12}f_2(\theta)u^{IV}\right)+f-\rho A\ddot{u}\right]\delta u dxdt$$
$$+\int_{t_0}^{t_1}\left[\left\{P-AE(f_1(\theta)u'+\frac{\eta^2}{12}f_2(\theta)u''')\right\}\delta u\right]_0^L dt$$
$$+\int_{t_0}^{t_1}\left[\left\{R-AE(-\frac{\eta^2}{12}f_2(\theta)u'')\right\}\delta u'\right]_0^L dt+\int_0^L[\rho A\dot{u}\delta u]_{t_0}^{t_1}dx=0 \qquad [3.21]$$

Thus, the first bracket of equation [3.21] represents the equation of motion of the rod:

$$AE\left[f_1(\theta)u''(x,t)+\frac{\eta^2}{12}f_2(\theta)u^{IV}(x,t)\right]+f=\rho A\ddot{u}(x,t) \qquad [3.22]$$

The initial conditions confirm the equation as follows:

$$\dot{u}(x,t_1)\delta u(x,t_1)-\dot{u}(x,t_0)\delta u(x,t_0)=0 \qquad [3.23]$$

The boundary conditions (classical and non-classical) are obtained as

$$\left[P(L,t)-AE\left[f_1(\theta)u'(L,t)+\frac{\eta^2}{12}f_2(\theta)u'''(L,t)\right]\right]\delta u(L,t)$$
$$-\left[P(0,t)-AE\left[f_1(\theta)u'(0,t)+\frac{\eta^2}{12}f_2(\theta)u'''(0,t)\right]\right]\delta u(0,t)$$
$$=0,$$

$$\left[R(L,t)-AE\left[-\frac{\eta^2}{12}f_2(\theta)u''(L,t)\right]\right]\delta u'(L,t)-$$
$$\left[R(0,t)-AE\left[-\frac{\eta^2}{12}f_2(\theta)u''(0,t)\right]\right]\delta u'(0,t)=0 \qquad [3.24]$$

Similar to the static case, ignoring the length scale parameter (η) in equations [3.22] and [3.24] gives the equations of motion of classical elasticity theory with classical boundary conditions.

3.3.3. *Static equilibrium equations of a nanobeam with periodic micro- and nanostructures*

The static governing equation and boundary conditions are found in a nanobeam with periodic nanostructures using the minimum potential energy principle:

$$\delta(U-W)=0 \qquad [3.25]$$

The strain energy of the beam can be defined as

$$U=\frac{1}{2}\int_0^L EI\left[f_1(\theta)(w'')^2-\frac{\eta^2}{12}f_2(\theta)(w''')^2\right]dx \qquad [3.26]$$

where I represents the moment of inertia of the beam cross-section. Then, we can define the first variation of the strain energy in the following form:

$$\delta U=\int_0^L EI\left(f_1(\theta)w^{IV}+\frac{\eta^2}{12}f_2(\theta)w^{VI}\right)\delta w dx$$
$$+\left[EI\left(f_1(\theta)w'''-\frac{\eta^2}{12}f_2(\theta)w^{V}\right)\delta u\right]_0^L$$
$$+\left[EI\left(\frac{\eta^2}{12}f_2(\theta)w^{VI}-f_1(\theta)w''\right)\delta w'\right]_0^L$$
$$-\left[EI\frac{\eta^2}{12}f_2(\theta)w'''\delta w''\right]_0^L \qquad [3.27]$$

Also, the first variation of the total work done by external forces is

$$\delta W=-\int_0^L f\delta w dx-[V\delta w]_0^L+[M\delta w']_0^L+[m\delta w'']_0^L \qquad [3.28]$$

where the shear force can be defined as $V=\partial M/\partial x$. Thus, equation [3.25] becomes:

$$\delta(U-W)=\int_0^L\left[EI\left(f_1(\theta)w^{IV}+\frac{\eta^2}{12}f_2(\theta)w^{VI}\right)+f\right]\delta w dx$$
$$+\left[\left\{V-EI\left(f_1(\theta)w'''+\frac{\eta^2}{12}f_2(\theta)w^{V}\right)\right\}\delta w\right]_0^L$$

$$-\left[\left\{M - EI\left(f_1(\theta)w'' + \tfrac{\eta^2}{12}f_2(\theta)w^{IV}\right)\right\}\delta w'\right]_0^L - \left[\left\{m - EI\left(-\tfrac{\eta^2}{12}f_2(\theta)w^{IV}\right)\right\}\delta w''\right]_0^L = 0 \quad [3.29]$$

Each term in square brackets in equation [3.29] must be zero. The first bracket of equation [3.29] gives the bending equation of the nanobeam with periodic nanostructures in DM theory.

$$EI\left[f_1(\theta)w^{IV}(x) + \tfrac{\eta^2}{12}f_2(\theta)w^{VI}(x)\right] + f = 0 \quad [3.30]$$

and the boundary conditions emerge from equation [3.29]:

$$\left[V(L) - EI\left[f_1(\theta)w'''(L) + \frac{\eta^2}{12}f_2(\theta)w^{V}(L)\right]\right]\delta w(L) - \left[V(0) - EI\left[f_1(\theta)w'''(0) + \frac{\eta^2}{12}f_2(\theta)w^{V}(0)\right]\right]\delta w(0) = 0,$$

$$\left[M(L) - EI\left[f_1(\theta)w''(L) + \frac{\eta^2}{12}f_2(\theta)w^{IV}(L)\right]\right]\delta w'(L) - \left[M(0) - EI\left[f_1(\theta)w''(0) + \frac{\eta^2}{12}f_2(\theta)w^{IV}(0)\right]\right]\delta w'(0) = 0,$$

$$\left[m(L) - EI\left[-\tfrac{\eta^2}{12}f_2(\theta)w'''(L)\right]\right]\delta w''(L) - \left[m(0) - EI\left[-\tfrac{\eta^2}{12}f_2(\theta)w'''(0)\right]\right]\delta w''(0) = 0 \quad [3.31]$$

where the displacement w, the rotation w', the shear force $V = EI(f_1(\theta)w''' + \frac{\eta^2}{12}f_2(\theta)w^{V})$ and the moment $M = EI(f_1(\theta)w'' + \frac{\eta^2}{12}f_2(\theta)w^{IV})$ represent the cases of modified classical boundary conditions, and w'' and the stress gradient moment $m = EI\frac{\eta^2}{12}f_2(\theta)w'''$ represent the cases of non-classical boundary conditions in the DM theory. It is seen that by ignoring the length scale parameter η in equations [3.30] and [3.31], the boundary conditions with the equation of motion can be found for classical elasticity theory.

3.3.4. *Equations of motion of a nanobeam with periodic micro- and nanostructures*

On the basis of the Hamilton principle, the equation of motion and the boundary conditions of the beam are defined as

$$\delta\left[\int_{t_0}^{t_1}(T - W - U)dt\right] = 0 \tag{3.32}$$

where U represents the strain energy defined in equation [3.26], W represents the total work done by external forces, and T represents the kinetic energy of the beam. The first variation of kinetic energy is

$$\delta T = \delta\left[\frac{1}{2}\int_0^L \rho A\dot{w}^2\, dx\right] \tag{3.33}$$

Upon substituting equations [3.27], [3.28] and [3.33] into equation [3.32], it leads to the variational equation as follows:

$$\int_{t_0}^{t_1}\int_0^L \left[EI\left(f_1(\theta)w^{IV} + \frac{\eta^2}{12}f_2(\theta)w^{VI}\right) + f + \rho A\ddot{w}\right]\delta w dx dt$$
$$+ \int_{t_0}^{t_1}\left[\left\{V - EI(f_1(\theta)w''' + \frac{\eta^2}{12}f_2(\theta)w^{V})\right\}\delta w\right]_0^L dt$$

$$+ \int_{t_0}^{t_1}\left[\left\{M + EI(f_1(\theta)w'' + \frac{\eta^2}{12}f_2(\theta)w^{IV})\right\}\delta w'\right]_0^L dt + \int_{t_0}^{t_1}\left[\left\{m - EI\frac{\eta^2}{12}f_2(\theta)w'''\right\}\delta w''\right]_0^L dt - \int_0^L [\rho A\dot{w}\delta w]_{t_0}^{t_1} dx = 0 \tag{3.34}$$

By equalizing the first bracket of equation [3.34] to zero, the equation of the motion of the beam with periodic nanostructures is obtained as

$$EI\left[f_1(\theta)w^{IV}(x) + \frac{\eta^2}{12}f_2(\theta)w^{VI}(x)\right] + \rho A\ddot{w} + f = 0 \tag{3.35}$$

The initial conditions satisfy the equation as follows:

$$[\rho A\dot{w}\delta w]_{t_0}^{t_1} = 0 \tag{3.36}$$

and the boundary conditions become:

$$\left[V(L,t) - EI\left[f_1(\theta)w'''(L,t) + \frac{\eta^2}{12}f_2(\theta)w^V(L,t)\right]\right]\delta w(L,t)$$
$$-\left[V(0,t) - EI\left[f_1(\theta)w'''(0,t) + \frac{\eta^2}{12}f_2(\theta)w^V(0,t)\right]\right]\delta w(0,t) = 0,$$

$$\left[M(L,t) - EI\left[f_1(\theta)w''(L,t) + \frac{\eta^2}{12}f_2(\theta)w^{IV}(L,t)\right]\right]\delta w'(L,t)$$
$$-\left[M(0,t) - EI\left[f_1(\theta)w''(0,t) + \frac{\eta^2}{12}f_2(\theta)w^{IV}(0,t)\right]\right]\delta w'(0,t) = 0,$$

$$\left[m(L,t) - EI\left[-\frac{\eta^2}{12}f_2(\theta)w'''(L,t)\right]\right]\delta w''(L,t) -$$
$$\left[m(0,t) - EI\left[-\frac{\eta^2}{12}f_2(\theta)w'''(0,t)\right]\right]\delta w''(0,t) = 0. \qquad [3.37]$$

Again, by neglecting the length scale parameter (η=0), the classical boundary conditions with the classical elasticity equation can be obtained.

3.4. Analytical solutions

3.4.1. *Axial deformation of nanorods with periodic nanostructures*

In this section, the axial deformation of a clamped–clamped nanorod with the periodic nanostructure model will be investigated. The static equilibrium equaticn of the nanorod was obtained in the previous section as

$$AE\left[f_1(\theta)u''(x) + \frac{\eta^2}{12}f_2(\theta)u^{IV}(x)\right] = -f \qquad [3.38]$$

This static equilibrium equation can be defined in the dimensionless form as follows:

$$f_1(\theta)\bar{u}'' + f_2(\theta)\beta^2\bar{u}^{IV} = \bar{f} \qquad [3.39]$$

The dimensionless parameters are

$$\bar{f} = \frac{-fL}{EA}, \beta^2 = \frac{\eta^2}{12L^2}, \bar{x} = \frac{x}{L}, \bar{u} = \frac{u}{L} \quad [3.40]$$

The axial load, $\bar{f}$, is known and can, therefore, be represented in the following Fourier-type sine-series form:

$$f(x) = \sum_{n=1}^{\infty} Q_n sin\frac{n\pi x}{L}, \quad F_n = \frac{2}{L}\int_0^L f(x) sin\frac{n\pi x}{L} \quad [3.41]$$

where the dimensionless coefficients $\bar{F}_n$ can be defined as

$$\bar{f}(\bar{x}) = \bar{f}_0, \ \bar{F}_n = \frac{4\bar{f}_0}{n\pi}, \qquad n = 1,3,5, ... \quad [3.42]$$

and the clamped–clamped boundary conditions are exactly satisfied by the following choice of the displacement field:

$$u(x) = \sum_{n=1}^{\infty} u_m sin\frac{n\pi x}{L} \quad [3.43]$$

The dimensionless form of the displacement field can be written as follows:

$$\bar{u}(\bar{x}) = \sum_{n=1}^{\infty} \bar{u}_m sin n\pi\bar{x} \quad [3.44]$$

The appropriate use of equations [3.39], [3.42] and [3.44], leads to the analytical solution of equation [3.39] in the following form:

$$\bar{u} = \sum_{n=1}^{\infty} \frac{4\bar{f}_0}{n\pi[f_1(\theta)(n\pi)^2 - f_2(\theta)\beta^2(n\pi)^4]} sin n\pi\bar{x} \quad [3.45]$$

It can be seen from equation [3.45] that setting β=0 gives the classical elasticity solution.

3.4.2. *Vibration analysis of nanorods with periodic nanostructures*

The equation of motion of a nanorod with periodic nanostructures can be defined without considering the axial load *f*:

$$AE\left[f_1(\theta)u''(x,t) + \frac{\eta^2}{12} f_2(\theta)u^{IV}(x,t)\right] = \rho A\ddot{u}(x,t) \quad [3.46]$$

For the present study, a Navier-type solution is presented for clamped–clamped boundary conditions in nanorods. Thus, the following choice of a displacement field is satisfied for clamped–clamped boundary conditions:

$$u(x) = \sum_{n=1}^{\infty} u_m cos \frac{n\pi x}{L} \cos \omega t \quad [3.47]$$

where u_m is the axial displacement with respect to the middle axis of the rod and ω is the natural frequency. Upon substituting equation [3.47] into equation [3.46], the axial frequency parameter can be obtained in the dimensionless form as

$$\lambda = \sqrt{n^2\pi^2 f_1(\theta) - f_2(\theta)\beta^2 n^4 \pi^4} \quad [3.48]$$

where λ^2 represents the dimensionless frequency parameter defined as

$$\lambda^2 = \frac{\rho\omega^2 L^2}{E} \quad [3.49]$$

3.4.3. *Axial wave propagation in nanorods with periodic nanostructures*

This section deals with axial wave propagation in a nanorod with periodic nanostructures. Axial displacement can be defined as follows:

$$u = ae^{i(kx-\omega t)} \quad [3.50]$$

where t represents the time, a represents the amplitude of the wave, and k represents the wavenumber. Substituting equation [3.50] into equation [3.46] leads to the wave characteristic equation, and the wave frequency in DM theory can be obtained in the following form:

$$\omega = c\sqrt{k^2 f_1(\theta) - \frac{\eta^2}{12} f_2(\theta) k^4} \quad [3.51]$$

where

$$c = \sqrt{\frac{E}{\rho}} \quad [3.52]$$

Then, the phase (V_p) and group (V_g) velocities of the nanorods can be obtained for the DM model as follows:

$$V_p = c\sqrt{f_1(\theta) - \frac{\eta^2}{12} f_2(\theta) k^2} \quad [3.53]$$

$$V_g = \frac{c}{2}\left(f_1(\theta)k^2 - \frac{\eta^2}{12}f_2(\theta)k^4\right)^{-1/2}\left(2f_1(\theta)k - 4f_2(\theta)\frac{\eta^2}{12}k^3\right) \quad [3.54]$$

It is clearly seen that phase velocities and group velocities take different values with different wavenumbers. This makes dispersive waves in nanorods. Furthermore, neglecting the length scale parameter (η) in equations [3.51], [3.53] and [3.54] generates the classical elasticity solution of nanorods with periodic nanostructures. The phase and group velocities of the nanorods are the same in this case, and non-dispersive waves in nanorods occur for the classical elasticity solution.

3.4.4. *Flexural deformation of nanobeams with periodic nanostructures*

The bending of a simply supported nanobeam with the periodic nanostructure model is investigated via DM theory. The static governing equation of the nanobeam was obtained in the previous section can be re-expressed as

$$EI\left[f_1(\theta)w^{IV}(x) + \frac{\eta^2}{12}f_2(\theta)w^{VI}(x)\right] = -f(x) \quad [3.55]$$

The distributed load, $f(x)$ can be represented in the following Fourier-type sine-series form:

$$f(x) = \sum_{n=1}^{\infty} F_n sin\frac{n\pi x}{L}, \quad F_n = \frac{2}{L}\int_0^L f(x)sin\frac{n\pi x}{L}dx \quad [3.56]$$

where the dimensionless coefficients $\bar{F}_n$ can be defined as

$$\bar{f}(\bar{x}) = \bar{f}_0, \ \bar{F}_n = \frac{4\bar{f}_0}{n\pi}, \qquad n = 1,3,5, ... \quad [3.57]$$

and the simply supported boundary conditions are satisfied exactly by the following choice of displacement field:

$$w(x) = \sum_{n=1}^{\infty} w_m sin\frac{n\pi x}{L} \quad [3.58]$$

The dimensionless form of the displacement field can be expressed as follows:

$$\bar{w}(\bar{x}) = \sum_{n=1}^{\infty} \bar{w}_m sinn\pi\bar{x} \quad [3.59]$$

Upon successively using equations [3.56], [3.57] and [3.59], the analytical solution of equation [3.55] can be obtained in the dimensionless form as follows:

$$\bar{w} = \sum_{n=1}^{\infty} \frac{4\bar{f}_0}{n\pi[f_1(\theta)(n\pi)^4 - f_2(\theta)\beta^2(n\pi)^6]} sinn\pi\bar{x} \quad [3.60]$$

where the dimensionless flexural displacement is defined as

$$\overline{w} = \frac{EI}{f_0 L^4} w \qquad [3.61]$$

It is noted that neglecting the length scale parameter in equation [3.60] leads to the classical elasticity solution.

3.4.5. *Buckling analysis of nanobeams with periodic nanostructures*

The buckling analysis of a simply supported nanobeam with length L is investigated computationally with respect to governing equation, which is written as follows:

$$EIf_1(\theta)w^{IV} + \frac{\eta^2}{12} f_2(\theta)EIw^{VI} + Nw'' = 0 \qquad [3.62]$$

where N represents the in-plane load that is subjected to both ends of the beam. The following solution of equation [3.62] is assumed to be valid:

$$w = \sum_{n=1}^{\infty} w_n \, sin\frac{n\pi x}{L} \qquad [3.63]$$

Substituting equation [3.63] into equation [3.62] leads to the critical buckling load in dimensionless form:

$$\overline{N} = f_1(\theta)(n\pi)^2 - \beta^2 f_2(\theta)(n\pi)^4 \qquad [3.64]$$

where the dimensionless critical buckling load, $\overline{N}$, can be defined as

$$\overline{N} = \frac{PL^2}{EI} \qquad [3.65]$$

From equation [3.64], it can be concluded that DM predicts lower critical buckling loads than the classical theory of elasticity.

3.4.6. *Vibration analysis of nanobeams with periodic nanostructures*

The equation of motion of a nanobeam with periodic nanostructures can be defined without considering the distributed load $f(x)$:

$$EIf_1(\theta)w^{IV} + \frac{\eta^2}{12} f_2(\theta)EIw^{VI} + \rho A\ddot{w} = 0 \qquad [3.66]$$

The basis of the Navier-type solution, following the choice of a displacement field, is satisfied for simply supported boundary conditions:

$$w(x,t) = \sum_{n=1}^{\infty} W_m \, sin\frac{n\pi x}{L} sin\omega t \quad [3.67]$$

where W_m is the flexural deflection with respect to the beam middle axis. Upon substituting equation [3.67] into equation [3.66], the axial frequency parameter can be obtained in the dimensionless form as:

$$\Omega = \sqrt{f_1(\theta)(n\pi)^4 - \beta^2 f_2(\theta)(n\pi)^6} \quad [3.68]$$

where the dimensionless frequency parameter, Ω, is

$$\Omega^2 = \frac{\rho A \omega^2 L^4}{EI} \quad [3.69]$$

It is obvious that DM predicts lower frequencies compared to the classical theory of elasticity (equation [3.68]).

3.4.7. *Flexural wave propagation in nanobeams with periodic nanostructures*

Flexural wave propagation analysis in a nanobeam with periodic nanostructures is investigated based on DM. The flexural displacement of the wave can be defined as follows:

$$w = ae^{i(kx-\omega t)} \quad [3.70]$$

By substituting equation [3.70] into equation [3.66], the wave characteristics equation and wave frequency in DM theory can be obtained as

$$\omega = d\sqrt{k^4 f_1(\theta) - \frac{\eta^2}{12} f_2(\theta) k^6} \quad [3.71]$$

where

$$d = \sqrt{\frac{EI}{\rho A}} \quad [3.72]$$

Then, the phase (V_p) and group (V_g) velocities of the nanobeams can be obtained for the DM model as follows:

$$V_p = dk\sqrt{f_1(\theta) - \frac{\eta^2}{12} f_2(\theta)k^2} \qquad [3.73]$$

$$V_g = \frac{d}{2}\left(f_1(\theta)k^4 - \frac{\eta^2}{12} f_2(\theta)k^6\right)^{-1/2}\left(4f_1(\theta)k^3 - 6f_2(\theta)\frac{\eta^2}{12}k^5\right) \qquad [3.74]$$

Similar to wave propagation in nanorods, dispersive waves in nanobeams are seen in the DM theory. It should be also noted that neglecting the length scale parameter (η) in equations [3.70], [3.73] and [3.74] leads to the classical elasticity solution of nanobeams with periodic nanostructures. In this way, non-dispersive waves in nanobeams are seen, and the geometric shape of the waves does not change during the propagation for classical elasticity theory.

3.5. Numerical results

All of the numerical results for different chiral angles (which take θ=0°, 15° and 30°, based on DM theory) are presented and discussed in this section. By virtue of the chiral angle, θ changes the mechanical behavior of periodic nanostructures. In this context, Figure 3.2 demonstrates the axial dimensionless displacement of the nanorod with different chiral angles (θ). The DM results are compared to the classical elasticity solution for L=1.5 nm in Figure 3.2. The stiffening and softening effects of DM depend on the chiral angle of the considered simple square crystal lattice. While for θ=0° and 15°, DM predicts lower displacement compared to classical elasticity theory. However, higher displacements are obtained by DM for θ=30°. This explains that atomic structure of the material directly affects the mechanical behavior of the material in DM theory. The difference between the two theories becomes maximum at the middle of the nanorod for all chiral angles. Figure 3.3 illustrates the dimensionless axial frequency parameter with different rod lengths for the first three modes. The nanorods become stiffer for θ=0° and 15°, and the difference between DM and classical theory is more apparent for higher mode numbers. This is because decreasing the wavelength ($2\pi/k$) increases the mode number, and the scale effects are more pronounced for smaller wavelengths. However, softening material behavior can be obtained in the case of θ=30°. The convergence of both theories is better when compared to the other values of chiral angles (θ=0° and 15°). Both theories converge with the increase in rod length, since the gradient effects disappear with increasing rod length. Vibration frequencies are presented with the different mode numbers in Figure 3.4 for L=3 nm. The difference between the two theories increases with the increase in the mode number, and in the case of the stiffening of the rod, both theories give very different frequency results.

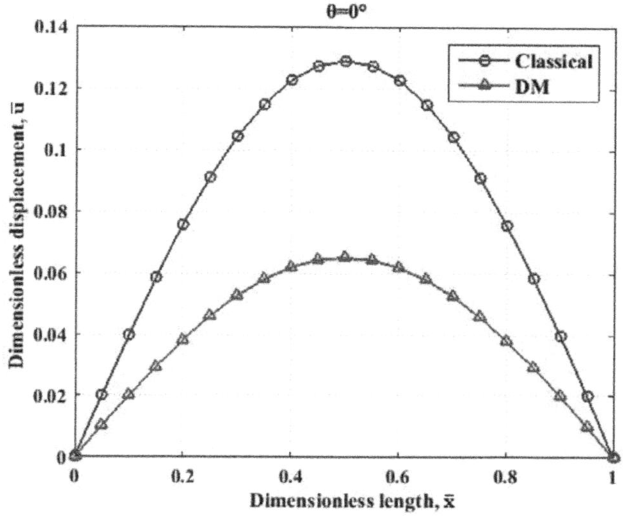

a) η=0.15 nm, L=1.5 nm

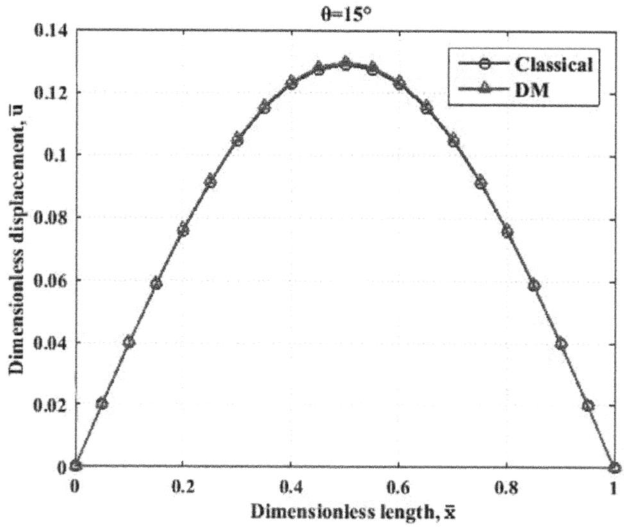

b) η=0.15 nm, L=1.5 nm

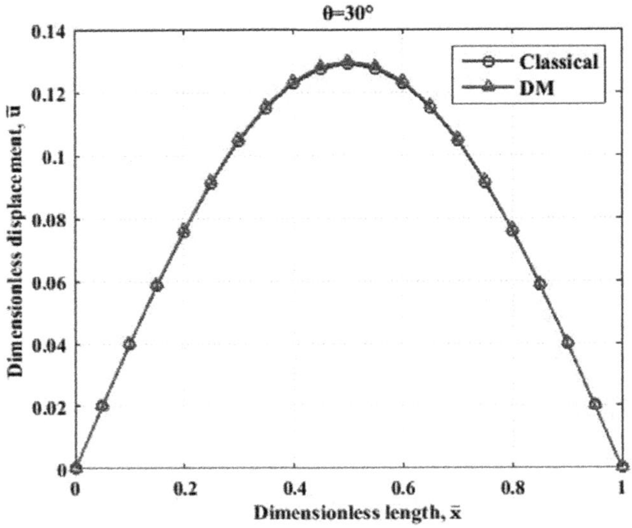

c) η=0.15 nm, L=1.5 nm

Figure 3.2. *Dimensionless displacement of nanorods with dimensionless rod length. For a color version of this figure, see www.iste.co.uk/challamel/mechanics2.zip*

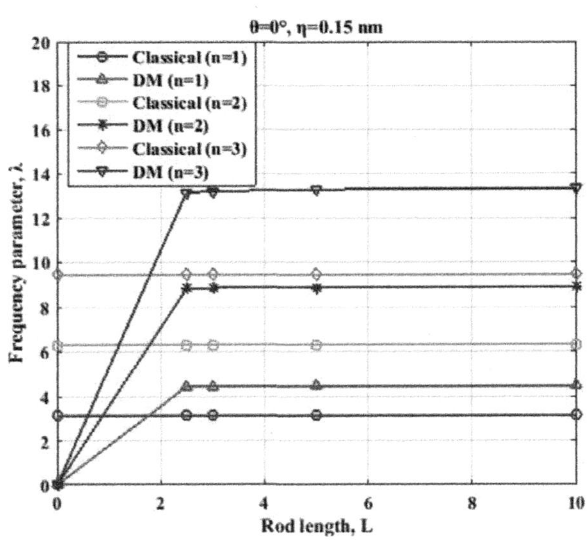

a)

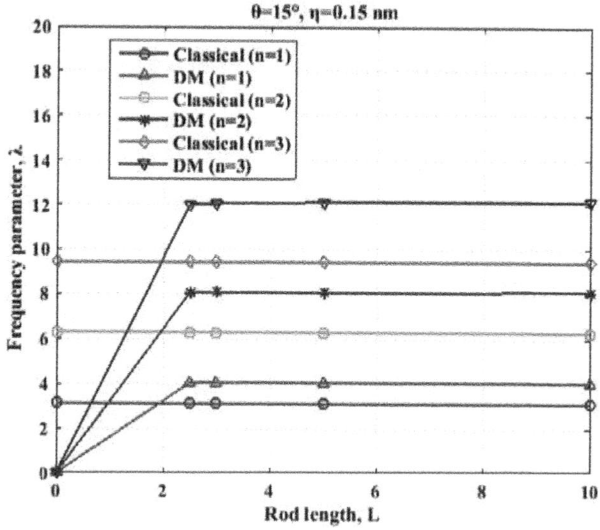

b)

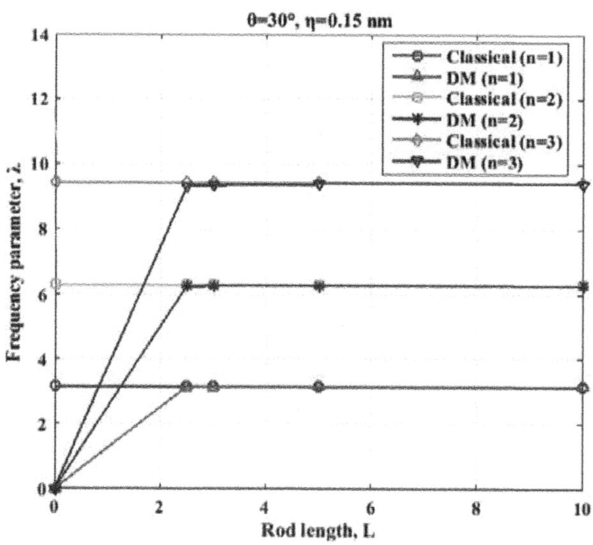

c)

Figure 3.3. *Frequency parameter of nanorods with rod length for the first three mode numbers. For a color version of this figure, see www.iste.co.uk/challamel/mechanics2.zip*

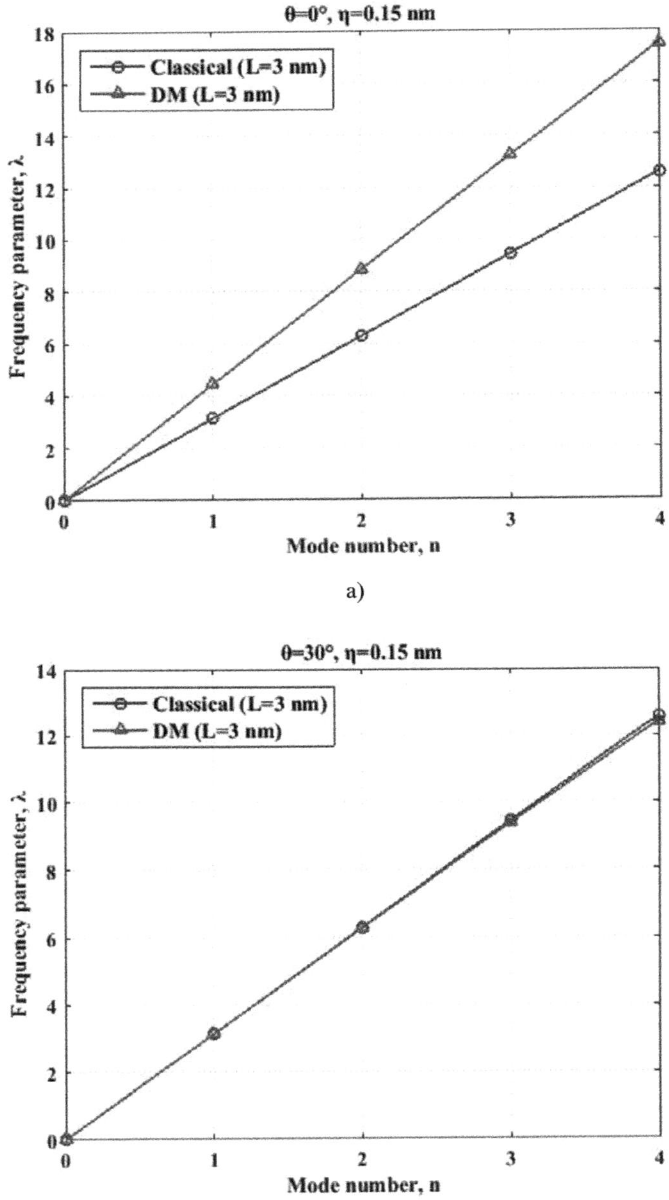

Figure 3.4. *Frequency parameter of nanorods with mode number. For a color version of this figure, see www.iste.co.uk/challamel/mechanics2.zip*

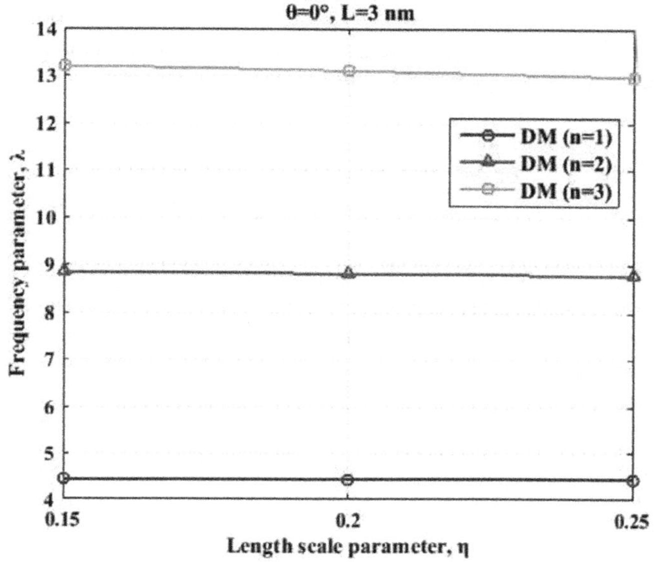

a)

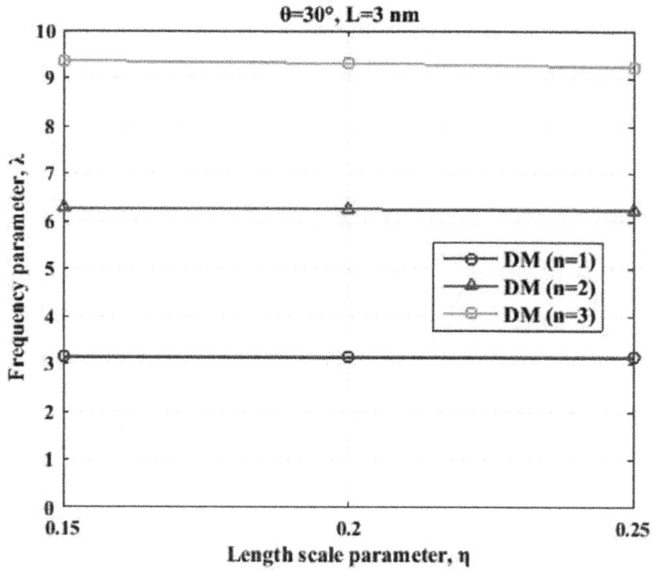

b)

Figure 3.5. *Frequency parameters of nanorods with doublet separation distance (η). For a color version of this figure, see www.iste.co.uk/challamel/mechanics2.zip*

In Figure 3.5, the variation of vibration frequencies with doublet separation distance, η is shown for three modes in DM theory. Axial vibration frequencies always decrease with the increase in the length scale parameter. This is due to the rise of inhomogeneity in the nanorods.

Another remarkable observation relates to wave dispersion relations of nanorods with a periodic nanostructure, as shown in Figures 3.6–3.8. In these figures, the density of the rod is equal to ρ=2300 kg/m^3 and the elasticity modulus is equal to E=1 TPa. Wave frequencies of the classical model linearly increase with the increase in the wavenumber, as shown in Figure 3.6. However, wave frequencies increase with limited wavenumber, and waves do not propagate after a certain wavenumber near to the first Brillouin zone (π/η) for DM and lattice dynamics (LD). In the softening case of the DM (θ=30°), wave frequencies are in perfect agreement with LD, and for while k=2x10^{10} 1/m, DM and LD give identical numerical results. In the stiffening cases of the DM (θ=0° and 15°), LD agrees well with the DM, especially for short wavelengths.

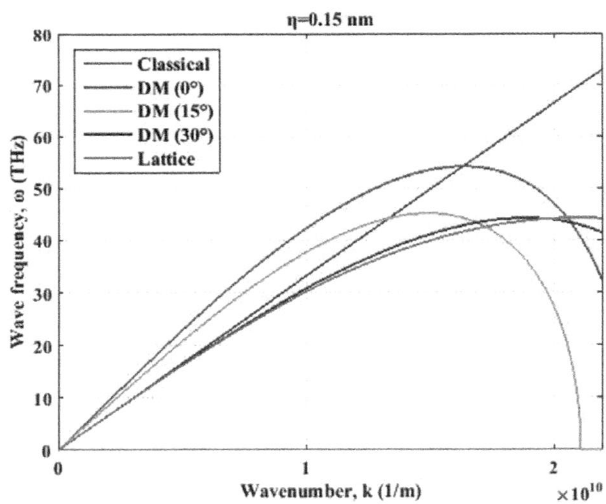

Figure 3.6. *Wave frequency of nanorods with wavenumber. For a color version of this figure, see www.iste.co.uk/challamel/mechanics2.zip*

Figures 3.7 and 3.8 illustrate the phase and group velocities when they go to zero, indicating localization and a stationary case. Similar to wave frequencies, the best agreement is obtained between the LD and the softening case of the DM (θ=30°) theory for η=0.15 nm. The accuracy of the DM theory with θ=15° for phase velocity is better than the DM theory with θ=0°, as LD is regarded for the given limit of the wavenumber. At the end of the first Brillouin zone, the softening case of the DM gives identical group velocity results to LD.

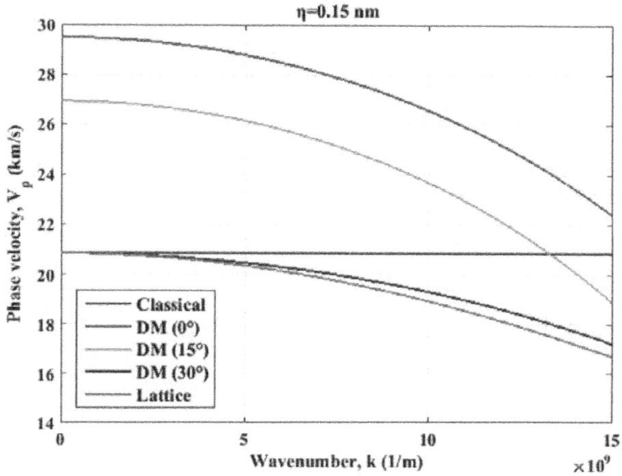

Figure 3.7. *Phase velocity of nanorods with wavenumber. For a color version of this figure, see www.iste.co.uk/challamel/mechanics2.zip*

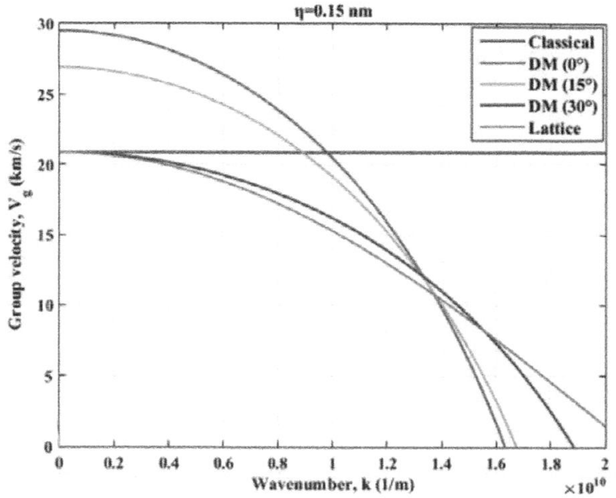

Figure 3.8. *Group velocity of nanorods with wavenumber. For a color version of this figure, see www.iste.co.uk/challamel/mechanics2.zip*

The similar static behaviors of nanobeams with nanorods presented in Figure 3.2 are obtained in Figure 3.9. It is seen here that the magnitude of the flexural displacement obtained by DM is lower than the classical elasticity model for θ=0° and θ=15°. This establishes the stiffening of nanobeams with respect to classical theory. However, the softening of nanobeams emerges for θ=30° in DM theory.

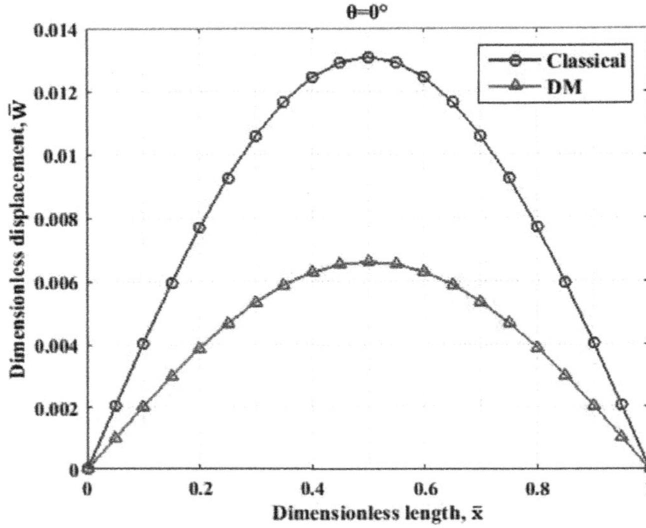

a) η=0.15 nm, L=1.5 nm

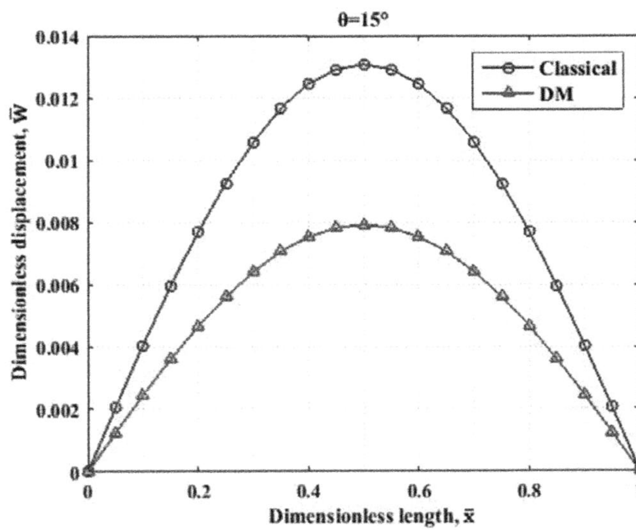

b) η=0.15 nm, L=1.5 nm

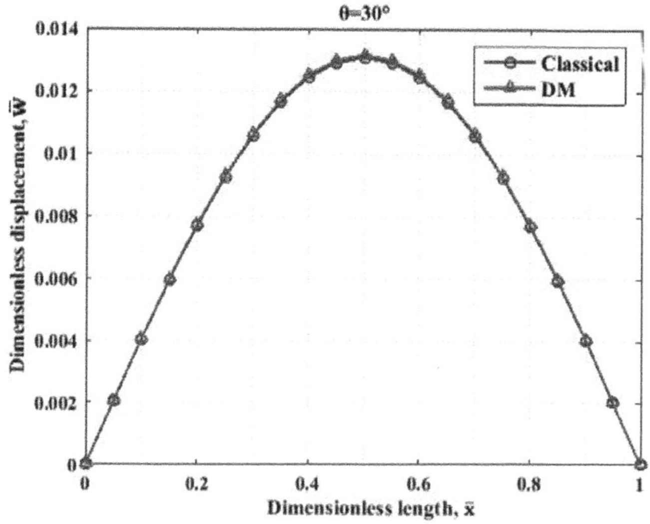

c) η=0.15 nm, L=1.5 nm

Figure 3.9. *Dimensionless displacement of nanobeams with dimensionless beam length. For a color version of this figure, see www.iste.co.uk/challamel/mechanics2.zip*

The variation of the critical buckling load with the beam length is illustrated in Figure 3.10. In the case of the softening of nanobeams (θ=30°), the DM curve converges to the classical elasticity model, with the increase in beam length. However, DM predicts a higher critical buckling load than classical elasticity for θ=0° and 15°, and, in this case, the difference between the two theories increases with the increase in beam length. Flexural vibration frequencies are shown for the first three modes in DM and classical elasticity theories in Figure 3.11. As it is similar to nanorod vibration, the stiffening of nanobeams occurs for θ=0°, whereas the softening of nanobeams occurs for θ=30° in DM theory. The difference between the two theories is more pronounced in the case of the stiffening of nanobeams and higher modes of vibration. However, in the case of the softening of nanobeams (θ=30°), this difference becomes small and DM results converge to classical theory with the increase in beam length. In this context, it is instructive for someone to observe that length scale parameter plays a more important role when θ=0° and 15°, compared to when θ=30° in the present DM theory.

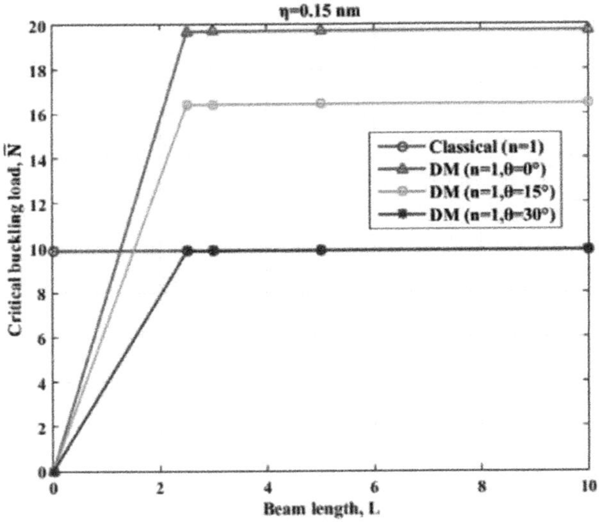

Figure 3.10. *Critical buckling load of nanobeams with beam length. For a color version of this figure, see www.iste.co.uk/challamel/mechanics2.zip*

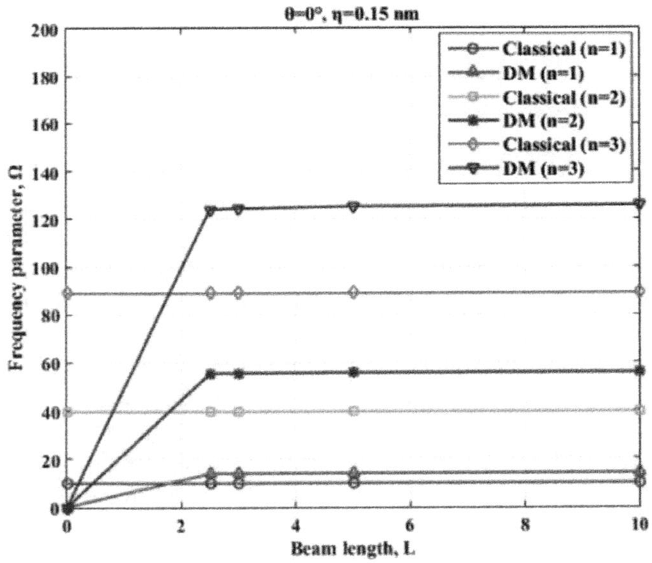

a)

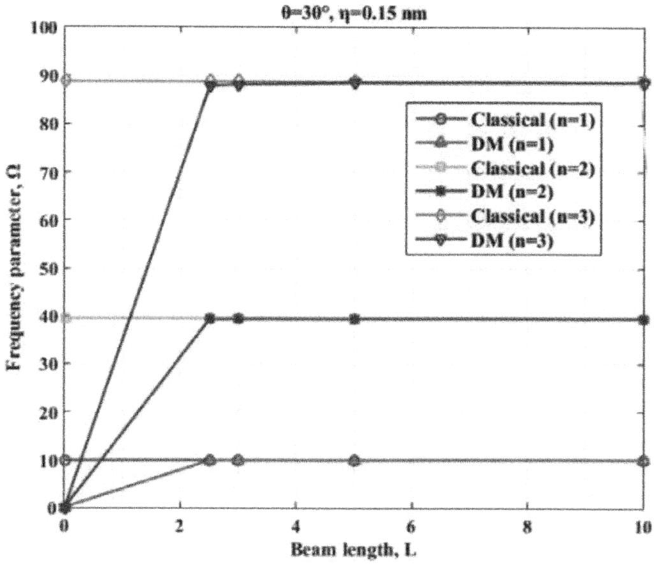

b)

Figure 3.11. *Frequency parameter of nanobeams with beam length for the first three mode numbers. For a color version of this figure, see www.iste.co.uk/challamel/mechanics2.zip*

Finally, the flexural wave dispersion relations of nanobeams are depicted in Figures 3.12–3.14. The material and geometrical properties are assumed to be: ρ=2300 kg/m^3, E=1 TPa, with the thickness of the nanobeam as t=0.34 nm, and the inner radius of the nanobeam as R_1=0.5 nm. Thus, the moment of inertia and the area of the cross-section of the beam can be computed as follows:

$$I = \frac{\pi(R_2^4 - R_1^4)}{4}, \; A = \pi(R_2^2 - R_1^2) \qquad [3.75]$$

where R_2 denotes the outer radius of the nanobeam. Similar trends are observed in flexural wave dispersion curves, with axial wave dispersion curves shown for DM theory. Accordingly, it is seen that wave frequencies obtained by DM tend to zero at the end of the first Brillouin zone (Figure 3.12). An exception to this trend is that flexural wave frequencies of the classical elasticity model increase with the increase in the wavenumber without any limit. The present flexural phase and group velocities are again, similar to their counterparts, are shown in Figures 3.7 and 3.8 for DM theory. The phase and group velocities converge to zero after a certain maximum wavenumber in DM theory for all chiral angles. However, these velocities increase with the increase in the wavenumber in the classical elasticity theory

(Figures 3.13 and 3.14). This behavior confirms that DM is a more physical theory, with Van Hove singularities.

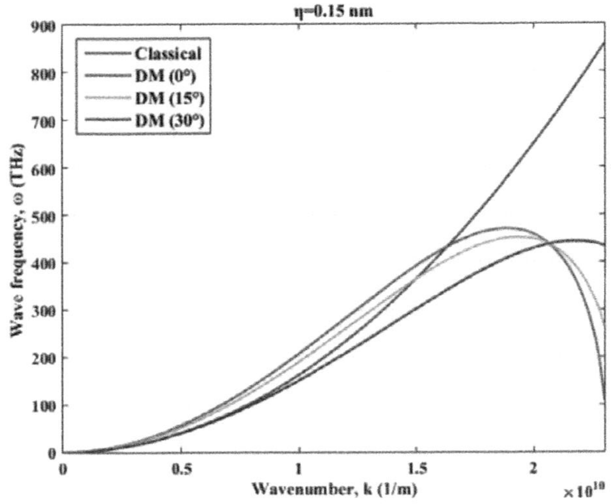

Figure 3.12. *Wave frequency of nanobeams with wavenumber*

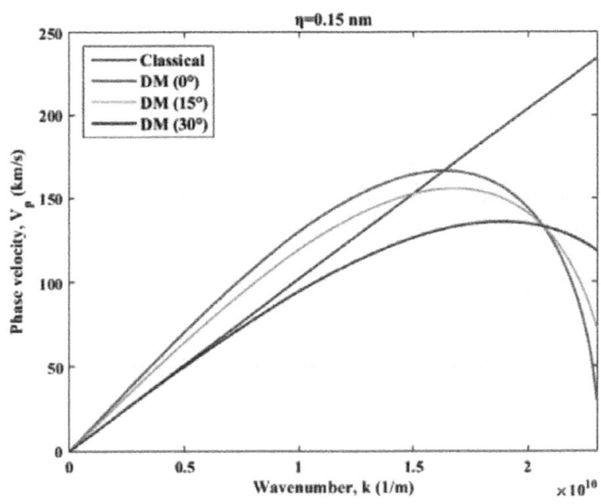

Figure 3.13. *Phase velocity of nanobeams with wavenumber*

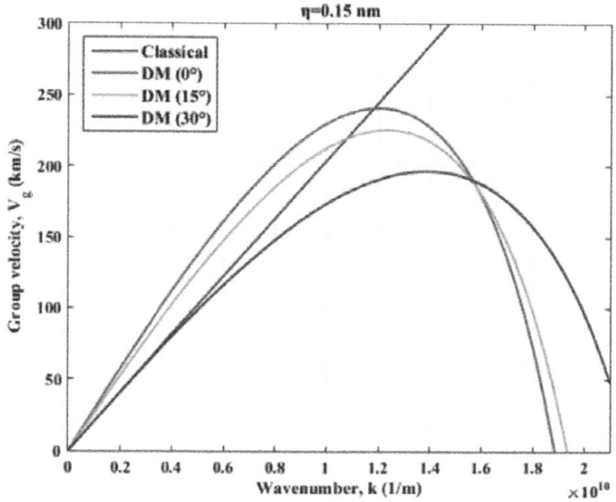

Figure 3.14. *Group velocity of nanobeams with wavenumber*

3.6. Conclusion

The presented DM formulation accounts for the static and dynamic analyses of nanorods and nanobeams with periodic nanostructures. After deriving the governing differential equations of a simple square crystal lattice of nanostructures, useful analytical solutions are obtained based on the DM model. By varying the chiral angle and the length scale parameter in DM, the stiffening and softening of nanorods and nanobeams occur compared to the classical elasticity theory. This proves that the mechanical behavior of nanomaterials is precisely related to the atomic structure of the considered material in size-dependent DM theory. This proof has made it easier to understand that the DM model is not phenomenological in contrast to other scale effect continuum theories like the nonlocal theory, stress gradient theories, peridynamics and so on. In this context, the value of the chiral angle of the simple square crystal lattice nanostructures plays the significant role of whether the nanostructures are stiff or soft. It is seen, in this context, that DM theory is in better agreement with LD results in the wave propagation of nanorods compared to the classical elasticity theory. This behavior can be seen in the case of the softening of the rod and beam for all the wavelengths, as well as in the case of the softening situation of DM for short wavelengths. The difference between the DM and classical elasticity results is more apparent for the higher modes of vibration and in the case of the stiffening of nanostructures. It is finally worth mentioning that an advantageous approach may be provided, by justifying the stiffening and softening effects of nanostructures.

3.7. References

Aifantis, E.C. (1992). On the role of gradients in the localization of deformation and fracture. *Int. J. Eng. Sci.*, 30, 1279–1299.

Altan, B.S. and Aifantis, E.C. (1997). On some aspects in the special theory of gradient elasticity. *J. Mech. Behav. Mater.*, 8, 231–282.

Apuzzo, A., Barretta, R., Luciano, R., Marotti de Sciarra, F., Penna, R. (2017). Free vibrations of Bernoulli–Euler nano-beams by the stress-driven nonlocal integral model. *Compos. Part B*, 123, 105–111.

Askes, H., Suiker, A.S.J., Sluys, L.J. (2002). A classification of higher-order strain-gradient models – Linear analysis. *Arch. Appl. Mech.*, 72, 171–188.

Aydogdu, M. and Gul, U. (2018a). Buckling analysis of double nanofibers embedded in an elastic medium using doublet mechanics theory. *Compos. Struct.*, 202, 355–363.

Aydogdu, M. and Gul, U. (2018b). Axial wave reflection and transmission in stepped nanorods using doublet mechanics theory. *MATEC Web of Conferences*, 148, 15002.

Barretta, R., Feo, L., Luciano, R., Marotti de Sciarra, F. (2016). Application of an enhanced version of the Eringen differential model to nanotechnology. *Compos. Part B*, 96, 274–280.

Challamel, N. and Wang, C.M. (2008). The small length scale effect for a non-local cantilever beam: A paradox solved. *Nanotechnology*, 19, 345703.

Chebakov, R., Kaplunov, J., Rogerson, G.A. (2016). Refined boundary conditions on the free surface of an elastic half-space taking into account non-local effects. *Proc. Roy. Soc. London A*, 472, 2186.

Chebakov, R., Kaplunov, J., Rogerson, G.A. (2017). A nonlocal asymptotic theory for thin elastic plates. *Proc. Roy. Soc. A*, 473, 20170249.

Cosserat, E. and Cosserat, F. (1909). *Sur la théorie des corps déformables*. Herman, Paris [in French].

Eringen, A.C. (1966). Linear theory of micropolar elasticity. *J. Math. Mech.*, 15(6), 909–923.

Eringen, A.C. (1972a). Nonlocal polar elastic continua. *Int. J. Eng. Sci.*, 10(1), 1–16.

Eringen, A.C. (1972b). Linear theory of nonlocal elasticity and dispersion of plane waves. *Int. J. Eng. Sci.*, 10(5), 425–435.

Eringen, A.C. (1983). On differential equations of nonlocal elasticity and solutions of screw dislocation and surface waves. *J. Appl. Phys.*, 54, 4703–4710.

Fernandez-Saez, J., Zaera, R., Loya, J.A., Reddy, J.N. (2016). Bending of Euler–Bernoulli beams using Eringen's integral formulation: A paradox resolved. *Int. J. Eng. Sci.*, 99, 107–116.

Ferrari, M., Granik, V.T., Imam, A., Nadeau, J. (1997). *Advances in Doublet Mechanics*. Springer, Berlin, Heidelberg.

Fleck, N.A., Muller, G.M., Ashby, M.F., Hutchinson, J.W. (1994). Strain gradient plasticity: Theory and experiment. *Acta Metall. et Mater.*, 42(2), 475–487.

Gentile, F., Sakamoto, J., Righetti, R., Decuzzi, P., Ferrari, M. (2011). A doublet mechanics model for the ultrasound characterization of malignant tissues. *J. Biomed. Sci. Eng.*, 4, 362–374.

Goldenveizer, A.L., Kaplunov, J.D., Nolde, E.V. (1993). On Timoshenko–Reissner type theories of plates and shells. *Int. J. Solids Struct.*, 30(5), 675–694.

Granik, V.T. (1978). Microstructural mechanics of granular media. Technique Report IM/MGU, Moscow State Univ. Mech. Eng., 78–241.

Granik, V.T. and Ferrari, M. (1993). Microstructural mechanics of granular media. *Mech. Mater.*, 15, 301–322.

Gul, U. and Aydogdu, M. (2017). Wave propagation in double walled carbon nanotubes by using doublet mechanics theory. *Phys. E*, 93, 345–357.

Gul, U. and Aydogdu, M. (2018a). Structural modelling of nanorods and nanobeams using doublet mechanics theory. *Int. J. Mech. Mater. Des.*, 14, 195–212.

Gul, U. and Aydogdu, M. (2018b). Noncoaxial vibration and buckling analysis of embedded double-walled carbon nanotubes by using doublet mechanics. *Compos. Part B*, 137, 60–73.

Gul, U. and Aydogdu, M. (2018c). Vibration analysis of Love nanorods using doublet mechanics theory. *J. Braz. Soc. Mech. Sci.*, 41, 351.

Gul, U., Aydogdu, M., Gaygusuzoglu, G. (2017). Axial dynamics of a nanorod embedded in an elastic medium using doublet mechanics. *Compos. Struct.*, 160, 1268–1278.

Gul, U., Aydogdu, M., Gaygusuzoglu, G. (2018). Vibration and buckling analysis of nanotubes (nanofibers) embedded in an elastic medium using doublet mechanics. *J. Eng. Math.*, 109, 85–111.

Kaplunov, J.D., Nolde, E.V., Rogerson, G.A. (2000). A low-frequency model for dynamic motion in pre-stressed incompressible elastic structures. *Proc. Roy. Soc. London A*, 456(2003), 2589–2610.

Kaplunov, J., Nolde, E., Rogerson, G.A. (2006). An asymptotic analysis of initial-value problems for thin elastic plates. *Proc. Roy. Soc. London A*, 462(2073), 2541–2561.

Kojic, M., Vlastelica, I., Decuzzi, P., Granik, V.T., Ferrari, M. (2011). A finite element formulation for the doublet mechanics modeling of microstructural materials. *Comput. Methods Appl. Mech. Eng.*, 200, 1446–1454.

Lam, D.C.C., Yang, F., Chong, A.C.M., Wang, J., Tong, P. (2003). Experiments and theory in strain gradient elasticity. *J. Mech. Phys. Solids*, 51, 1477–1508.

Li, C., Yao, L., Chen, W., Li, S. (2015). Comments on nonlocal effects in nano-cantilever beams. *Int. J. Eng. Sci.*, 87, 47–57.

Lin, S.S. and Shen, Y.C. (2005). Stress fields of a half-plane caused by moving loads-resolved using doublet mechanics. *Soil Dynam. Earthquake Eng.*, 25, 893–904.

Mindlin, R.D. and Eshel, N.N. (1968). On first strain-gradient theories in linear elasticity. *Int. J. Solids Struct.*, 4, 109–124.

Mindlin, R.D. and Tiersten, H.F. (1964). Micro-structure in linear elasticity. *Arch. Ration. Mech. Analy*, 16, 51–78.

Romano, G. and Barretta, R. (2016). Comment on the paper "Exact solution of Eringen's nonlocal integral model for bending of Euler–Bernoulli and Timoshenko beams" by Meral Tuna & Mesut Kirca. *Int. J. Eng. Sci.*, 109, 240–242.

Romano, G. and Barretta, R. (2017). Stress-driven versus strain-driven nonlocal integral model for elastic nano-beams. *Compos. Part B*, 114, 184–188.

Romano, G., Barretta, R., Diaco, M. (2017). On nonlocal integral models for elastic nano-beams. *Int. J. Mech. Sci.*, 131–132, 490–499.

Peddieson, J., Buchanan, G.R., McNitt, R.P. (2003). Application of nonlocal continuum models to nanotechnology. *Int. J. Eng. Sci.*, 41(3–5), 305–312.

Silling, S.A. (2000). Reformulation of elasticity theory for discontinuities and long-range forces. *J. Mech. Phys. Solids*, 48(1), 175–209.

Sudak, L.J. (2003). Column buckling of multiwalled carbon nanotubes using nonlocal continuum mechanics. *J. Appl. Phys.*, 94, 7281.

Vajari, A.F. (2018). A new method for evaluating the natural frequency in radial breathing like mode vibration of double-walled carbon nanotubes. *ZAMM-J. Appl. Math. Mech.*, 98(2), 255.

Vajari, A.F. and Imam, A. (2016a). Axial vibration of single-walled carbon nanotubes using doublet mechanics. *Indian J. Phys.*, 90(4), 447–455.

Vajari, A.F. and Imam, A. (2016b). Torsional vibration of single-walled carbon nanotubes using doublet mechanics. *ZAMP-J. Appl. Math. Phys.*, 67, 81.

Vajari, A.F. and Imam, A. (2016c). Analysis of radial breathing mode of vibration of single-walled carbon nanotubes via doublet mechanics. *ZAMM-J. Appl. Math. Mech.*, 96(9), 1020–1032.

4

Free Vibration of Micro-Beams and Frameworks Using the Dynamic Stiffness Method and Modified Couple Stress Theory

There are many practical engineering applications of micro-beams and micro-frames. For example, the micro-beam theory plays a very important role in the design and development of micro-electromechanical systems. In such applications, the usually adopted classical beam theory that does not account for the small-scale effect is inadequate. This inadequacy becomes even more pronounced when predicting the dynamic behavior of micro-beams and micro-frames. This has motivated researchers to investigate the static, dynamic and buckling characteristics and other performances of micro-beams. Over the past two decades, considerable efforts have been made to investigate the free vibration behavior of micro-beams and micro-frames using various continuum models. Foremost among these models are those that focus on non-local elasticity, classical couple stress, gradient elasticity and modified couple stress theories. The work described in this chapter deals with the free vibration problems of micro-beams and frameworks based on the dynamic stiffness method through the implementation of the modified couple stress theory (MCST). The advantages of the MCST are well known. An important aspect of the MCST is that, unlike other theories, it uses only one material length scale parameter to account for the smallness of the structure. Essentially, the MCST is the underlying scheme based on which the dynamic stiffness theory is developed for a micro-beam in this chapter. This is principally achieved by first deriving the governing differential equations of motion of the micro-beam in free vibration using Hamilton's principle. The differential equations are then solved in closed analytical

Chapter written by J.R. BANERJEE.

form. Finally, the dynamic stiffness matrix of the micro-beam is formulated by relating the amplitudes of the forces to those of the corresponding displacements at the ends of the beam. The theory is applied in conjunction with the Wittrick–Williams algorithm as a solution technique to investigate the free vibration characteristics of micro-beams and frameworks. Results for the natural frequencies and mode shapes of several examples of micro-beams and micro-frames for different boundary conditions are presented and discussed. The method is validated by some comparative results in the literature. The effects of the length scale parameter on the free vibration characteristics of micro-beams and micro-frames are demonstrated with significant conclusions.

4.1. Introduction

Micro-beams and micro-frames have abundant applications right across the board, particularly in the design and development of micro-electromechanical systems for which the microstructure-dependent size effect is an important parameter to consider (see, for example, Attia *et al.* (1998); Li *et al.* (2003); Pei *et al.* (2004); Akgoz and Civalek (2012); Moeenfard and Ahmadian (2013)). There is plenty of experimental evidence that substantiates the importance of the size effect in microstructures, as evident from the literature (Nix 1989; Fleck *et al.* 1994; Ma and Clarke 1995; Stolken and Evans 1998; Chong and Lam 1999; Chong *et al.* 2001; Li *et al.* 2018). Clearly, conventional classical structural theories cannot be meaningfully applied to analyze microstructures because they do not account for the small-scale size effect. Obviously, the application of classical theories to microstructures may lead to unacceptably inaccurate and unreliable results. This is particularly true when predicting the dynamic behavior of microstructures. From an engineering point of view and within the confines of continuum mechanics, a literature survey has shown that for well over a century, research in the area of structural behavior and other performances of microstructures is constantly evolving, and publications have continued to grow. In the context of the present study on micro-beams and micro-frames, a small but carefully selected sample of published literature (Anthoine 2000; Park and Gao 2006; Challamel and Wang 2008; Kong *et al.* 2008; Challamel *et al.* 2015; Chebakov *et al.* 2017; Hache *et al.* 2019a, 2019b) is included here, but for further reading, interested readers are referred to a survey paper (Thai *et al.* 2017) that gives widespread coverage, as well as numerous cross-references on the subject. From a historical perspective, the pioneering contributions of Cosserat and Cosserat (1909), Mindlin and Tiersten (1962), Toupin (1962), Mindlin (1963), Koiter (1964) and Eringen (1966, 1972, 1983) towards the development of micro-continuum theories have been well documented.

Now, referring to the present work, it should be noted that researchers have carried out investigations on the free vibration behavior of micro-beams using various continuum theories of elasticity. Foremost among these are the non-local elasticity theory first proposed by Eringen (1972, 1983) and then further developed by Reddy (2007); the classical couple stress theory introduced by Toupin (1962), Mindlin and Tiersten (1962) and Koiter (1964); the gradient elasticity theory presented by Mindlin (1964, 1965), Lam *et al.* (2003), Kong *et al.* (2009) and Askes and Aifantis (2011); and the modified couple stress theory developed by Yang *et al.* (2002). In recent years, an asymptotic approach, through the application of non-local elasticity, has been developed by Hache *et al.* (2019a) for beam models and by Chebakov *et al.* (2017) and Hache *et al.* (2019b) for plate models. However, the modified couple stress theory (MCST) of Yang *et al.* (2002) in relation to micro-beams is chosen in the present work. This is mainly because, unlike other theories, the MCST uses only one material length scale parameter to account for the smallness of the structure. In essence, the MCST modifies the classical couple stress theory of Mindlin and Tiersten (1962), Toupin (1962) and Koiter (1964) in such a way that it reduces the number of length scale parameters to only one. This particularly useful aspect of the MCST (Yang *et al.* 2002) is exploited in the current research. The main contribution made is the development of the dynamic stiffness method (DSM) for a micro-beam using the MCST and then to extend it to analyze frameworks. However, it should be acknowledged that the proposed dynamic stiffness theory is based on the Bernoulli–Euler hypothesis, and thus it does not account for the effects of shear deformation and rotatory inertia. In this respect, it is worth noting that one of the earliest contributions on micro-beam analysis using the MCST based on the Bernoulli–Euler theory was made by Park and Gao (2006). Although the investigation in their paper was focused on static analysis of a micro-beam using the MCST, they nevertheless successfully demonstrated the size effect on the deflection of a cantilever micro-beam and highlighted the inadequacy of the classical theory. A couple of years later, Kong *et al.* (2008) studied the free vibration behavior of micro-beams using the Bernoulli–Euler theory in conjunction with the MCST. On the contrary, Noori *et al.* (2016) developed a higher-order model for free vibration analysis of micro-beams, and they compared their results computed from the Bernoulli–Euler, Timoshenko and higher-order theories. Mustapha (2020) appears to be the only author who extended the free vibration investigation of individual micro-beams to micro-frames. Investigation using a nonlinear MCST model is outside the scope of this work, but interested readers are referred to the publications of Farokhi *et al.* (2013), Ghayesh *et al.* (2013), Wang et al. (2013) and Simek (2014).

To the best of the author's knowledge, the DSM has not been applied earlier to investigate the free vibration behavior of micro-beams and frameworks. Therefore, the main purpose of this chapter is to develop the dynamic stiffness matrix of a micro-beam and then carry out its free vibration analysis, and finally extend the theory to the free vibration analysis of micro-frames. The investigation is carried out in the following steps. First, the potential and kinetic energies of a micro-beam are formulated using the MCST that accounts for the material length scale parameter. Using the expressions for the kinetic and potential energies, the governing differential equations and the associated natural boundary conditions are derived by applying Hamilton's principle. For harmonic oscillation, the governing differential equations are solved in closed analytical form. Explicit algebraic expressions for the amplitudes of axial displacement, bending displacement, bending rotation, as well as those for the corresponding amplitudes of axial force, shear force and bending moment are obtained from the resulting solution. Next, the dynamic stiffness matrix of the micro-beam is formulated by relating the amplitudes of forces and moments to those of the displacements and rotation at the ends of the micro-beam element. Finally, the well-established Wittrick–Williams algorithm (Wittrick and Williams 1971) is applied to the resulting dynamic stiffness matrix, in order to compute the natural frequencies and mode shapes of some carefully chosen examples that include both individual micro-beams and micro-frames. The effects of the small length scale parameter on the results are demonstrated, and some results are validated using published results.

Readers who are not acquainted with the DSM, but are familiar with the finite element method (FEM), may find the following account of the DSM and its similarities and differences with the FEM useful.

As a universal tool, the FEM is generally used to solve complex structural engineering problems (Rao 2017). When carrying out the free vibration analysis, the element stiffness and mass matrices are assembled in the FEM to form the overall master stiffness and mass matrices of the final structure. Following this procedure, a linear eigenvalue problem is generally set up, yielding the natural frequencies and mode shapes of the structure. Predictably, the number of natural frequencies that can be meaningfully computed in the FEM is limited by the order of the mass and stiffness matrices of the structure. Of course, the inaccuracy will build up when computing higher-order natural frequencies and mode shapes. To overcome this restriction, there is an alternative to the FEM to address the free vibration problem more accurately and efficiently. The proposed method is not restrictive, but at the same time, it always provides accurate results regardless of the order of the natural frequencies and mode shapes. An alternative method is that of the DSM, which is robust and undoubtedly accurate, as demonstrated by Williams and Wittrick (1983), Williams (1993) and Banerjee (2015). When investigating the free vibration behavior of complex structures, the DSM can be used in a wider context like the FEM. Although the DSM is different from the FEM, it nevertheless shares many common features with the FEM. For example, the assembly of the structural properties of individual elements in the DSM is similar to the FEM. However, there are some major differences between the DSM and the FEM. One important difference is that the former is not affected by the number of elements used in the analysis as it always gives exact results, regardless of the number of elements and nodes, whereas the latter is significantly mesh dependent. Thus, the computational accuracy in the FEM depends very much on the number of elements used in the analysis, whereas in the DSM, the accuracy of results is independent of the number of elements. A single structural element can be used in the DSM to compute any number of natural frequencies without losing any accuracy, which is clearly not possible in the FEM. The uncompromising accuracy of results in the DSM originates from the fact that the shape function used to derive the element dynamic stiffness matrix of a structural element is obtained from the exact solution of the governing differential equation of motion of the element undergoing free natural vibration. The element dynamic stiffness matrix is always frequency dependent, comprising both the mass and stiffness properties of the element. This is very different from the FEM for which the mass and stiffness matrices are always separate entities and are frequency independent. A systematic procedure illustrating the formulation of the dynamic stiffness matrix of a structural element can be found in the work of Banerjee (1997). The overall frequency-dependent dynamic stiffness matrix of the final structure is assembled from the dynamic stiffness matrices of all individual elements in the structure. However, the resulting overall dynamic stiffness matrix leads to a nonlinear eigenvalue problem, usually solved by the well-known Wittrick–Williams algorithm (Wittrick and William 1971) to extract the natural frequencies and mode shapes of the final structure. The DSM is called an *exact* method because all assumptions in the method are within the limits of the governing differential equations of motion. Thus, the DSM is most suitable for free vibration analysis in all frequency ranges.

4.2. Formulation of the potential and kinetic energies

Figure 4.1 shows, in a rectangular right-handed Cartesian coordinate system, a uniform micro-beam of length L at a greatly enlarged scale with the X-axis that characterizes the direction of the axial displacement coinciding with the centroidal axis of the beam, whereas the Y-axis that is upwards and perpendicular to the X-axis represents the direction of the bending or transverse displacement of the beam. The Z-axis, which is not shown, but is perpendicular to the plane of the paper, represents the neutral axis of the beam.

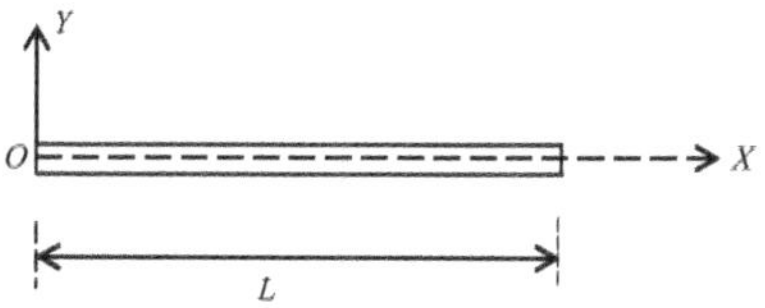

Figure 4.1. *Coordinate system of a micro-beam*

The MCST used here was first developed by Yang *et al.* (2002) with the fundamental premise that the strain energy is a function of both the strain tensor, which is conjugated to the stress tensor and, importantly, the curvature tensor, which is conjugated to the couple stress tensor. Based on this postulate, the strain or potential energy U of a linearly deformed elastic body, like the micro-beam shown in Figure 4.1, can be written in tensorial notation as (Park and Gao 2006; Kong *et al.* 2008; Reddy 2011)

$$U = \frac{1}{2}\int_V (\boldsymbol{\sigma}:\boldsymbol{\varepsilon} + \mathbf{m}:\boldsymbol{\chi})\, dv \Leftrightarrow \frac{1}{2}\int_V \left(\sigma_{ij}\varepsilon_{ij} + m_{ij}\chi_{ij}\right) dv \tag{4.1}$$

where σ_{ij} and ε_{ij} $(i, j = 1, 2, 3)$ are the components of the stress ($\boldsymbol{\sigma}$) and strain ($\boldsymbol{\varepsilon}$) tensors, and m_{ij} and χ_{ij} signify the components of the deviatoric part of the symmetric couple stress tensor ($\mathbf{m}$) and the symmetric curvature tensor ($\boldsymbol{\chi}$), respectively, and the integration is carried out over the entire volume of the elastic body.

The expressions for σ_{ij}, ε_{ij}, m_{ij} and χ_{ij} in the usual notations are given by (see Kong *et al.* 2008; Ma *et al.* 2008)

$$\boldsymbol{\sigma} = \lambda tr(\boldsymbol{\varepsilon})\mathbf{I} + 2\mu\boldsymbol{\varepsilon} \Leftrightarrow \sigma_{ij} = \lambda tr(\varepsilon_{ij})\delta_{ij} + 2\mu\varepsilon_{ij} \tag{4.2}$$

$$\boldsymbol{\varepsilon} = \frac{1}{2}[\nabla\mathbf{u} + (\nabla\mathbf{u})^T] \Leftrightarrow \varepsilon_{ij} = \frac{1}{2}\left(\frac{\partial u_i}{\partial x_j} + \frac{\partial u_j}{\partial x_i}\right) \tag{4.3}$$

$$\mathbf{m} = 2\mu l^2\boldsymbol{\chi} \Leftrightarrow m_{ij} = 2\mu l^2\chi_{ij} \tag{4.4}$$

$$\boldsymbol{\chi} = \frac{1}{2}[\nabla\boldsymbol{\theta} + (\nabla\boldsymbol{\theta})^T] \Leftrightarrow \chi_{ij} = \frac{1}{2}\left(\frac{\partial \theta_i}{\partial x_j} + \frac{\partial \theta_j}{\partial x_i}\right) \tag{4.5}$$

where λ and μ (also denoted by G) are Lame's constant, $\mathbf{I}$ is the unit or identity matrix (so that its elements δ_{ij} are 1 when $i = j$ and 0 when $i \neq j$), u_i, u_j are the components of the displacement vector $\mathbf{u}$, θ_i and θ_j are the components of the rotation vector $\boldsymbol{\theta}$, and l is the material length scale parameter that identifies the effect of the couple stress as described by Yang *et al.* (2002). It should be noted that Chong *et al.* (2001) successfully determined the small length scale parameter l by experiment. For a homogeneous and isotropic material, Lame's constants λ and μ, given in equations [4.2] and [4.4], can be found in standard textbooks on elasticity. These are given by

$$\lambda = \frac{E\nu}{(1+\nu)(1-2\nu)}, \quad \mu = \frac{E}{2(1+\nu)} \tag{4.6}$$

where E is the Young modulus, ν is the Poisson ratio, and μ is effectively the shear modulus or the modulus of rigidity of the material, often denoted by G.

For the micro-beam problem in two dimensions, as shown in Figure 4.1, the tensorial notation used in equations [4.1]–[4.5] simplifies as follows:

$$u_1 = u(x,t) - y\phi(x,t),\ u_2 = v(x,t),\quad u_3 = 0 \qquad [4.7]$$

where u_1, u_2 and u_3 are the components of the displacements in X, Y and Z directions, respectively, and $\phi(x,t)$ is the angle of rotation of the cross-section about the Z-axis which, according to the Bernoulli–Euler hypothesis, is given by

$$\phi(x,t) = \frac{\partial v(x,t)}{\partial x} \qquad [4.8]$$

From equations [4.7] and [4.8], equation [4.3] gives the expression for the strains as

$$\varepsilon_{11} = \varepsilon_{xx} = \frac{\partial u}{\partial x} - y\frac{\partial^2 v}{\partial x^2},\ \varepsilon_{22} = \varepsilon_{yy} = \varepsilon_{12} = \varepsilon_{xy} = \varepsilon_{23} = \varepsilon_{yz} = 0 \qquad [4.9]$$

The components of rotations based on the displacement field defined in equation [4.7] are given by

$$\theta_1 = \theta_x = \theta_2 = \theta_y = 0,\ \theta_3 = \theta_z = \frac{\partial v}{\partial x} \qquad [4.10]$$

From equations [4.5] and [4.10], the components of the curvature tensor $\boldsymbol{\chi}$ are obtained as

$$\chi_{13} = \chi_{xz} = \frac{1}{2}\frac{\partial^2 v}{\partial x^2},\quad \chi_{11} = \chi_{22} = \chi_{33} = \chi_{12} = \chi_{23} = 0 \qquad [4.11]$$

The components of the classical stress tensor $\boldsymbol{\sigma}$ and the deviatoric part of the couple stress tensor $\mathbf{m}$ from equations [4.2] and [4.4] are, respectively, given by

$$\sigma_{11} = \sigma_{xx} = E\left(\frac{\partial u}{\partial x} - y\frac{\partial^2 v}{\partial x^2}\right),\ \sigma_{22} = \sigma_{33} = \sigma_{12} = \sigma_{23} = \sigma_{13} = 0 \qquad [4.12]$$

$$m_{13} = 2\mu l^2 \chi_{13} = \mu l^2 \frac{\partial^2 v}{\partial x^2},\ m_{11} = m_{22} = m_{33} = m_{12} = m_{23} = 0 \qquad [4.13]$$

Now, the following expression for the potential energy U of the micro-beam can be obtained by substituting equations [4.9]–[4.13] into equation [4.1] with the assumption of the linear small deflection theory:

$$U = \frac{1}{2}\int_0^L \int_A (\varepsilon_{11}\sigma_{11} + 2\chi_{13}m_{13})\, dA dx$$

$$= \frac{1}{2}\int_0^L \left[EA\left(\frac{\partial u}{\partial x}\right)^2 + (EI + \mu A l^2)\left(\frac{\partial^2 v}{\partial x^2}\right)^2\right] dx \qquad [4.14]$$

where A is the area and I is the second moment of area of the micro-beam cross-section.

The expression of the kinetic energy (T) is given by

$$T = \frac{1}{2}\int_0^L \int_A \rho\left[\left(\frac{\partial u_1}{\partial t}\right)^2 + \left(\frac{\partial u_2}{\partial t}\right)^2 + \left(\frac{\partial u_3}{\partial t}\right)^2\right] dAdx \qquad [4.15]$$

where ρ is the density of the micro-beam material.

The substitution of equation [4.7] into equation [4.15] gives, when neglecting the rotatory inertia term,

$$T = \frac{1}{2}\int_0^L \int_A \rho\left[\left(\frac{\partial u}{\partial t}\right)^2 + \left(\frac{\partial v}{\partial t}\right)^2\right] dAdx = \frac{1}{2}\rho A\int_0^L \left[\left(\frac{\partial u}{\partial t}\right)^2 + \left(\frac{\partial v}{\partial t}\right)^2\right] dx \qquad [4.16]$$

4.3. Derivation of the governing differential equations

The expressions for the kinetic (T) and potential (U) energies developed above can now be used to derive the governing differential equations of motion of a micro-beam undergoing free natural vibration. In the context of the present work, this can be best accomplished by applying Hamilton's principle that has the added advantage of generating natural boundary conditions as required in the dynamic stiffness formulation.

Hamilton's principle states that

$$\delta \int_{t_1}^{t_2} (T - U)\, dt = 0 \qquad [4.17]$$

where t_1 and t_2 are the time intervals in the dynamic trajectory, and δ is the usual variational operator.

Substituting T and U from equations [4.16] and [4.14] into equation [4.17] leads to

$$\delta \int_{t_1}^{t_2} \int_0^L \frac{1}{2}[\rho A(\dot{u}^2 + \dot{v}^2) - EA(u')^2 - (EI + \mu A l^2)(v'')^2] dxdt = 0 \qquad [4.18]$$

where an over dot and a prime denote partial differentiation with respect to time t and distance x, respectively.

Using the δ operator in equation [4.18], and then integrating by parts and collecting terms lead to

$$\int_{t_1}^{t_2}\int_0^L\{(EAu'' - \rho A\ddot{u})\delta u\}dxdt - \int_{t_1}^{t_2}\int_0^L\{(EI + \mu AL^2)v'''' + \rho A\ddot{v}\}\delta v dxdt - \int_{t_1}^{t_2}[EAu'\delta u]_0^L\,dt - \int_{t_1}^{t_2}[(EI + \mu Al^2)v''\delta v']_0^L\,dt + \int_{t_1}^{t_2}[(EI + \mu Al^2)v'''\delta v]_0^L + \int_0^L[\rho A\dot{u}\,\delta u]_{t_1}^{t_2}\,dx + \int_0^L[\rho A\dot{v}\,\delta v]_{t_1}^{t_2}\,dx = 0 \quad [4.19]$$

Since δu and δv are completely arbitrary, the governing differential equations of motion and the natural boundary conditions giving the expressions for the axial force, the transverse shear force and the bending moment follow from equation [4.19] as:

Governing differential equations:

$$EAu'' - \rho A\ddot{u} = 0 \quad [4.20]$$

$$(EI + \mu Al^2)v'''' + \rho A\ddot{v} = 0 \quad [4.21]$$

Natural boundary conditions:

Axial force: $f(x,t) = -EA\dfrac{\partial u}{\partial x}$ [4.22]

Shear force: $s(x,t) = (EI + \mu Al^2)v'''$ [4.23]

Bending moment: $m(x,t) = -(EI + \mu Al^2)v''$ [4.24]

The axial force *f*, the shear force *s* and the bending moment *m* resulting from the Hamiltonian formulation, and given in equations [4.22]-[4.24], correspond to the positive sign convention shown in Figure 4.2.

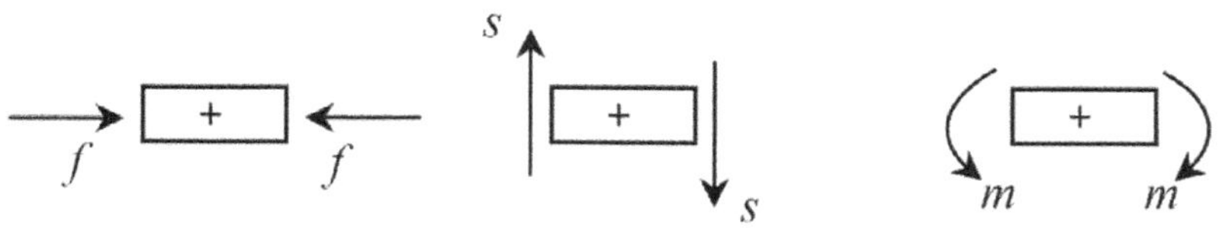

Figure 4.2. *Sign convention for the positive axial force* f, *the shear force* s *and the bending moment* m

4.4. Development of the dynamic stiffness matrix

The dynamic stiffness matrix of a structural element relates the amplitudes of forces to those of the corresponding displacements at the nodes of the elements. The first step in the derivation of the dynamic stiffness matrix is to solve the governing differential equations of motion of the structural element exhibiting free natural vibration.

For harmonic oscillation with the angular or circular frequency ω, the displacements $u(x, t)$ and $v(x, t)$ given in equations [4.20] and [4.21] can be written as

$$u(x,t) = \bar{U}e^{i\omega t}, \ v(x,t) = \bar{V}e^{i\omega t} \tag{4.25}$$

where $\bar{U}$ and $\bar{V}$ are the amplitudes of the axial and bending displacements of the micro-beam, respectively.

Substituting equation [4.25] into equations [4.20] and [4.21] gives the following ordinary differential equations:

$$EA\frac{d^2\bar{U}}{dx^2} + \rho A\omega^2\bar{U} = 0 \tag{4.26}$$

$$(EI + \mu A l^2)\frac{d^4\bar{V}}{dx^4} - \rho A\omega^2\bar{V} = 0 \tag{4.27}$$

By introducing the independent variable ξ and non-dimensional parameters α and β, as defined in equation [4.28], equations [4.26] and [4.27] can now be transformed into equations [4.29] and [4.30] as follows:

$$\xi = \frac{x}{L}, \ \eta = \frac{\mu A l^2}{EI}, \ \alpha^2 = \frac{\rho A\omega^2 L^2}{EA}, \ \beta^4 = \frac{\rho A\omega^2 L^4}{EI(1+\eta)} \tag{4.28}$$

$$\frac{d^2\bar{U}}{d\xi^2} + \alpha^2\bar{U} = 0 \tag{4.29}$$

$$\frac{d^4\bar{V}}{d\xi^4} - \beta^4\bar{V} = 0 \tag{4.30}$$

The solutions of equations [4.29] and [4.30] give, respectively, the amplitudes of the axial and bending displacements $\bar{U}$ and $\bar{V}$ as

$$\bar{U} = C_1\cos\alpha\,\xi + C_2\sin\alpha\xi \tag{4.31}$$

$$\bar{V} = C_3\cos\beta\xi + C_4\sin\beta\xi + C_5\cosh\beta\xi + \sinh\beta\xi \tag{4.32}$$

The expression for the amplitude of the bending rotation Φ is given by

$$\Phi(x) = \Phi(\xi) = \frac{1}{L}\frac{d\bar{V}}{d\xi} = \frac{\beta}{L}(-C_3 \sin\beta\xi + C_4 \cos\beta\xi + C_5 \sinh\beta\xi + C_6 \cosh\beta\xi) \quad [4.33]$$

The solutions given in equations [4.31] and [4.32] using equations [4.22]–[4.24] and [4.28] give the expressions for the amplitudes of the axial force (F), the shear force (S) and the bending moment (M) as follows:

$$F(x) = F(\xi) = -EA\frac{d\bar{U}}{dx} = -\frac{EA}{L}\frac{d\bar{U}}{d\xi} = \frac{EA}{L}\alpha(C_1 \sin\alpha\xi - C_2 \cos\alpha\xi) \quad [4.34]$$

$$S(x) = S(\xi) = \frac{EI}{L^3}(1+\eta)\frac{d^3\bar{V}}{d\xi^3}$$

$$= \frac{EI}{L^3}(1+\eta)\beta^3(C_3 \sin\beta\xi - C_4 \cos\beta\xi + C_5 \sinh\beta\xi + C_6 \cosh\beta\xi) \quad [4.35]$$

$$M(x) = M(\xi) = -\frac{EI(1+\eta)}{L^2}\frac{d^2\bar{V}}{d\xi^2}$$

$$= \frac{EI(1+\eta)}{L^2}\beta^2(C_3 \cos\beta\xi + C_4 \sin\beta\xi - C_5 \cosh\beta\xi - C_6 \sinh\beta\xi) \quad [4.36]$$

As the equations of motion in axial and bending vibration are not coupled (see equations [4.26] and [4.27]), the dynamic stiffness matrix of the micro-beam can now be developed by separately considering axial and bending stiffnesses, which can be combined later to perform the free vibration analysis in a general context.

4.4.1. *Axial stiffnesses*

Figure 4.3 shows the boundary conditions for axial displacements and forces at two ends (i.e. nodes 1 and 2) of the micro-beam, which can be related to formulate the dynamic stiffness matrix in axial motion. Essentially, the boundary conditions are:

$$\text{At } x = 0 \text{ (i.e. } \xi = 0\text{)}, \ \bar{U} = \bar{U}_1 \text{ and } F = F_1 \quad [4.37]$$

$$\text{At } x = L \text{ (i.e. } \xi = 1\text{)}, \ \bar{U} = \bar{U}_2 \text{ and } F = -F_2 \quad [4.38]$$

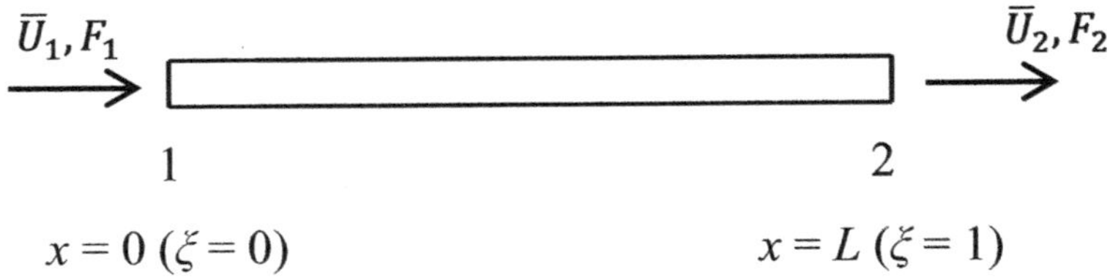

Figure 4.3. *Boundary conditions for the axial displacement and the axial force*

Substituting equations [4.37] and [4.38] into equations [4.31] and [4.34], the following matrix relationships can be obtained:

$$\begin{bmatrix} \bar{U}_1 \\ \bar{U}_2 \end{bmatrix} = \begin{bmatrix} 1 & 0 \\ \cos\alpha & \sin\alpha \end{bmatrix} \begin{bmatrix} C_1 \\ C_2 \end{bmatrix} \qquad [4.39]$$

and

$$\begin{bmatrix} F_1 \\ F_2 \end{bmatrix} = \alpha \begin{bmatrix} 0 & -1 \\ -\sin\alpha & \cos\alpha \end{bmatrix} \begin{bmatrix} C_1 \\ C_2 \end{bmatrix} \qquad [4.40]$$

The constants C_1 and C_2 can now be eliminated from equations [4.39] and [4.40] to give the dynamic stiffness matrix of the axially vibrating micro-beam relating the amplitudes of the forces to those of the displacements at its ends as follows:

$$\begin{bmatrix} F_1 \\ F_2 \end{bmatrix} = \begin{bmatrix} a_1 & a_2 \\ a_2 & a_1 \end{bmatrix} \begin{bmatrix} \bar{U}_1 \\ \bar{U}_2 \end{bmatrix} \qquad [4.41]$$

where the frequency-dependent elements of the above 2×2 dynamic stiffness matrix are given by

$$a_1 = \frac{EA}{L}\alpha\cot\alpha, \quad a_2 = -\frac{EA}{L}\alpha\,\mathrm{cosec}\,\alpha \qquad [4.42]$$

4.4.2. *Bending stiffnesses*

Now referring to Figure 4.4, the boundary conditions for the bending displacement, the bending rotation, the shear force and the bending moment can be applied as follows:

At $x = 0$ (i.e. $\xi = 0$), $\bar{V} = \bar{V}_1,\ \Phi = \Phi_1,\ S = S_1$ and $M = M_1$ [4.43]

At $x = L$ (i.e. $\xi = 1$), $\bar{V} = \bar{V}_2,\ \Phi = \Phi_2,\ S = -S_2$ and $M = -M_2$ [4.44]

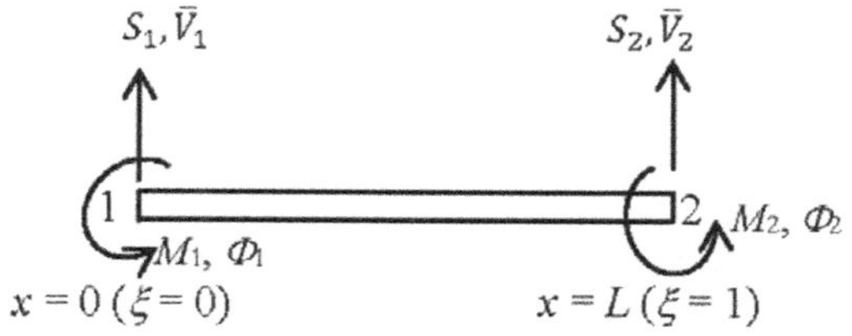

Figure 4.4. *Boundary conditions for the bending displacement, the bending rotation, the shear force and the bending moment*

Substituting equations [4.43] and [4.44] into equations [4.32], [4.33], [4.35] and [4.36], the following two matrix equations can be obtained for displacements and forces, respectively, in terms of the constants C_3–C_6:

$$\begin{bmatrix} \bar{V}_1 \\ \Phi_1 \\ \bar{V}_2 \\ \Phi_2 \end{bmatrix} = \begin{bmatrix} 1 & 0 & 1 & 0 \\ 0 & \beta/L & 0 & \beta/L \\ C_\beta & S_\beta & C_{h\beta} & S_{h\beta} \\ -\beta S_\beta/L & \beta C_\beta/L & \beta S_{h\beta}/L & \beta C_{h\beta}/L \end{bmatrix} \begin{bmatrix} C_3 \\ C_4 \\ C_5 \\ C_6 \end{bmatrix} \quad [4.45]$$

and

$$\begin{bmatrix} S_1 \\ M_1 \\ S_2 \\ M_2 \end{bmatrix} = \begin{bmatrix} 0 & -W_3\beta^3 & 0 & W_3\beta^3 \\ W_2\beta^2 & 0 & -W_2\beta^2 & 0 \\ -W_3\beta^3 S_\beta & W_3\beta^3 C_\beta & -W_3\beta^3 S_{h\beta} & -W_3\beta^3 C_{h\beta} \\ -W_2\beta^2 C_\beta & -W_2\beta^2 S_\beta & W_2\beta^2 C_{h\beta} & W_2\beta^2 S_{h\beta} \end{bmatrix} \begin{bmatrix} C_3 \\ C_4 \\ C_5 \\ C_6 \end{bmatrix} \quad [4.46]$$

where

$$C_\beta = \cos\beta, \quad S_\beta = \sin\beta, \quad C_{h\beta} = \cosh\beta, \quad S_{h\beta} = \sinh\beta \quad [4.47]$$

and

$$W_2 = \frac{EI(1+\eta)}{L^2}, \quad W_3 = \frac{EI(1+\eta)}{L^3} \quad [4.48]$$

The constants C_3–C_6 can now be eliminated from equations [4.45] and [4.46] to give the 4×4 dynamic stiffness matrix of the micro-beam in bending vibration as follows:

$$\begin{bmatrix} S_1 \\ M_1 \\ S_2 \\ M_2 \end{bmatrix} = \begin{bmatrix} d_1 & d_2 & d_4 & d_5 \\ d_2 & d_3 & -d_5 & d_6 \\ d_4 & -d_5 & d_1 & -d_2 \\ d_5 & d_6 & -d_2 & d_3 \end{bmatrix} \begin{bmatrix} \bar{V}_1 \\ \bar{\Phi}_1 \\ \bar{V}_2 \\ \bar{\Phi}_2 \end{bmatrix} \quad [4.49]$$

where

$$d_1 = \frac{W_3\beta^3(S_\beta C_{h\beta}+C_\beta S_{h\beta})}{\Delta}, \quad d_2 = \frac{W_2\beta^2 S_\beta S_{h\beta}}{\Delta}, \quad d_3 = \frac{W_1\beta(S_\beta C_{h\beta}-C_\beta S_{h\beta})}{\Delta} \qquad [4.50]$$

$$d_4 = -\frac{W_3\beta^3(S_\beta+S_{h\beta})}{\Delta}, \quad d_5 = \frac{W_2\beta^2(C_{h\beta}-C_\beta)}{\Delta}, \quad d_6 = \frac{W_1\beta(S_{h\beta}-S_\beta)}{\Delta} \qquad [4.51]$$

with

$$W_1 = \frac{EI(1+\eta)}{L}, \qquad \Delta = 1 - C_\beta C_{h\beta} \qquad [4.52]$$

4.4.3. *Combination of axial and bending stiffnesses*

A simple superposition is now possible to combine the axial and bending dynamic stiffnesses to express the force–displacement relationship of the micro-beam that can be applied to frameworks. Superposing Figures 4.3 and 4.4 will result in Figure 4.5

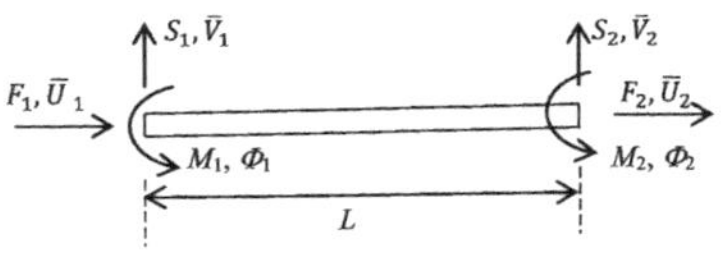

Figure 4.5. *Amplitudes of displacements and forces at the ends of the micro-beam*

Using Figure 4.5, equations [4.41] and [4.49] give the dynamic stiffness matrix of the micro-beam in the following matrix form:

$$\begin{bmatrix} F_1 \\ S_1 \\ M_1 \\ F_2 \\ S_2 \\ M_2 \end{bmatrix} = \begin{bmatrix} a_1 & 0 & 0 & a_2 & 0 & 0 \\ 0 & d_1 & d_2 & 0 & d_4 & d_5 \\ 0 & d_2 & d_3 & 0 & -d_5 & d_6 \\ a_2 & 0 & 0 & a_1 & 0 & 0 \\ 0 & d_4 & -d_5 & 0 & d_1 & -d_2 \\ 0 & d_5 & d_6 & 0 & -d_2 & d_3 \end{bmatrix} \begin{bmatrix} \bar{U}_1 \\ \bar{V}_1 \\ \Phi_1 \\ \bar{U}_2 \\ \bar{V}_2 \\ \Phi_2 \end{bmatrix} \qquad [4.53]$$

or

$$\mathbf{P} = \mathbf{K}\,\boldsymbol{\delta} \qquad [4.54]$$

where **P** and **d** are, respectively, the force and displacement vectors, and **K** is the 6×6 frequency-dependent dynamic stiffness matrix of the micro-beam whose elements have already been defined in equations [4.42], [4.50] and [4.51]. Note that **K** is symmetric, as expected.

4.4.4. *Transformation matrix*

The dynamic stiffness **K** derived above corresponds to local coordinates, i.e. the coordinates that are local to the micro-beam element. For the application of the dynamic stiffness matrix to frameworks, **K** needs to be transformed into global (or datum) coordinates. In the local coordinate system, the centroidal axis of the beam coincides with the X-axis, and the Y-axis is perpendicular in the anticlockwise direction as shown in the right-handed coordinate system in Figure 4.6, which also shows a set of global or datum coordinates $\bar{X}$ and $\bar{Y}$. The angle between the global (or datum) coordinate and the local coordinate systems is θ, as shown in the figure. Note that θ is measured from the global (or datum) $\bar{X}$ axis to the local X axis and is positive in the anticlockwise direction. The dynamic stiffness **K** in local coordinates needs to be transformed into the global (or datum) coordinates, in order to make it possible to perform the free vibration analysis of frameworks.

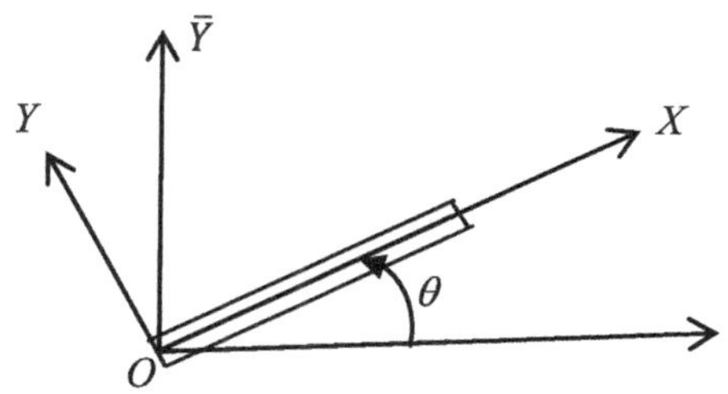

Figure. 4.6. *Alignment of the local* (OXY) *and global* (O$\bar{X}\bar{Y}$) *coordinate systems*

The transformation matrix **T** needed to transform the stiffness matrix from the local to global (or datum) coordinate system is given by (Przemieniecki 1985; Rao 2017)

$$\mathbf{T} = \begin{bmatrix} C & S & 0 & 0 & 0 & 0 \\ -S & C & 0 & 0 & 0 & 0 \\ 0 & 0 & 1 & 0 & 0 & 0 \\ 0 & 0 & 0 & C & S & 0 \\ 0 & 0 & 0 & -S & C & 0 \\ 0 & 0 & 0 & 0 & 0 & 1 \end{bmatrix} \qquad [4.55]$$

where

$$C = \cos\theta\,, \quad S = \sin\theta \qquad [4.56]$$

Using the transformation matrix **T** in equation [4.55], the dynamic stiffness matrix in the global (datum) coordinates $\bar{\mathbf{K}}$ can be obtained as

$$\bar{\mathbf{K}} = \mathbf{T}^t\mathbf{K}\mathbf{T} \tag{4.57}$$

where $\mathbf{T}^t$ is the transpose of **T**.

Using equation [4.57], the element dynamic stiffness matrices in a framework can be assembled in the global (datum) coordinates using the standard procedure to form the overall dynamic stiffness matrix of the final structure.

4.5. Application of the Wittrick–Williams algorithm

The dynamic stiffness matrix derived above can now be used to compute the natural frequencies and mode shapes of either an individual micro-beam or an assembly of micro-beams for different boundary conditions. A non-uniform micro-beam can be investigated for its free vibration characteristics by modeling it as an assemblage of many uniform micro-beams. A reliable method to compute the natural frequencies and mode shapes of a structure using the DSM is to apply the Wittrick–Williams algorithm (Wittrick and Williams 1971) which has been explained in numerous papers (Williams and Wittrick 1983; Williams 1993). Before the algorithm can be applied, the dynamic stiffness matrices of all individual elements in the structure should be assembled to form the overall dynamic stiffness matrix $\mathbf{K}_f$ of the final structure which may of course, as a special case, consist of a single element. Essentially, the algorithm monitors the Sturm sequence property of $\mathbf{K}_f$ in a secure way so that no natural frequency of the structure is missed. The algorithm, unlike its proof, is simple to apply, but its basic working principle is summarized as follows.

If ω represents an arbitrary circular (or angular) frequency of a freely vibrating structure, then the Wittrick–Williams algorithm (Wittrick and Williams 1971) states that j, the number of natural frequencies passed, as ω is increased from zero to ω^*, is

$$j = j_0 + s\{\mathbf{K}_f\} \tag{4.58}$$

where $\mathbf{K}_f$, the overall dynamic stiffness matrix of the final structure whose elements all depend on ω, is evaluated at $\omega = \omega^*$. The term $s\{\mathbf{K}_f\}$ in equation [4.58] is the number of negative elements on the leading diagonal of $\mathbf{K}_f^\Delta$, where $\boldsymbol{K}_f^\Delta$ is the upper triangular matrix obtained by applying Gauss elimination to $\mathbf{K}_f$. In equation [4.58], j_0 is the number of natural frequencies of the structure which can still lie between $\omega = 0$ and $\omega = \omega^*$ when the displacement components to which $\mathbf{K}_f$ corresponds are all zeros. (It should be noted that the structure can still have natural frequencies when all its nodes are fully clamped. This is because the exact element equation allows each individual element to have displacements between nodes with an infinite number of degrees of freedom, and hence the infinite number of natural frequencies between nodes). Thus

$$j_0 = \sum j_m \qquad [4.59]$$

where j_m is the number of natural frequencies between $\omega = 0$ and $\omega = \omega^*$ for an individual component member with its ends fully clamped. The summation in equation [4.59] extends over all individual members of the structure. Thus, with the knowledge of equations [4.58] and [4.59], it is possible to determine how many natural frequencies of a structure exist below an arbitrarily chosen trial frequency. As successive trial frequencies can be chosen by the user, it is possible to establish the upper and lower bounds of a natural frequency and finally converge on any required natural frequency to any desired accuracy.

4.6. Numerical results and discussion

The first set of results was computed for an individual micro-beam of length L with a rectangular cross-section of width b and depth (or thickness) h, for three different boundary conditions, namely the simply supported–simply supported (S–S), clamped–free (C–F) and clamped–clamped (C–C). The data used in the analysis are taken from Reddy (2011) and Mustapha (2020), which are as follows:

Young's modulus, E=1.44 GPa; density, ρ = 1220 kg/m^3; Poisson's ratio, $\nu = 0.38$; $h = 17.6\times10^{-6}$ m; $b = 2h$; and $L = 20h$.

The first five non-dimensional natural frequencies $\lambda_i = \omega_i L^2\sqrt{\rho A/EI}$ (i = 1, 2, … 5) of the micro-beam for the S–S, C–F and C–C boundary conditions are given in Tables 4.1–4.3 for various values of the size-dependent non-dimensional material length scale parameter l/h. (Note that the axial displacement is prevented at the simple support in the S–S case.)

Comparative results for the first three natural frequencies of the micro-beam for the S–S boundary condition were published by Reddy (2011), which are given in Table 4.1. Clearly, the results computed from the present theory are in excellent agreement with those of Reddy (2011) (see his Table 2).

As can be seen in the results shown in Tables 4.1–4.3, the natural frequencies increase when the l/h ratio increases. This is expected because the material length scale parameter l essentially increases the bending rigidity EI of the beam by a factor $(1+\eta)$, where $\eta = \frac{\mu A l^2}{EI}$ (see equation [4.28]) so that the natural frequencies increase by a factor $\sqrt{1+\eta}$. Note that the results presented in Tables 4.1–4.3 exclude the axial natural frequencies that are not affected by the material length scale parameter (l). Thus, Tables 4.1–4.3 only present the bending natural frequencies. The results for the case when l/h = 0 agree completely with those obtained from the classical Bernoulli–Euler theory (Rao 2019), as expected.

l/h	Non-dimensional natural frequency $\lambda_i = \omega_i L^2 \sqrt{\rho A/EI}$							
	λ_1		λ_2		λ_3		λ_4	λ_5
	Present theory	Reddy (2011)	Present theory	Reddy (2011)	Present theory	Reddy (2011)	Present theory	Present theory
0.0	9.8696	9.86	39.478	39.32	88.827	88.02	157.91	246.74
0.2	10.693	10.68	42.773	42.60	96.240	95.36	171.09	267.35
0.4	12.852	12.84	51.409	51.20	115.67	114.61	205.63	321.31
0.6	15.808	15.79	63.229	62.97	142.27	140.97	252.91	395.18
0.8	19.195	19.18	76.781	76.47	172.76	171.18	307.12	479.88
1.0	22.824	22.80	91.295	90.92	205.41	203.54	365.17	570.58

Table 4.1. *Natural frequencies of a simply supported (S–S) micro-beam*

l/h	Non-dimensional natural frequency $\lambda_i = \omega_i L^2 \sqrt{\rho A/EI}$				
	λ_1	λ_2	λ_3	λ_4	λ_5
0.0	3.5159	22.035	61.698	120.90	199.86
0.2	3.8095	23.875	66.848	130.99	216.54
0.4	4.5786	28.692	80.341	157.44	260.25
0.6	5.6313	35.291	98.817	193.64	320.10
0.8	6.8383	42.854	119.99	235.13	388.70
1.0	8.1308	50.955	142.68	279.58	462.19

Table 4.2. *Natural frequencies of a cantilever (C–F) micro-beam*

l/h	Non-dimensional natural frequency $\lambda_i = \omega_i L^2 \sqrt{\rho A/EI}$				
	λ_1	λ_2	λ_3	λ_4	λ_5
0.0	22.373	61.674	120.90	199.86	298.55
0.2	24.241	66.821	131.00	216.54	323.49
0.4	29.134	80.309	157.44	260.25	388.76
0.6	35.835	98.777	193.64	320.10	478.17
0.8	43.514	119.95	235.13	388.70	580.66
1.0	51.739	142.62	279.60	462.19	690.43

Table 4.3. *Natural frequencies of a clamped–clamped (C–C) micro-beam*

The mode shapes corresponding to the size-dependent non-dimensional material length scale parameter $l/h = 1$ for the S–S, C–F (cantilever) and C–C boundary conditions are shown in Figures 4.7–4.9, respectively. Understandably, the mode shapes have the same patterns as those generally found using the classical theory because all that happens when applying the MCST instead is that the stiffening effect due to the material length scale parameter increases the bending rigidity of the micro-beam by a constant factor $(1+\eta)$, which will have no influence on normalized mode shapes.

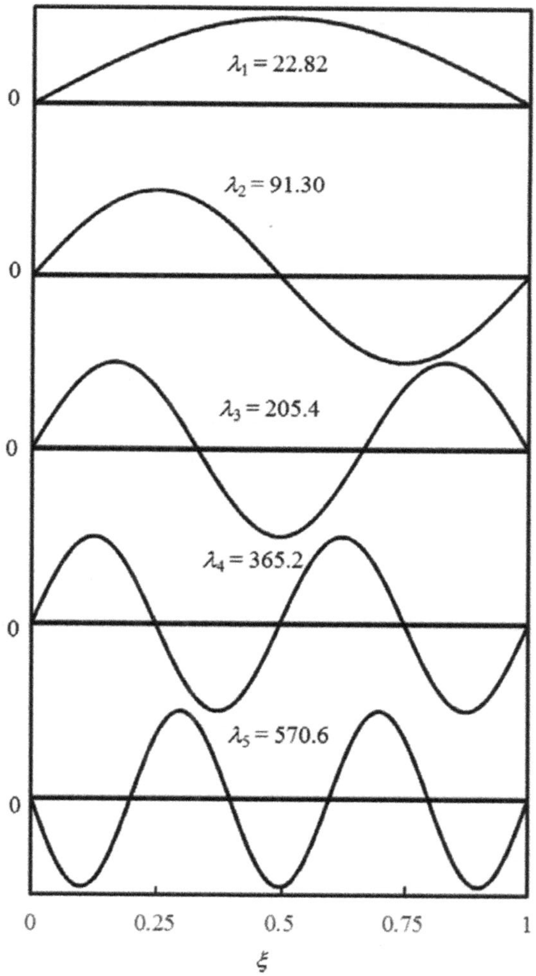

Figure 4.7. *Natural frequencies and mode shapes of a micro-beam with the simply supported (S–S) boundary condition for* l/h = *1*

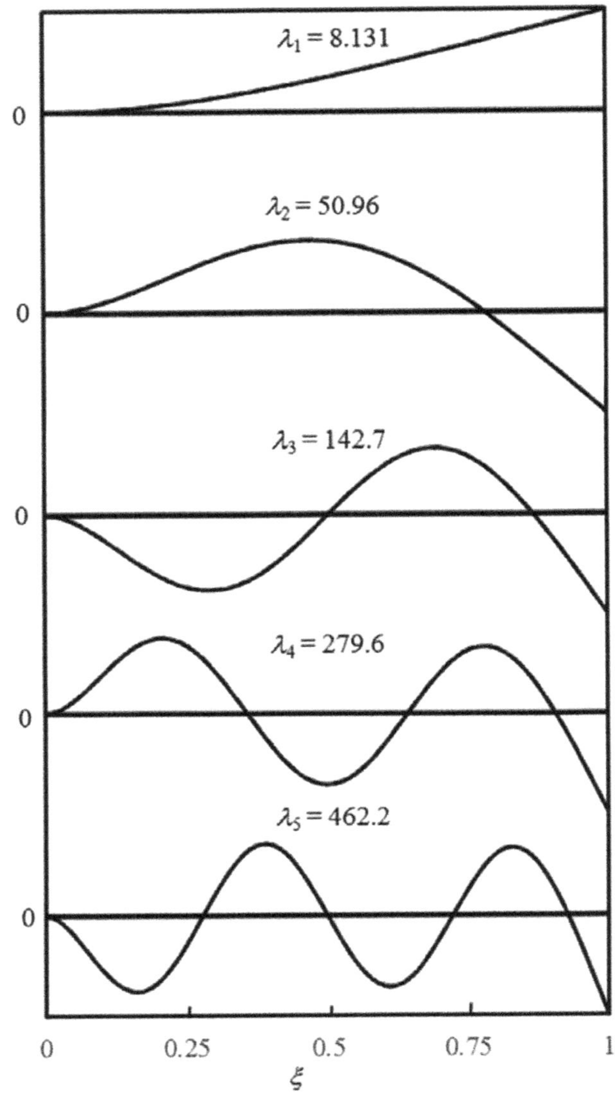

Figure 4.8. *Natural frequencies and mode shapes of a micro-beam with the cantilever (C–F) boundary condition for* l/h = 1

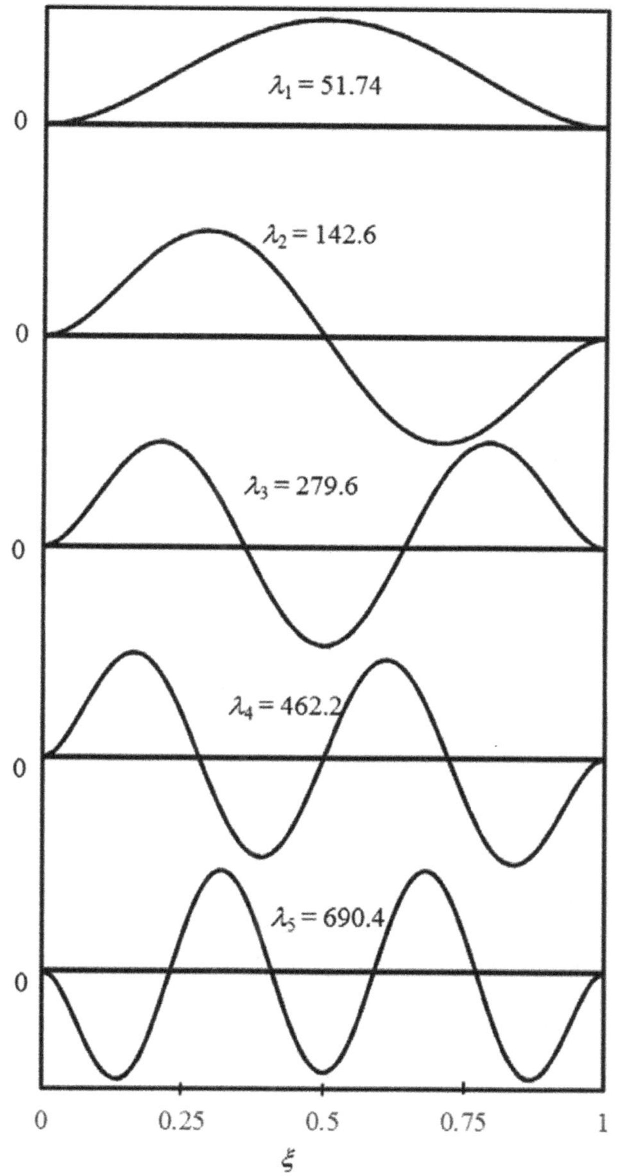

Figure 4.9. *Natural frequencies and mode shapes of a micro-beam with the clamped–clamped (C–C) boundary condition for* l/h = *1*

The effect of the material length scale parameter *l/h* on the fundamental natural frequency λ_1 of the micro-beam for the S–S, C–F and C–C boundary conditions is shown in Figure 4.10. Clearly, the natural frequency increases with the increasing values of the material length scale parameter, as expected.

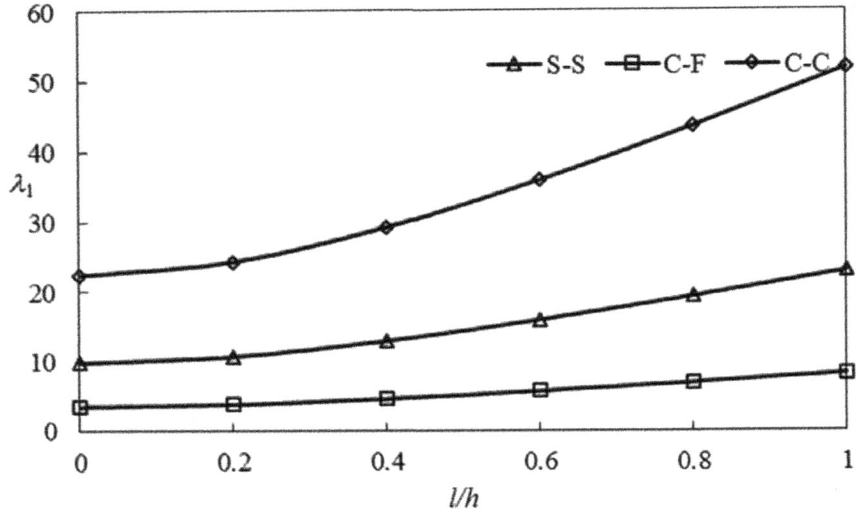

Figure 4.10. *Effect of the material length scale parameter* l/h *on the fundamental natural frequency* λ_1 *of a micro-beam for various boundary conditions*

The next set of results was computed for an L-shaped micro-frame shown in Figure 4.11. This example problem was investigated earlier by Mei (2012) using the classical Bernoulli–Euler theory and very recently by Mustapha (2020) in the context of micro-beam analysis using the MCST. The data and other properties used for the two-constituent individual micro-beams of the frame are taken from Reddy (2011) and Mustapha (2020), which are essentially the same as those of the previous example. The first four natural frequencies of the L-shaped frame are given in Table 4.4 in non-dimensional form, alongside the published results given in parentheses and reported by Mustapha (2020) who applied the FEM and modeled the L-shaped frame using 20–100 elements, as opposed to the current DSM that used only two elements, as it is an exact method. In the comparative result given in parentheses, the number of elements used in the FEM was 100. The agreement between the results from the current theory and the finite element theory (Mustapha 2020) is very good. The difference in results for the first three natural frequencies is less than 3%, whereas in the fourth frequency, it is less than 6.5% for all values of the l/h ratios. The natural frequencies computed by Mustapha (2020) are slightly lower than those computed from the present theory. The reason for this can be attributed to the fact that his method was based on the Rayleigh–Love and Timoshenko beam theories that account for transverse inertia in axial vibration, and shear deformation and rotatory inertia in bending vibration of the micro-beam.

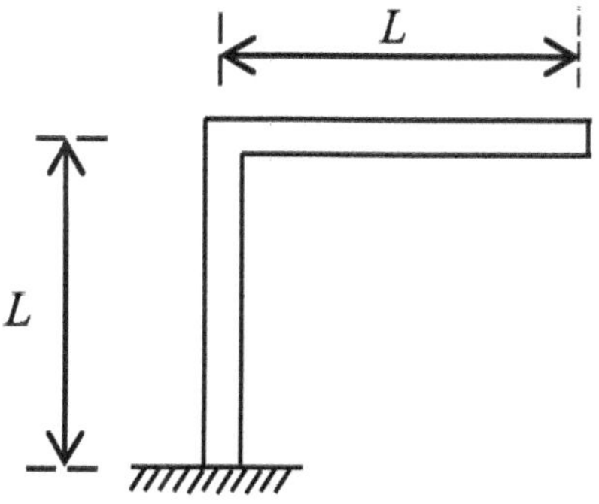

Figure. 4.11. *An L-shaped micro-frame (L=20h, b=2h, h=17.6 μm)*

l/h	Non-dimensional natural frequency $\lambda_i = \omega_i L^2 \sqrt{\rho A/EI}$			
	λ_1	λ_2	λ_3	λ_4
0.0	1.1717 (1.1710)	3.1895 (3.1796)	15.713 (15.589)	22.968 (22.619)
0.25	1.3213 (1.3204)	3.5963 (3.5826)	17.706 (17.541)	25.861 (25.381)
0.5	1.6925 (1.6908)	4.6053 (4.5772)	22.631 (22.322)	32.973 (32.037)
1.0	2.7087 (2.7029)	7.3609 (7.2546)	35.859 (34.829)	51.620 (48.450)

Table 4.4. *Natural frequencies of an L-shaped micro-frame*

The mode shapes corresponding to the first four natural frequencies of the L-shaped micro-beam computed from the present theory for l/h = 1 are shown in Figure 4.12, where the solid and dashed lines represent the undeformed and deformed frames, respectively. The mode shapes are in good agreement with those presented by Mustapha (2020).

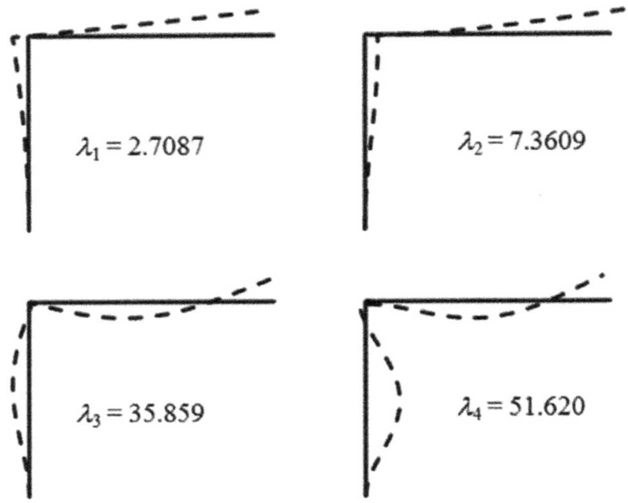

Figure 4.12. *Mode shapes of an L-shaped micro-frame for* l/h = 1

The final set of results was computed for a portal micro-frame shown in Figure 4.13. The properties used for the micro-beam element are the same as those of the previous example.

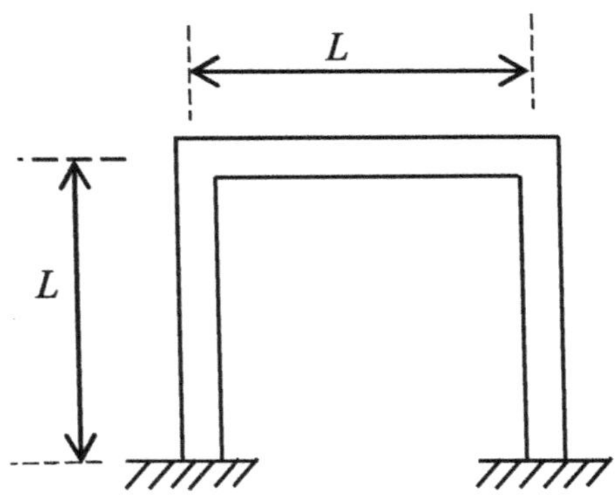

Figure. 4.13. *A micro portal frame (L=20h, b=2h, h=17.6 μm)*

The first four natural frequencies of the micro portal frame computed from the present theory are given in Table 4.5, together with those reported by Mustapha (2020) which are shown in parentheses. The quoted results of Mustapha (2020) are based on 100-element idealization in the FEM, whereas the results from the current DSM used only three elements to represent the portal frame. An excellent agreement between the two sets of results is evident. The natural frequencies computed by Mustapha (2020) are slightly lower than those computed by the present theory because he used the Rayleigh–Love and Timoshenko theories in his formulation, which diminished the natural frequencies, as expected.

l/h	Non-dimensional natural frequency $\lambda_i = \omega_i L^2 \sqrt{\rho A/EI}$			
	λ_1	λ_2	λ_3	λ_4
0.0	3.2028 (3.1904)	12.591 (12.498)	20.617 (20.349)	22.168 (21.795)
0.25	3.6113 (3.5939)	14.181 (14.053)	23.246 (22.874)	24.934 (24.416)
0.5	4.6239 (4.5880)	18.097 (17.847)	29.765 (29.027)	31.689 (30.671)
1.0	7.3880 (7.2510)	28.503 (27.629)	47.554 (44.886)	48.995 (45.665)

Table 4.5. *Natural frequencies of a micro portal frame*

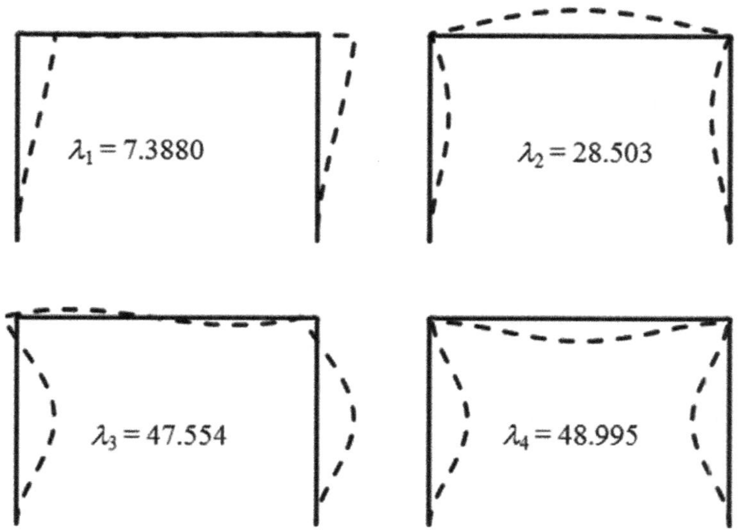

Figure 4.14. *Mode shapes of a micro portal frame for* l/h = *1*

The mode shapes corresponding to the first four natural frequencies of the micro portal frame computed from the present theory are shown in Figure 4.14, where the solid and dashed lines represent the undeformed and deformed shape of the portal frame, respectively. The mode shapes are in good agreement with those presented by Mustapha (2020).

4.7. Conclusion

The dynamic stiffness matrix of a micro-beam was developed for the first time by applying the MCST in conjunction with the classical Bernoulli–Euler theory. The investigation was carried out first by deriving the governing differential equations of motion of the micro-beam using Hamilton's principle and accounting for the small length material scale parameter in the strain energy formulation. The resulting differential equations were then solved in an exact sense by presenting them in explicit algebraic form. The expressions for the axial force, the shear force and the bending moment generated by the natural boundary condition of the Hamiltonian formulation were also expressed in explicit analytical form. Next, the dynamic stiffness matrix was developed by relating the amplitudes of the forces to those of the corresponding displacements at the ends of the freely vibrating micro-beam. Illustrative examples were given for individual micro-beams with classical boundary conditions, as well as for two frameworks, of which one is L-shaped and the other is portal. A substantial amount of the results computed from the current theory were compared with published results, which showed excellent agreement. As the developed method gives exact results that are independent of the number of elements used in the analysis, its usefulness becomes apparent in the free vibration analysis of micro-beam structures in all frequency ranges. The proposed theory can be used as an aid to validate the FEM and other approximate methods.

4.8. Acknowledgments

The author benefited from related earlier projects on the dynamic stiffness method funded by the Engineering and Physical Sciences Research Council, UK and a current project funded by the Leverhulme Trust, UK (grant reference EM-2019-061). The author also wishes to thank Dr Ajandan Ananthapuvirajah, his former PhD student, for his help in the preparation of the manuscript.

4.9. References

Akgoz, B. and Civalek, O. (2012). Analysis of micro-sized beams for various boundary conditions based on the strain gradient elasticity theory. *Arch. Appl. Mech.*, 82(3), 423–443.

Anthoine, A. (2000). Effect of couple-stresses on the elastic bending of beams. *Int. J. Solids Struct.*, 37(7), 1003–1018.

Askes, H. and Aifantis, E.C. (2011). Gradient elasticity in statics and dynamics: An overview of formulations, length scale identification procedures, finite element implementations and new results. *Int. J. Solids Struct.*, 48(13), 1962–1990.

Attia, M.A. and Emam, S.A. (2018). Electrostatic nonlinear bending, buckling and free vibrations of viscoelastic microbeams based on the modified couple stress theory. *Acta Mech.*, 229(8), 3235–3255.

Attia, P., Tremblay, G., Laval, R., Hesto, P. (1998). Characterisation of a low-voltage actuated gold microswitch. *Mater. Sci. Eng. B*, 51(1–3), 263–266.

Banerjee, J.R. (1997). Dynamic stiffness formulation for structural elements: A general approach. *Comput. Struct.*, 63(1), 101–103.

Banerjee, J.R. (2015). The dynamic stiffness method: Theory, practice and promise. *Comput. Technol. Rev.*, 11, 31–57.

Challamel, N. and Wang, C.M. (2008). The small length scale effect for a non-local cantilever beam: A paradox solved. *Nanotechnology*, 19(34), 345703.

Challamel, N., Picandet, V., Elishakoff, I., Wang, C.M., Collet, B., Michelitsch, T. (2015). On nonlocal computation of eigenfrequencies of beams using finite difference and finite element methods. *Int. J. Struct. Stab. Dynam.*, 15(7), 1540008.

Chebakov, R., Kaplunov, J., Rogerson, G.A. (2017). A non-local asymptotic theory for thin elastic plates. *Proc. R. Soc. A*, 473(2203), 20170249.

Chong, A.C.M. and Lam, D.C.C. (1999). Strain gradient plasticity effect in indentation hardness of polymers. *J. Mater. Res.*, 14(10), 4103–4110.

Chong, A.C.M., Yang, F., Lam. D.C.C., Tong, P. (2001). Torsion and bending of micron-scaled structures. *J. Mater. Res.*, 16(4), 1052–1058.

Cosserat, E. and Cosserat, F. (1909). *Theory of Deformable Bodies*. Hermann et Fils, Paris.

Eringen, A.C. (1966). Linear theory of micropolar elasticity. *J. Math. Mech.*, 15(6), 909–923.

Eringen, A.C. (1972). Nonlocal polar elastic continua. *Int. J. Eng. Sci.*, 10(1), 1–16.

Eringen, A.C. (1983). On differential equations of nonlocal elasticity and solutions of screw dislocation and surface waves. *J. Appl. Phys.*, 54(9), 4703–4710.

Farokhi, H., Ghayesh, M.H., Amabili, M. (2013). Nonlinear dynamics of a geometrically imperfect microbeam based on the modified couple stress theory. *Int. J. Eng. Sci.*, 68, 11–23.

Fleck, N.A., Muller, G.M., Ashby, M.F., Hutchinson, J.W. (1994). Strain gradient plasticity: Theory and experiment. *Acta Metallur. Mater.*, 42(2), 475–487.

Ghayesh, M.H., Farokhi, H., Amabili, M. (2013). Nonlinear dynamics of a microscale beam based on the modified couple stress theory. *Compos., Part B. Eng.*, 50, 318–324.

Hache, F., Challamel, N., Elishakoff, I. (2019a). Asymptotic derivation of nonlocal beam models from two-dimensional nonlocal elasticity. *Math. Mech. Solids*, 24(8), 2425–2443.

Hache, F., Challamel, N., Elishakoff, I. (2019b). Asymptotic derivation of nonlocal plate models from three-dimensional stress gradient elasticity. *Continuum Mech. Thermo.*, 31(1), 47–70.

Koiter, W.T. (1964). Couple-stresses in the theory of elasticity: I and II. *Proc. K. Ned Akad. Wet. B*, 67(1), 17–44.

Kong, S., Zhou, S., Nie, Z., Wang, K. (2008). The size-dependent natural frequency of Bernoulli–Euler micro-beams. *Int. J. Eng. Sci.*, 46(5), 427–437.

Kong, S., Zhou, S., Nie, Z., Wang, K. (2009). Static and dynamic analysis of microbeams based on strain gradient elasticity theory. *Int. J. Eng. Sci.*, 47(4), 487–498.

Lam, D.C.C., Yang, F., Chong, A.C.M., Wong, J., Tong, P. (2003). Experiments and theory in strain gradient elasticity. *J. Mech. Phys. Solids.*, 51(8), 1477–1508.

Li, X., Bhushan, B., Takashima, K., Baek, C.W., Kim, Y.K. (2003). Mechanical characterization of micro/nanoscale structures for MEMS/NEMS applications using nanoindentation techniques. *Ultramicroscopy*, 97(1–4), 481–494.

Li, Z., He, Y., Lei, J., Guo, S., Liu, D., Wang, L. (2018). A standard experimental method for determining the material length scale based on modified couple stress theory. *Int. J. Mech. Sci.*, 141, 198–205.

Ma, Q. and Clarke, D.R. (1995). Size dependent hardness of silver single crystals. *J. Mater. Res.*, 10, 853–863.

Ma, H.M., Gao, X.L., Reddy, J.N. (2008). A microstructure-dependent Timoshenko beam model based on a modified couple stress theory. *J. Mech. Phys. Solids*, 56, 3379–3391.

Mei, C. (2012). Wave analysis of in-plane vibrations of L-shaped and portal frame structures. *J. Vibr. Acous.*, 134(2).

Mindlin, R.D. (1963). Influence of couple-stresses on stress concentrations. *Exp. Mech.*, 3, 1–7.

Mindlin, R.D. (1964). Micro-structure in linear elasticity. *Arch. Rat. Mech. Anal.*, 16, 51–78.

Mindlin, R.D. (1965). Second gradient of strain and surface-tension in linear elasticity. *Int J. Solids Struct.*, 1, 417–438.

Mindlin, R.D. and Tiersten, H.F. (1962). Effects of couple-stresses in linear elasticity. *Arch. Rational Mech. Anal.*, 11, 415–448.

Moeenfard, H. and Ahmadian, M.T. (2013). Analytical modelling of bending effect on the torsional response of electrostatically actuated micromirrors. *Optik.*, 124(12), 1278–1286.

Mustapha, K.B. (2020). Free vibration of microscale frameworks using modified couple stress and a combination of Rayleigh–Love and Timoshenko theories. *J. Vib. Contr.*, 26(13–14), 1285–1310.

Nix, W.D. (1989). Mechanical properties of thin films. *Metallu. Trans. A*, 20(11), 2217.

Noori, J., Fariborz, S.J., Vafa, J.P. (2016). A higher-order micro-beam model with application to free vibration. *Mech. Adv. Mater. Struct.*, 23(4), 443–450.

Park, S.K. and Gao, X.L. (2006). Bernoulli–Euler beam model based on a modified couple stress theory. *J. Micromech. Microeng.*, 16(11), 2355–2359.

Pei, J., Tian, F., Thundat, T. (2004). Glucose biosensor based on the microcantilever. *Analy. Chem.*, 76(2), 292–297.

Przemieniecki, J.S. (1985). *Theory of Matrix Structural Analysis*. Dover Publications, New York.

Rao, S.S. (2017). *The Finite Element Method in Engineering*, 6th edition. Butterworth-Heinemann, Oxford.

Rao, S.S. (2019). *Vibration of Continuous Systems*, 2nd edition. John Wiley & Sons, New York.

Reddy, J.N. (2007). Nonlocal theories for bending, buckling and vibration of beams. *Int. J. Eng. Sci.*, 45(2–8), 288–307.

Reddy, J.N. (2011). Microstructure-dependent couple stress theories of functionally graded beams. *J. Mech. Phy. Solids*, 59(11), 2382–2399.

Simsek, M. (2014). Nonlinear static and free vibration analysis of microbeams based on the nonlinear elastic foundation using modified couple stress theory and He's variational method. *Compos. Struct.*, 112, 264–272.

Stolken, J.S. and Evans, A.G. (1998). A microbend test method for measuring the plasticity length scale. *Acta. Mater.*, 46(14), 5109–5115.

Thai, H.T., Vo, T.P., Nguyen, T.K., Kim, S.E. (2017). A review of continuum mechanics models for size-dependent analysis of beams and plates. *Compos. Struct.*, 177, 196–219.

Toupin, R.A. (1962). Elastic materials with couple-stresses. *Arch. Rational Mech. Anal.*, 11, 385–414.

Wang, Y.G., Lin, W.H., Liu, N. (2013). Nonlinear free vibration of a microscale beam based on modified couple stress theory. *Phys. E: Low-dimensional Sys. Nanostruct.*, 47, 80–85.

Williams, F.W. (1993). Review of exact buckling and frequency calculations with optional multi-level substructuring. *Comput. Struct.*, 48(3), 547–552.

Williams, F.W. and Wittrick, W.H. (1983). Exact buckling and frequency calculations surveyed. *J. Struct. Eng.*, 109(1), 169–187.

Wittrick, W.H. and Williams, F.W. (1971). A general algorithm for computing natural frequencies of elastic structures. *Q. J. Mech. Appl. Math.*, 24(3), 263–284.

Yang, F., Chong, A.C.M., Lam, D.C.C., Tong, P. (2002). Couple stress based strain gradient theory for elasticity. *Int. J. Solids Struct.*, 39(10), 2731–2743.

5

On the Geometric Nonlinearities in the Dynamics of a Planar Timoshenko Beam

In the past, several simplified models have been developed for planar nonlinear dynamics of beams based on specific geometrical assumptions and negligibility of some terms. Thus, the question arises about the reliability of these models, i.e. whether the neglected terms can really be overlooked. Recently, the authors have considered in a series of papers a geometrically exact model in which all the nonlinearities are taken into account. A Timoshenko beam is considered, whose equations of motion up to the third order contain several nonlinear terms. The question is: are all these terms important, or do any of them play a minor role and can thus be neglected? To answer these questions, in this chapter, a detailed comparison of the role of nonlinear terms in the geometrically exact equations is given. The outcome is a simplified model having only the most important nonlinear terms.

5.1. Introduction

According to Villaggio (1997, section 20), "The first complete theory of large displacement but small strain in initially straight elastic rods was proposed by Kirchhoff (1859), who extended an early result of Euler (1744)". Limiting to the planar case, the Kirchhoff theory assumes that (i) the axis is inextensible; (ii) the bending moment depends linearly on the curvature; (iii) there is no shear deformation between the cross-section and the axis; and (iv) the cross-section is undeformable in its plane. This is what is today known as the *Euler–Bernoulli beam*, both in linear and nonlinear regimes (Hodges 1980), even if the axial inextensibility is relaxed.

Chapter written by Stefano LENCI and Giuseppe REGA.

When rotatory inertia is considered, sometimes it is called the Rayleigh beam (Rao 2007), but when the beam is loaded only by concentrated forces and moments at one end, it is called *elastica* (Love 1944; Villaggio 1997).

Early contributions by Galileo, Mariotte, Jacob Bernoulli, Euler and Saint-Venant were reported by Timoshenko (1983), while the contribution by da Vinci was highlighted by Ballarini (2003). Moreover, Love (1944) addressed the problem, considering the contributions of Kirchhoff and Clebsch (Clebsch 1862) to the initially curved beams. Other interesting historical hints and perspectives can be found in Nayfeh and Pai (2004), Antman (2005), Rao (2007) and Lacarbonara (2013), who have also reported on some more recent important contributions. Historical surveys, quoting more recent – and less known – and not only "modern" papers, can be found in the introduction of Hamdan and Shabneh (1997) and in Sathyamoorthy (1982) and Marur (2001).

Relaxing the unshearability hypothesis, we fall in the realm of *Timoshenko beams*. For a detailed history of this model in the linear case, including its variants, we refer to Elishakoff *et al.* (2015) and Elishakoff (2019). Early works on the nonlinear oscillations of Timoshenko beams are those by Rao *et al.* (1976) and Luongo *et al.* (1987). The kinematically exact equations of motion for the planar nonlinear Timoshenko beam are given in Lenci and Rega (2016a) and Lenci *et al.* (2016) and will be summarized in section 5.2.

The "full" exact model, having axial, shear and flexural deformation, is also known as the Cosserat model (Villaggio 1997; Antman 2005; Lacarbonara 2013), although sometimes this reference is more related to the use of the director theory to obtain the model rather than the mechanical characteristics of the model itself.

In the case of the elastica, the exact solution is known, in terms of Jacobian elliptic functions, when the loads are concentrated forces at the end of the beam (Love 1944; Villaggio 1997). However, when distributed loads are present, and/or axial deformations cannot be neglected, the solution is no longer available, which thus triggers the development of approximate beam models, liable to having simpler solutions while keeping the main relevant mechanical behaviors. Lacarbonara (2013) states that approximate models "are usually based on *ad hoc* mechanical models which account for geometric and inertial nonlinearities by resorting to variational formulations based on truncated kinematic models".

Among the many models that have been developed, we first mention the model developed by Mettler (1962), since it is well known and has been used extensively in the past. It is valid within the Euler–Bernoulli framework, when moderate displacements are considered, and only for axially restraining boundary conditions. It is based on the main *ad hoc* assumptions that (i) curvature nonlinearities can be neglected; (ii) the axial strain can be approximated by $\varepsilon = W' + 1/2U'^2$

(i.e. quadratic terms in the axial displacement W are neglected, which is similar to the von Kármán hypothesis for plates); and (iii) the axial and rotatory inertia can be neglected. The latter, in particular, allows us to perform the so-called static (or kinematic) condensation, in which the axial displacement W is expressed as a (nonlinear) function of the transversal displacement U, thus becoming a "slave" unknown. The ensuing governing equation, without damping and excitation, is (Lacarbonara 2013)

$$\rho A \ddot{U} + EJ\,U^{IV} - \frac{EA}{L}U''[W(L,T) - W(0,T)] - \frac{EA}{2L}U'' \int_0^L U'^2 dz = 0, \quad [5.1]$$

where ρA is the mass per unit length, EA and EJ are the axial and flexural stiffnesses, respectively, and L is the length of the beam. The dot denotes the derivative with respect to time T, and the prime denotes the derivative with respect to the spatial coordinate Z.

One of the motivations of the success of this model relies on the fact that the nonlinearities are of integral form, a circumstance that greatly simplifies the solution, and that for certain boundary conditions allows us to easily obtain the nonlinear exact solution.

A simplified model developed by the Russian school is the Bolotin (1956) model:

$$\begin{aligned} &EA\,W'' - \rho A \ddot{W} = \rho A \int_0^Z [U'\ddot{U}' + (\dot{U}')^2] d\xi, \\ &EJ\,U^{IV} + \rho A \ddot{U} + EA\,(U'W')' = 0, \end{aligned} \quad [5.2]$$

which also takes care of the inertial nonlinearity in an approximate way (see the integral in the first equation), together with the axial strain nonlinearity (in the second equation). Again, no curvature nonlinearities are considered.

A model that instead considers the curvature nonlinearities is the one developed by Takahashi (1979) (note that, in the original work, the nonlinear terms have an incorrect sign, as highlighted by Hamdan and Shabneh (1997)):

$$\rho A \ddot{U} + EJ\,U^{IV} + EJ\,(2U'U''U''' + U''^2 + 2U'^3U''U''' + 3U''^3U'^2), \quad [5.3]$$

which assumes that the beam is inextensible (allowing us to obtain W as a function of U) and is a fifth-order approximation of a model previously obtained by Wagner (1965).

Lacarbonara (2013) highlighted that "there seems to be a lack of internal consistency in the way the truncated models of nonlinear beams are obtained by means of *ad hoc* kinematic approximations". In Figures 5.14 and 5.15, he addressed the problem by comparing the results of the exact shearable, exact unshearable and Mettler models, in terms of frequency response curves for forced vibrations with immovable boundary conditions. He concluded that the three models provide the same results for slender beams (Figure 5.15) and different results for thick beams (Figure 5.14), as somehow expected, although a detailed discussion has not been reported.

A comparison of different approximate models was performed by Singh *et al.* (1990), who discussed seven different formulations (see their Table 2), also paying attention to some mechanical items such as the effect of the axial displacement (see their Table 1).

Although limited to the linear regime, an interesting comparison between the Euler–Bernoulli, Timoshenko and two-dimensional elasticity models was made by Labuschagne *et al.* (2009), while a control-oriented comparison was studied by Morris and Vidyasagar (1990).

Here, still with the (first) aim of checking the reliability of approximate models, a different approach is proposed. No *a priori* (and *ad hoc*) mechanical assumptions are made to neglect some terms, but *a posteriori* we compare the importance of all nonlinear terms appearing in the kinematically exact equations of motion, based on an approximate solution obtained by the Poincaré–Lindstedt method. Thus, the negligibility of some terms is not assumed, but checked, thus allowing us to select only the most important nonlinear terms, which is the second goal of this work. Contrary to Singh *et al.* (1990) and Lacarbonara (2013), who provided only an overall comparison of different models, here we perform a detailed comparison of each nonlinear term in the geometrically exact equations of motion, highlighting the most important and negligible ones.

In fact, this work, aimed at checking the relevance of nonlinear terms, is a continuation of a series of papers by the authors and co-workers on the nonlinear Timoshenko beam hinged on one side and simply supported on the other side with a linear boundary spring in the axial direction. The stiffness of the spring can range from $\kappa = 0$, meaning that a simply supported (axially movable) boundary condition is considered, up to $\kappa \to \infty$, representing a hinged (axially immovable) constraint.

These two limit cases are mechanically very different: the former is (mildly) softening, and here the axial deformation is negligible while the axial inertia is very important; the latter is (strongly) hardening, and the axial deformation is important while the axial inertia can be neglected (Lenci *et al.* 2016). By varying the spring stiffness, we can easily switch between these two opposite nonlinear behaviors.

In Lenci and Rega (2016a), the equations of motion were obtained by considering geometrically exact kinematics; axial, shear and bending deformations; axial and rotatory inertia; and linear elastic constitutive behavior. The Poincaré–Lindstedt method was applied to obtain the third-order approximate solution for free nonlinear oscillations.

A detailed investigation of the effects of the various mechanical parameters, in particular the stiffness of the boundary spring, on the frequency nonlinear correction coefficient (which summarizes the nonlinear effects) was performed by Lenci *et al.* (2016). The negligibility of axial and rotational inertia was also discussed.

Previous analytical findings were compared with numerical simulations by Clementi *et al.* (2017) to check their reliability or better the amplitude extent to which they are accurate.

A comparison between two different definitions of curvature, the geometrical and the mechanical one, was made by Lenci *et al.* (2017a), showing that the difference was minimal for thick beams and negligible for slender structures. This topic was further explored by Lenci *et al.* (2017b), where the differences between the two curvatures were addressed from a theoretical point of view.

The main contribution by Lenci and Rega (2016b) was to consider both axial and transversal displacements (the former being neglected in previous works) starting from the first order. It was then shown that, depending on the slenderness of the beam and the stiffness of the boundary spring, six different cases could exist: (i)–(ii) only transversal uncoupled solutions exist and are hardening (softening); (iii)–(iv) axial–transversal coupled and transversal uncoupled solutions coexist, and are both hardening (softening); and (v)–(vi) axial–transversal coupled and transversal uncoupled solutions coexist, the former (latter) being softening and the latter (former) hardening. No internal resonances between axial and transversal linear modes were considered.

While, in previous works, only free nonlinear oscillations were considered, Kloda *et al.* (2018) studied the effect of excitation and damping. The multiple time-scale method was used instead of the Poincaré–Lindstedt method to obtain the frequency response curves. The analytical results were compared with numerical (FEM) simulations, showing a good agreement up to moderate amplitudes of the displacements. The numerical simulations further highlighted the presence of superharmonic resonances in the neighborhood of the principal resonance for certain values of the spring stiffness.

Higher-order resonances were considered by Kloda *et al.* (2019), while the comparison between the numerical simulation and preliminary experimental results was made by Kloda *et al.* (2020). It should be noted that the experimental results can be reproduced very well, even for large excitations, by assuming an amplitude-dependent damping.

5.2. The geometrically exact planar Timoshenko beam

The geometrically exact equations of the free motions of a planar, initially straight, linearly elastic, shearable (Timoshenko) beam are (Lenci and Rega 2016a, 2016b; Lenci *et al.* 2016; Clementi *et al.* 2017):

$$\begin{aligned}&\Bigg\{EA[\sqrt{(1+W')^2+U'^2}-1]\frac{1+W'}{\sqrt{(1+W')^2+U'^2}}\\&+GA\left[\theta-\arctan\left(\frac{U'}{1+W'}\right)\right]\frac{U'}{\sqrt{(1+W')^2+U'^2}}\Bigg\}'=\rho A\ddot{W},\end{aligned} \qquad [5.4]$$

$$\begin{aligned}&\Bigg\{EA[\sqrt{(1+W')^2+U'^2}-1]\frac{U'}{\sqrt{(1+W')^2+U'^2}}\\&-GA\left[\theta-\arctan\left(\frac{U'}{1+W'}\right)\right]\frac{1+W'}{\sqrt{(1+W')^2+U'^2}}\Bigg\}'=\rho A\ddot{U},\end{aligned} \qquad [5.5]$$

$$\begin{aligned}&\left[EJ\frac{\theta'}{\sqrt{(1+W')^2+U'^2}}\right]'\\&-GA\left[\theta-\arctan\left(\frac{U'}{1+W'}\right)\right]\sqrt{(1+W')^2+U'^2}=\rho J\ddot{\theta}.\end{aligned} \qquad [5.6]$$

Here, $W(Z,T)$ and $U(Z,T)$ are the displacements along the Z and X axes (i.e. along and perpendicular to the undeformed configuration), $\theta(Z,T)$ is the rotation of the cross-section, while EA, GA and EJ are the axial, shear and bending stiffnesses, respectively, and ρA and ρJ are the translational and rotational inertia, respectively; they are constant since the beam is assumed to be homogeneous. The geometrical curvature is used (Lenci *et al.* 2017a).

The boundary conditions are

$$\begin{aligned}&U(0,T)=0, \quad M(0,T)=0, \quad W(0,T)=0,\\&U(L,T)=0, \quad M(L,T)=0, \quad H_o(L,T)+\kappa W(L,T)=0,\end{aligned} \qquad [5.7]$$

where M is the bending moment and H_o is the internal force in the Z direction (which is the axial force only in the linear approximation) (see Lenci *et al.* (2016) for more details).

Although equations [5.4]–[5.6] are valid for arbitrarily large displacements and rotations, their mathematical third-order approximations are reported since asymptotic methods (up to the third order) are used:

$$EA\left(W' + \frac{1}{2}U'^2 - U'^2W'\right)' + GA(U'\theta - U'^2 + 2U'^2W' - U'W'\theta)' = \rho A\ddot{W}, \quad [5.8]$$

$$EA\left(U'W' + \frac{1}{2}U'^3 - U'W'^2\right)' + GA\left(U' - \theta - U'W' + \frac{1}{2}U'^2\theta - \frac{5}{6}U'^3 + U'W'^2\right)' = \rho A\ddot{U}, \quad [5.9]$$

$$EJ\left(\theta' - W'\theta' + W'^2\theta' - \frac{1}{2}U'^2\theta'\right)' + GA\left(U' - \theta - W'\theta - \frac{1}{2}U'^2\theta + \frac{1}{6}U'^3\right) = \rho J\ddot{\theta}. \quad [5.10]$$

To this order, we also have

$$M = EJ\theta'\left(1 - W' + W'^2 - \frac{1}{2}U'^2\right), \quad [5.11]$$

$$H_o = EA\left(W' + \frac{1}{2}U'^2 - U'^2W'\right) + GA(\theta U' - U'^2 + 2U'^2W' - \theta U'W'). \quad [5.12]$$

Assuming that $W(Z,T)$ and $U(Z,T)$ are of the same order, which is the case considered by Lenci and Rega (2016b), we can collect terms with the same order on the left-hand side of each equation:

$$\begin{aligned} \text{first:}\quad & +EA\,W'' \\ \text{second:}\quad & +EA\,U'U'' + GA\,(U''\theta + U'\theta' - 2U'U'') \\ \text{third:}\quad & -EA\,(U'^2W'' + 2\,U'U''W') \\ & +GA\,(2\,U'^2W'' + 4\,U'U''W' - U''W'\theta - U'W''\theta - U'W'\theta') \\ & = \rho A\ddot{W}, \end{aligned} \quad [5.13]$$

$$\begin{aligned}
\text{first:}\quad & + GA\,(U'' - \theta') \\
\text{second:}\quad & + EA\,(U''W' + U'W'') - GA\,(U''W' + U'W'') \\
\text{third:}\quad & + EA\,(3/2\,U'^2U'' - U''W'^2 - 2\,U'W'W'') \\
& + GA\,(U'U''\theta + 1/2\,U'^2\theta' - 5/2\,U'^2U'' + U''W'^2 + 2\,U'W'W'') \\
& = \rho A\ddot{U},
\end{aligned} \qquad [5.14]$$

$$\begin{aligned}
\text{first:}\quad & + EJ\,\theta'' + GA\,(U' - \theta) \\
\text{second:}\quad & - EJ\,(W''\theta' + W'\theta'') - GA\,W'\theta \\
\text{third:}\quad & + EJ\,(W'^2\theta'' + 2\,W'W''\theta' - 1/2\,U'^2\theta'' - U'U''\theta') \\
& + GA\,(-1/2\,U'^2\theta + 1/6\,U'^3) \\
& = \rho J\,\ddot{\theta}.
\end{aligned} \qquad [5.15]$$

If, on the other hand, we assume that the axial displacement W is smaller than the transversal displacements U and θ, for example U and θ are of the first order and W is of the second order, which is the case considered by Lenci and Rega (2016a), Lenci *et al.* (2016) and Clementi *et al.* (2017), the previous equations become

$$\begin{aligned}
\text{first:}\quad & \text{none} \\
\text{second:}\quad & + EA\,(W'' + U'U'') + GA\,(U''\theta + U'\theta' - 2U'U'') \\
\text{third:}\quad & \text{none} \\
& = \rho A\ddot{W},
\end{aligned} \qquad [5.16]$$

$$\begin{aligned}
\text{first:}\quad & + GA\,(U'' - \theta') \\
\text{second:}\quad & \text{none} \\
\text{third:}\quad & + EA\,(U''W' + U'W'' + 3/2\,U'^2U'') \\
& + GA\,(-U''W' - U'W'' + U'U''\theta + 1/2\,U'^2\theta' - 5/2\,U'^2U'') \\
& = \rho A\ddot{U},
\end{aligned} \qquad [5.17]$$

$$\begin{aligned}
\text{first:}\quad & + EJ\,\theta'' + GA\,(U' - \theta) \\
\text{second:}\quad & \text{none} \\
\text{third:}\quad & + EJ\,(-W''\theta' - W'\theta'' - 1/2\,U'^2\theta'' - U'U''\theta') \\
& + GA\,(-W'\theta - 1/2\,U'^2\theta + 1/6\,U'^3) \\
& = \rho J\,\ddot{\theta}.
\end{aligned} \qquad [5.18]$$

The first conclusion is that the following terms, which are of order higher than the third, can be neglected:

$$\begin{aligned} \text{1st equation:} \quad & -EA\,(U'^2W'' + 2\,U'U''W') \\ & + GA\,(2\,U'^2W'' + 4\,U'U''W' \\ & - U''W'\theta - U'W''\theta - U'W'\theta'), \end{aligned} \qquad [5.19]$$

$$\begin{aligned} \text{2nd equation:} \quad & -EA\,(U''W'^2 + 2\,U'W'W'') \\ & + GA\,(U''W'^2 + 2\,U'W'W''), \end{aligned} \qquad [5.20]$$

$$\text{3rd equation:} \quad + EJ\,(W'^2\theta'' + 2\,W'W''\theta'), \qquad [5.21]$$

This is the case considered in this chapter, which is further aimed at studying the relative role played by the five, eight and seven nonlinear terms, respectively, in equations [5.16]–[5.18].

5.3. The asymptotic solution

The asymptotic solution of [5.4]–[5.6] up to the third order, obtained by the Poincaré–Lindstedt method, is sought after in the form (Lenci and Rega 2016a; Lenci *et al.* 2016; Clementi *et al.* 2017)

$$W(Z,t) = \varepsilon^2 W_2(Z,t) + \varepsilon^3 W_3(Z,t) + ..., \qquad [5.22]$$

$$U(Z,t) = \varepsilon U_1(Z,t) + \varepsilon^2 U_2(Z,t) + \varepsilon^3 U_3(Z,t) + ..., \qquad [5.23]$$

$$\theta(Z,t) = \varepsilon \theta_1(Z,t) + \varepsilon^2 \theta_2(Z,t) + \varepsilon^3 \theta_3(Z,t) + ..., \qquad [5.24]$$

$$\omega = \omega_0 + \varepsilon^2 \omega_2 + ..., \qquad [5.25]$$

$$t = \omega T, \qquad [5.26]$$

where ω is the nonlinear frequency. Up to the second order, the solution is given by

$$U_1(Z,t) = U_{1a}(Z)\sin(t), \qquad [5.27]$$

$$\theta_1(Z,t) = \theta_{1a}(Z)\sin(t), \qquad [5.28]$$

$$W_2(Z,t) = W_{2a}(Z) + W_{2b}(Z)\cos(2t), \qquad [5.29]$$

$$U_2(Z,t) = 0, \qquad [5.30]$$

$$\theta_2(Z,t) = 0, \qquad [5.31]$$

where

$$\frac{U_{1a}(Z)}{U_a} = \sin\left(\frac{n\pi Z}{L}\right), \tag{5.32}$$

$$\frac{\theta_{1a}(Z)}{U_a} = \alpha_1 \frac{n\pi}{L} \cos\left(\frac{n\pi Z}{L}\right), \tag{5.33}$$

$$\alpha_1 = \left(1 - \frac{\rho J}{GA}\omega_0^2 + \frac{EJ}{GA}\frac{n^2\pi^2}{L^2}\right)^{-1}, \tag{5.34}$$

$$\omega_0^2 = \frac{n^2\pi^2}{2L^2}\left(\frac{EJ}{\rho J} + \frac{GA}{\rho A}\right) + \frac{GA}{2\rho J} - \sqrt{\frac{n^4\pi^4}{4L^4}\left(\frac{EJ}{\rho J} - \frac{GA}{\rho A}\right)^2 + \frac{n^2\pi^2 GA}{2L^2\rho J}\left(\frac{EJ}{\rho J} + \frac{GA}{\rho A}\right) + \left(\frac{GA}{2\rho J}\right)^2}, \tag{5.35}$$

$$\frac{W_{2a}(Z)}{U_a^2} = -\frac{n\pi}{16L}\left[1 + 2\frac{GA}{EA}(\alpha_1 - 1)\right]\sin\left(2\frac{n\pi Z}{L}\right) + \frac{c_1}{L}\frac{Z}{L}, \tag{5.36}$$

$$\frac{W_{2b}(Z)}{U_a^2} = \frac{n^3\pi^3}{16L^3}\frac{1 + 2\frac{GA}{EA}(\alpha_1 - 1)}{\frac{n^2\pi^2}{L^2} - \frac{\rho A}{EA}\omega_0^2}\sin\left(2\frac{n\pi Z}{L}\right) + \frac{c_2}{L}\sin\left(\frac{2\omega_0\sqrt{\rho A}}{\sqrt{EA}}Z\right), \tag{5.37}$$

$$c_1 = -n^2\pi^2\frac{\frac{1}{8} + \frac{GA}{4EA}(\alpha_1 - 1)}{1 + \frac{\kappa L}{EA}}, \tag{5.38}$$

$$c_2 = \frac{n^2\pi^2}{8}\frac{\dfrac{\left[1 + 2\frac{GA}{EA}(\alpha_1 - 1)\right]\left(2\omega_0^2\frac{\rho A}{EA}L^2 - n^2\pi^2\right)}{\omega_0^2\frac{\rho A}{EA}L^2 - n^2\pi^2}}{\frac{\kappa L}{EA}\sin\left(\frac{2\omega_0\sqrt{\rho A}}{\sqrt{EA}}L\right) + 2L\omega_0\sqrt{\frac{\rho A}{EA}}\cos\left(\frac{2\omega_0\sqrt{\rho A}}{\sqrt{EA}}L\right)}, \tag{5.39}$$

and the boundary conditions [5.7] are used. Details on the derivation of the previous solution can be found in Lenci and Rega (2016a) and Lenci *et al.* (2016). For different boundary conditions, the analytical expressions are more involved, but can be detected by a symbolic software manipulator.

The dimensionless slenderness (l), the spring stiffness (κ_h) and the shear stiffness ($z = [2(1+\nu)\chi]^{-1}$, where ν is the Poisson coefficient and χ is the shear correction coefficient) are conveniently introduced through

$$EA = \frac{EJ\,l^2}{L^2}, \quad \rho J = \frac{\rho A\,L^2}{l^2}, \quad GA = \frac{EJ\,l^2}{L^2} z, \quad \kappa = \frac{EJ\,\kappa_h}{L^3}. \qquad [5.40]$$

In the following, we assume $\nu = 0.3$ and $\chi = 1.2$, which gives $z = 0.3205$. Only l and κ_h (and the order of resonance n) are left free to vary for parametric analyses. Furthermore, we will use the dimensionless amplitude $a = \varepsilon U_a/L$. Note that axial/rotational inertias and strains are *not* neglected.

For the expression of ω_2, which is important but not relevant for this work, we refer to Lenci and Rega (2016a) and Lenci *et al.* (2016).

5.4. The importance of nonlinear terms

5.4.1. *An initial case*

To begin with, let us consider the case $l = 100$ (i.e. a quite slender beam), $\kappa_h = 0$ (no end spring, i.e. a simply supported – or movable – boundary) and $n = 1$ (first resonance). In this case, we have (where $s = Z/L$ is the dimensionless space variable)

$$\frac{U_{1a}}{L\,a} = \sin(\pi s), \qquad [5.41]$$

$$\frac{\theta_{1a}}{a} = 3.1320\cos(\pi s), \qquad [5.42]$$

$$\frac{W_{2a}}{L\,a^2} = -0.1960\sin(6.2832s) - 1.2313s, \qquad [5.43]$$

$$\frac{W_{2b}}{L\,a^2} = 0.1962\sin(6.2833s) + 6.3673\sin(0.1970s), \qquad [5.44]$$

$$\omega L^2\sqrt{\frac{\rho A}{EJ}} = 9.8498 - 17.611\,a^2 + ... \qquad [5.45]$$

The constant nonlinear terms in the *axial* equation of motion [5.16], which are related to W_{2a}, are then

$$\textbf{(1)} = EAW''\frac{L^3}{EJ\,a^2} \to 77363.3\sin(2\pi s) + 0.000001\sin(0.1970s), \quad [5.46]$$

$$\textbf{(2)} = EAU'U''\frac{L^3}{EJ\,a^2} \to -77515.7\sin(2\pi s), \qquad [5.47]$$

$$\textbf{(3)} = GAU''\theta\frac{L^3}{EJ\,a^2} \to GAU'\theta'\frac{L^3}{EJ\,a^2} = -24767.6\sin(2\pi s), \qquad [5.48]$$

$$\textbf{(4)} = -2GAU'U''\frac{L^3}{EJ\,a^2} \to 49687.6\sin(2\pi s), \qquad [5.49]$$

and are shown in Figure 5.1a.

The nonlinear terms proportional to $\cos(2t)$ in [5.16], which are related to W_{2b}, are instead

$$\textbf{(1)} = EAW''\frac{L^3}{EJ\,a^2} \to -77439.4\sin(2\pi s) - 2471.0\sin(0.1970s), \qquad [5.50]$$

$$\textbf{(2)} = EAU'U''\frac{L^3}{EJ\,a^2} \to 77515.7\sin(2\pi s), \qquad [5.51]$$

$$\textbf{(3)} = GAU''\theta\frac{L^3}{EJ\,a^2} \to GAU'\theta'\frac{L^3}{EJ\,a^2} = 24767.6\sin(2\pi s), \qquad [5.52]$$

$$\textbf{(4)} = -2GAU'U''\frac{L^3}{EJ\,a^2} \to -49687.6\sin(2\pi s), \qquad [5.53]$$

and are shown in Figure 5.1b.

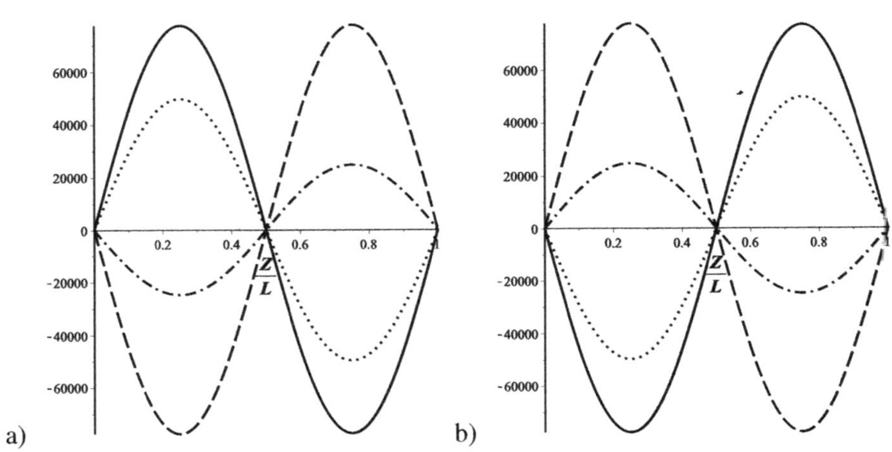

Figure 5.1. *The nonlinear terms in the axial direction. a) Constant terms; b) terms proportional to* $\cos(2t)$. ***(1)*** *= continuous;* ***(2)*** *= dashed;* ***(3)*** *= dashdot;* ***(4)*** *= dot.* $l = 100$, $n = 1$, $\kappa_h = 0$

Figure 5.1 shows that all nonlinear terms are important in this case. To have a *quantitative* comparison between the different terms, we compute their L^2 norm ($|f|_2 = \sqrt{\int_0^L f(Z)^2 dZ}$), and then their percentage with respect to the L^2 norm of all nonlinear terms. The results are reported in Table 5.1, and confirm that all terms are comparable between each other, and thus none can be neglected. Note that the two columns practically provide the same results.

	Constant term	Proportional to $\cos(2t)$
(1)	30.45%	30.42%
(2)	30.51%	30.52%
(3)	19.49%	19.50%
(4)	19.55%	19.56%

Table 5.1. *The L^2 norm percentage of the various nonlinear terms in the axial equation. $l = 100$, $n = 1$, $\kappa_h = 0$*

We now address the transversal [5.17] and rotational [5.18] equations. Here, there are also linear terms, which are not discussed since we are only interested in comparing different nonlinear terms.

The nonlinear terms proportional to $\sin(t)$ in the *transversal* equation of motion [5.17] are

$$\textbf{(1)} = EAU''W'\frac{L^3}{EJ\,a^3} \rightarrow 91171.4\sin(3\pi s) + 30350.6\sin(\pi s) + 30949.2[\sin(3.3386s) + \sin(2.9446s)], \quad [5.54]$$

$$\textbf{(2)} = EAU'W''\frac{L^3}{EJ\,a^3} \rightarrow 182342.8[\sin(3\pi s) + \sin(\pi s)] + 1940.7[\sin(3.3386s) - \sin(2.9446s)], \quad [5.55]$$

$$\textbf{(3)} = \frac{3}{2}EAU'^2U''\frac{L^3}{EJ\,a^3} \rightarrow -273963.1[\sin(3\pi s) + \sin(\pi s)], \quad [5.56]$$

$$\textbf{(4)} = -GAU''W'\frac{L^3}{EJ\,a^3} \rightarrow -29220.4\sin(3\pi s) - 9727.4\sin(\pi s) - 9919.2[\sin(3.3386s) + \sin(2.9446s)], \quad [5.57]$$

$$\textbf{(5)} = -GAU'W''\frac{L^3}{EJ\,a^3} \rightarrow -58440.9[\sin(3\pi s) + \sin(\pi s)] + 621.99[-\sin(3.3386s) + \sin(2.9446s)], \quad [5.58]$$

$$\textbf{(6)} = GAU'U''\theta\frac{L^3}{EJ\,a^3} \to -58357.2[\sin(3\pi s) + \sin(\pi s)], \qquad [5.59]$$

$$\textbf{(7)} = \frac{1}{2}GAU'^2\theta'\frac{L^3}{EJ\,a^3} \to -29178.6[\sin(3\pi s) + \sin(\pi s)], \qquad [5.60]$$

$$\textbf{(8)} = -\frac{5}{2}GAU'^2U''\frac{L^3}{EJ\,a^3} \to 146341.9[\sin(3\pi s) + \sin(\pi s)], \qquad [5.61]$$

and are shown in Figure 5.2a.

The nonlinear terms proportional to $\sin(3t)$ in [5.17] are instead

$$\textbf{(1)} = EAU''W'\frac{L^3}{EJ\,a^3} \to 30410.4[-\sin(3\pi s) + \sin(\pi s)]$$
$$- 30949.2[\sin(3.3386s) + \sin(2.9446s)], \qquad [5.62]$$

$$\textbf{(2)} = EAU'W''\frac{L^3}{EJ\,a^3} \to -60820.8[\sin(3\pi s) + \sin(\pi s)]$$
$$+ 1940.7[-\sin(3.3386s) + \sin(2.9446s)], \qquad [5.63]$$

$$\textbf{(3)} = \frac{3}{2}EAU'^2U''\frac{L^3}{EJ\,a^3} \to 91321.0[\sin(3\pi s) + \sin(\pi s)], \qquad [5.64]$$

$$\textbf{(4)} = -GAU''W'\frac{L^3}{EJ\,a^3} \to 9746.5[\sin(3\pi s) - \sin(\pi s)]$$
$$+ 9919.2[\sin(3.3386s) + \sin(2.9446s)], \qquad [5.65]$$

$$\textbf{(5)} = -GAU'W''\frac{L^3}{EJ\,a^3} \to 19493.0[\sin(3\pi s) + \sin(\pi s)]$$
$$+ 621.99[\sin(3.3386s) - \sin(2.9446s)], \qquad [5.66]$$

$$\textbf{(6)} = GAU'U''\theta\frac{L^3}{EJ\,a^3} \to 19452.4[\sin(3\pi s) + \sin(\pi s)], \qquad [5.67]$$

$$\textbf{(7)} = \frac{1}{2}GAU'^2\theta'\frac{L^3}{EJ\,a^3} \to 9726.2[\sin(3\pi s) + \sin(\pi s)], \qquad [5.68]$$

$$\textbf{(8)} = -\frac{5}{2} GAU'^2 U'' \frac{L^3}{EJ\,a^3} \to -48780.6[\sin(3\pi s) + \sin(\pi s)], \qquad [5.69]$$

and are shown in Figure 5.2b.

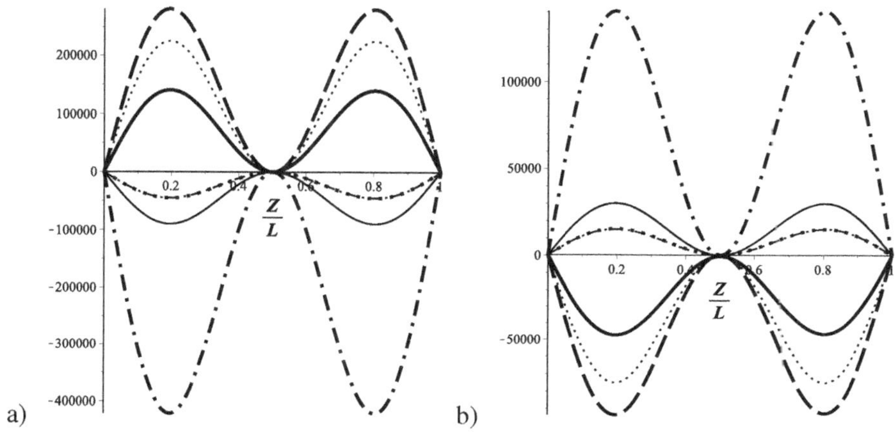

Figure 5.2. *The nonlinear terms in the transversal direction. a) Terms proportional to* $\sin(t)$*; b) terms proportional to* $\sin(3t)$*.* ***(1)*** *= Thick continuous;* ***(2)*** *= thick dashed;* ***(3)*** *= thick dashdot;* ***(4)*** *= thick dot;* ***(5)*** *= thin continuous;* ***(6)*** *= thin dashed;* ***(7)*** *= thin dashdot;* ***(8)*** *= thin dot.* $l = 100$, $n = 1$, $\kappa_h = 0$

Again, a quantitative comparison is obtained by percentages of the L^2 norm of each term, which are reported in Table 5.2. We note that, in this case, the terms $GAU''W'$ and $\frac{1}{2}GAU'^2\theta'$ are smaller than the others and could be neglected in the equation with a reasonable (but not infinitesimal) error.

	Proportional to $\sin(t)$	Proportional to $\sin(3t)$
(1)	10.53%	10.61%
(2)	20.96%	20.92%
(3)	31.52%	31.49%
(4)	3.37%	3.40%
(5)	6.72%	6.70%
(6)	6.71%	6.71%
(7)	3.36%	3.35%
(8)	16.83%	16.82%

Table 5.2. *The* L^2 *norm percentage of the various nonlinear terms in the transversal equation.* $l = 100$, $n = 1$, $\kappa_h = 0$

The nonlinear terms proportional to $\sin(t)$ in the *rotational* equation of motion [5.18] are

$$\begin{aligned}(\mathbf{1}) &= -EJW''\theta'\frac{L^2}{EJ\,a^3} \to 57.109[-\cos(3\pi s)+\cos(\pi s)]\\ &+0.6078[-\cos(3.3386s)+\cos(2.9446s)],\end{aligned} \quad [5.70]$$

$$\begin{aligned}(\mathbf{2}) &= -EJW'\theta''\frac{L^2}{EJ\,a^3} \to -28.554\cos(3\pi s)-66.615\cos(\pi s)\\ &-9.6931[\cos(3.3386s)+\cos(2.9446s)],\end{aligned} \quad [5.71]$$

$$(\mathbf{3}) = -EJU'U''\theta'\frac{L^2}{EJ\,a^3} \to 57.203[\cos(3\pi s)-\cos(\pi s)], \quad [5.72]$$

$$(\mathbf{4}) = -\frac{1}{2}EJU'^2\theta''\frac{L^2}{EJ\,a^3} \to 28.601\cos(3\pi s)+85.804\cos(\pi s), \quad [5.73]$$

$$\begin{aligned}(\mathbf{5}) &= -GAW'\theta\frac{L^2}{EJ\,a^3} \to 9272.6\cos(3\pi s)+21632.1\cos(\pi s)\\ &+3147.71[\cos(3.3386s)+\cos(2.9446s)],\end{aligned} \quad [5.74]$$

$$(\mathbf{6}) = -\frac{1}{2}GAU'^2\theta\frac{L^2}{EJ\,a^3} \to -9287.8\cos(3\pi s)-27863.5\cos(\pi s), \quad [5.75]$$

$$(\mathbf{7}) = \frac{1}{6}GAU'^3\frac{L^2}{EJ\,a^3} \to 3105.5\cos(3\pi s)+9316.4\cos(\pi s), \quad [5.76]$$

and are shown in Figure 5.3a.

The nonlinear terms proportional to $\sin(3t)$ in [5.18] are instead

$$\begin{aligned}(\mathbf{1}) &= -EJW''\theta'\frac{L^2}{EJ\,a^3} \to 19.049[\cos(3\pi s)-\cos(\pi s)]\\ &+0.6078[\cos(3.3386s)-\cos(2.9446s)],\end{aligned} \quad [5.77]$$

$$\begin{aligned}(\mathbf{2}) &= -EJW'\theta''\frac{L^2}{EJ\,a^3} \to 9.5244[\cos(3\pi s)+\cos(\pi s)]\\ &+9.6931[\cos(3.3386s)+\cos(2.9446s)],\end{aligned} \quad [5.78]$$

$$\textbf{(3)} = -EJU'U''\theta' \frac{L^2}{EJ\,a^3} \to 19.067[-\cos(3\pi s) + \cos(\pi s)], \qquad [5.79]$$

$$\textbf{(4)} = -\frac{1}{2}EJU'^2\theta'' \frac{L^2}{EJ\,a^3} \to -9.5338\cos(3\pi s) - 28.601\cos(\pi s), \qquad [5.80]$$

$$\textbf{(5)} = -GAW'\theta \frac{L^2}{EJ\,a^3} \to -3092.9[\cos(3\pi s) + \cos(\pi s)]$$
$$- 3147.7[\cos(3.3386s) + \cos(2.9446s)], \qquad [5.81]$$

$$\textbf{(6)} = -\frac{1}{2}GAU'^2\theta \frac{L^2}{EJ\,a^3} \to 3095.9\cos(3\pi s) + 9287.8\cos(\pi s), \qquad [5.82]$$

$$\textbf{(7)} = \frac{1}{6}GAU'^3 \frac{L^2}{EJ\,a^3} \to -1035.2\cos(3\pi s) - 3105.5\cos(\pi s), \qquad [5.83]$$

and are shown in Figure 5.3b.

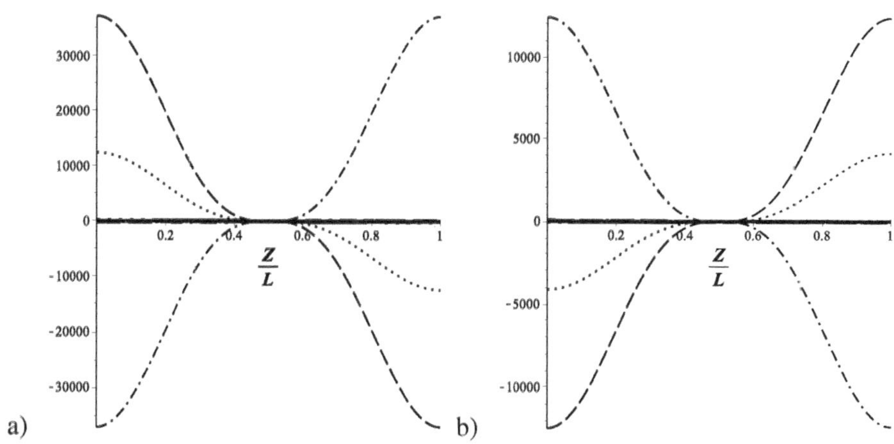

Figure 5.3. *The nonlinear terms in the rotational direction. a) Terms proportional to* $\sin(t)$*; b) terms proportional to* $\sin(3t)$*.* ***(1)*** *= Thick continuous;* ***(2)*** *= thick dashed;* ***(3)*** *= thick dashdot;* ***(4)*** *= thick dot;* ***(5)*** *= thin continuous;* ***(6)*** *= thin dashed;* ***(7)*** *= thin dashdot.* $l = 100$, $n = 1$, $\kappa_h = 0$

The quantitative comparison obtained by percentage of the L^2 norm of each term is reported in Table 5.3. We note that, in this case, the terms $EJW''\theta'$, $EJW'\theta''$, $EJU'U''\theta'$ and $\frac{1}{2}EJU'^2\theta''$ (i.e. the terms $EJ(W'\theta')'$ and $\frac{1}{2}EJ(U'^2\theta')'$) can certainly be neglected.

	Proportional to $\sin(t)$	Proportional to $\sin(3t)$
(1)	0.12%	0.12%
(2)	0.13%	0.13%
(3)	0.12%	0.12%
(4)	0.13%	0.13%
(5)	42.63%	42.74%
(6)	42.62%	42.54%
(7)	14.25%	14.22%

Table 5.3. *The L^2 norm percentage of the various nonlinear terms in the rotational equation.* $l = 100$, $n = 1$, $\kappa_h = 0$

5.4.2. *The effect of the slenderness*

To examine the effect of the slenderness, we keep $n = 1$ and $\kappa_h = 0$ and vary l from $l = 5$ (going below this threshold is unphysical – for example, for a rectangular cross-section, for which $l = \sqrt{12}(L/H)$, which means that the length L should be lesser than 1.44 times the height H) to $l = 100$ (above this threshold, all values are practically independent of l. See the following figures).

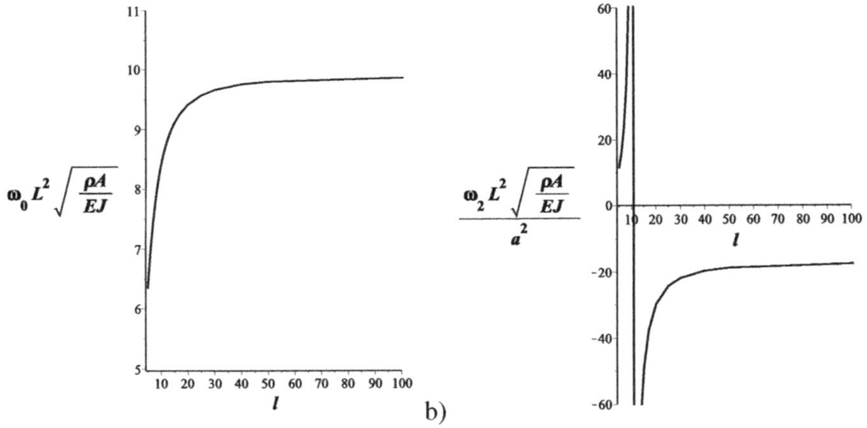

Figure 5.4. *a) ω_0 and b) ω_2 for varying l.* $n = 1$, $\kappa_h = 0$

First we report in Figure 5.4 the linear natural frequency (ω_0) and the nonlinear correction coefficient (ω_2). We see that for low values of the slenderness $\omega_2 > 0$, i.e. the beam has a hardening nonlinear behavior, that becomes softening ($\omega_2 < 0$) for increasing l. The transition is through a singular point at $l \cong 10.917$, where ω_2 tends to infinity. This behavior, which has been already reported in previous papers (Lenci and Rega 2016a; Lenci *et al.* 2016; Clementi *et al.* 2017), involves internal resonances and requires a more detailed analysis in its neighborhood. This is left for future works, so that the results of this chapter are valid out of a neighborhood of the singular point $l \cong 10.917$.

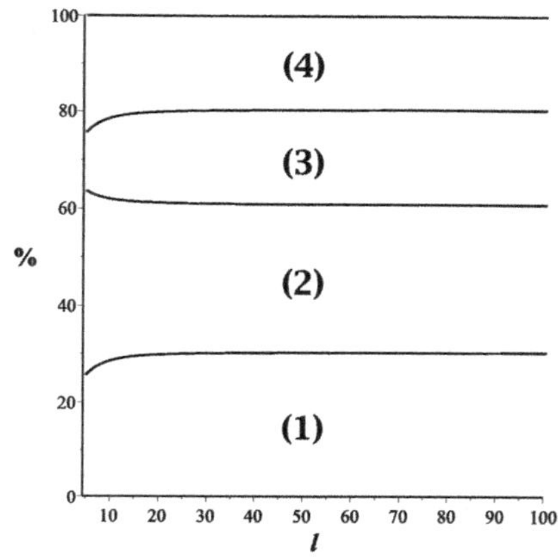

Figure 5.5. *The L^2 norm percentage of the various nonlinear terms in the constant part of the axial equation, for varying l. $n = 1$, $\kappa_h = 0$*

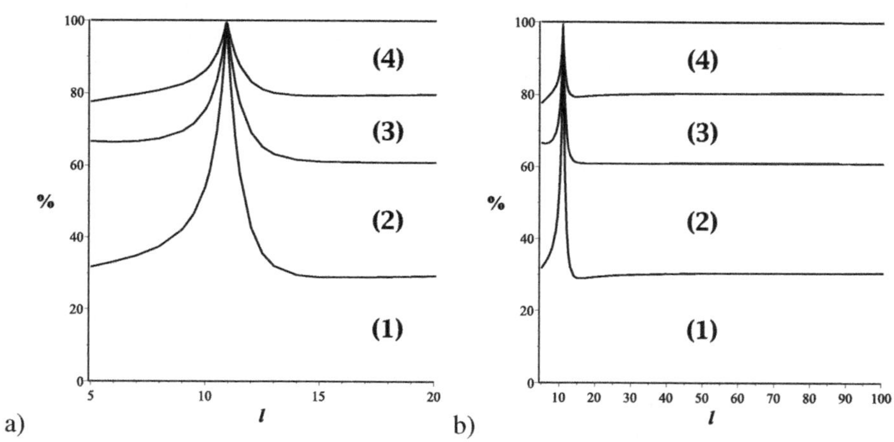

Figure 5.6. *The L^2 norm percentage of the various nonlinear terms in the part proportional to $\cos(2t)$ of the axial equation, for varying l. $n = 1$, $\kappa_h = 0$. a) Zoom around the singular point; b) large view*

The comparisons of the various nonlinear terms, in each equation, are shown in Figures 5.5–5.10, which extend Tables 5.1–5.3.

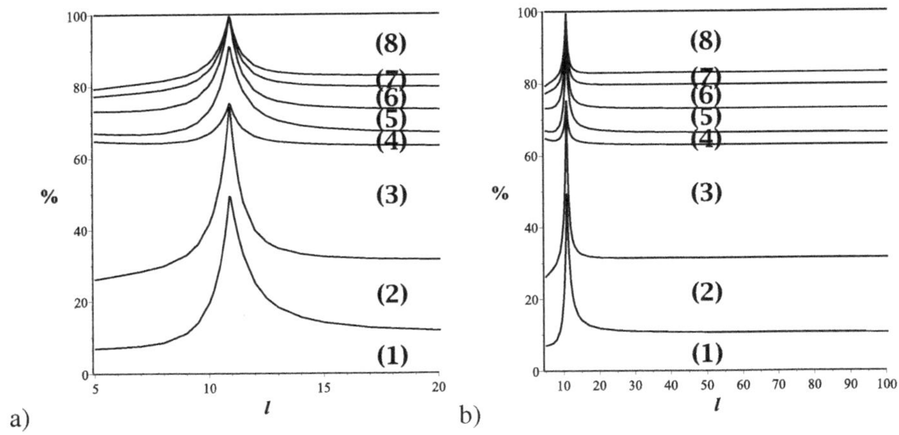

Figure 5.7. *The L^2 norm percentage of the various nonlinear terms in the part proportional to $\sin(t)$ of the transversal equation, for varying l. $n = 1$, $\kappa_h = 0$. a) Zoom around the singular point; b) large view*

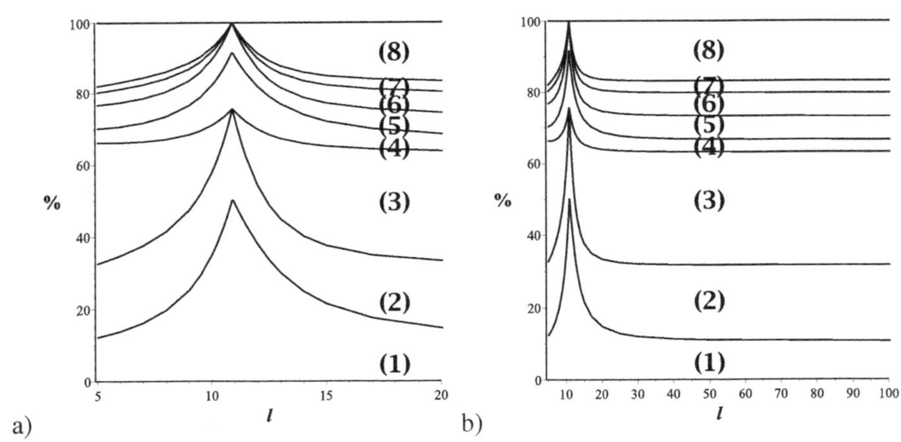

Figure 5.8. *The L^2 norm percentage of the various nonlinear terms in the part proportional to $\sin(3t)$ of the transversal equation, for varying l. $n = 1$, $\kappa_h = 0$. a) Zoom around the singular point; b) large view*

We initially note that only the case of Figure 5.5 is not affected by the singularity.

From Figures 5.5 and 5.6, we can conclude that no nonlinear term can be neglected in the axial equation, thus extending to the whole l-range the conclusion previously drawn for $l = 100$. The dominant behavior of the **(1)** term around the singular point needs a deeper analysis in order to be (eventually) confirmed.

Figures 5.7 and 5.8 confirm that terms **(4)** and **(7)** can be neglected in the transversal equation of motion, although they are not infinitesimal.

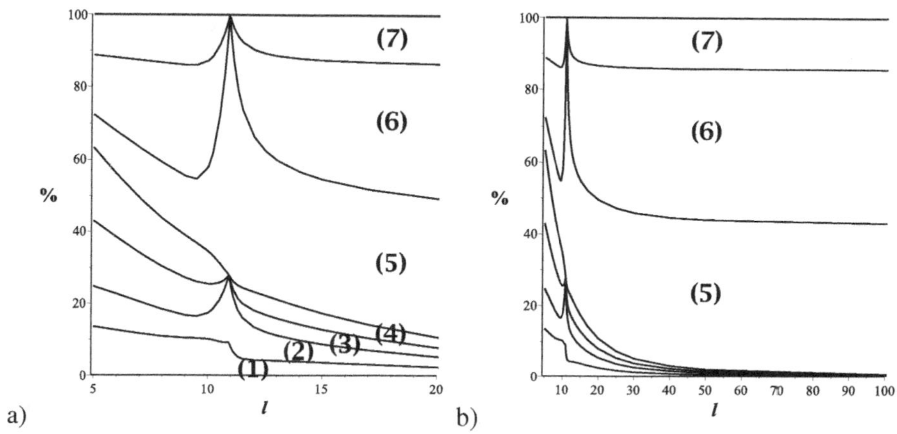

Figure 5.9. *The L^2 norm percentage of the various nonlinear terms in the part proportional to $\sin(t)$ of the rotational equation, for varying l. $n = 1$, $\kappa_h = 0$. a) Zoom around the singular point; b) large view*

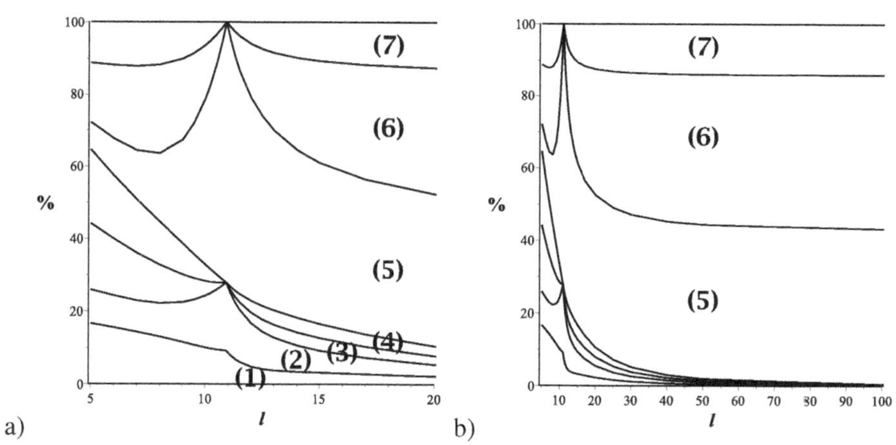

Figure 5.10. *The L^2 norm percentage of the various nonlinear terms in the part proportional to $\sin(3t)$ of the rotational equation, for varying l. $n = 1$, $\kappa_h = 0$. a) Zoom around the singular point; b) large view*

Finally, from Figures 5.9 and 5.10, as seen for the case $l = 100$, we deduce that terms **(1)**, **(2)**, **(3)** and **(4)** can be neglected in the rotational equation for slender beams. However, for thick beams (say, for $l < 20$ – for a rectangular cross-section

$L < 5.77H$), they are no longer negligible, and they become dominant for very thick beams; see, for example, the case $l = 5$ where all these terms give more than 60% of the whole nonlinearity. Again, the behavior around the singular point needs further analysis.

5.4.3. *The effect of the end spring*

To detect the effect of the end spring stiffness, we keep $n = 1$ and consider $\kappa_h = 10^{20}$ (which practically corresponds to $\kappa \to \infty$ and thus to an immovable, or hinged, boundary), and again vary l from $l = 5$ to $l = 100$.

In this case, we first note that there is no longer the singularity of ω_2, and thus the curves are regular. The beam is always hardening, and $\omega_2 > 0$ increases monotonically with l (see (Lenci *et al.* 2016)).

The results for the axial equation are shown in Figure 5.11. The conclusion is that, also for this boundary condition, no nonlinear terms can be neglected.

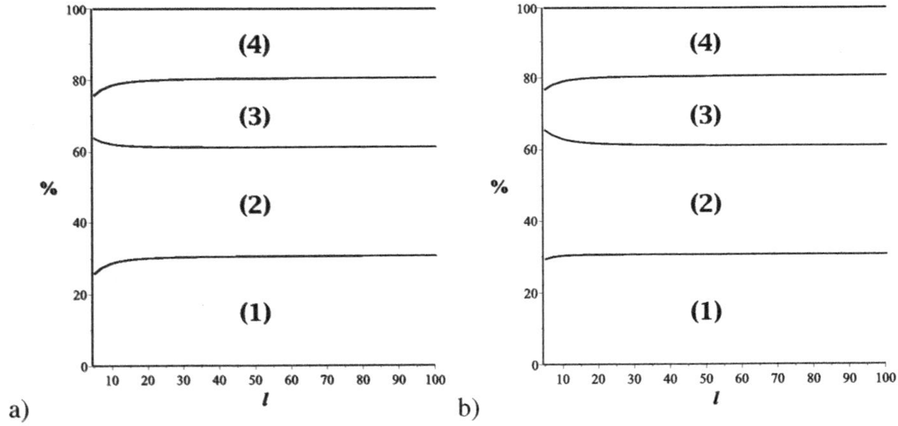

Figure 5.11. *The L^2 norm percentage of the various nonlinear terms in a) the constant part; and b) the part proportional to $\cos(2t)$ of the axial equation, for varying l. $n = 1$, $\kappa_h = 10^{20}$*

The results for the transversal equation are shown in Figure 5.12. It is confirmed that the nonlinear terms **(4)** and **(7)** can be neglected, although their contribution is not infinitesimal.

The results for the rotational equation are shown in Figure 5.13. Also for the hinged case, the nonlinear terms **(1)**, **(2)**, **(3)** and **(4)** can be neglected for slender beams, while they are dominating for very thick beams.

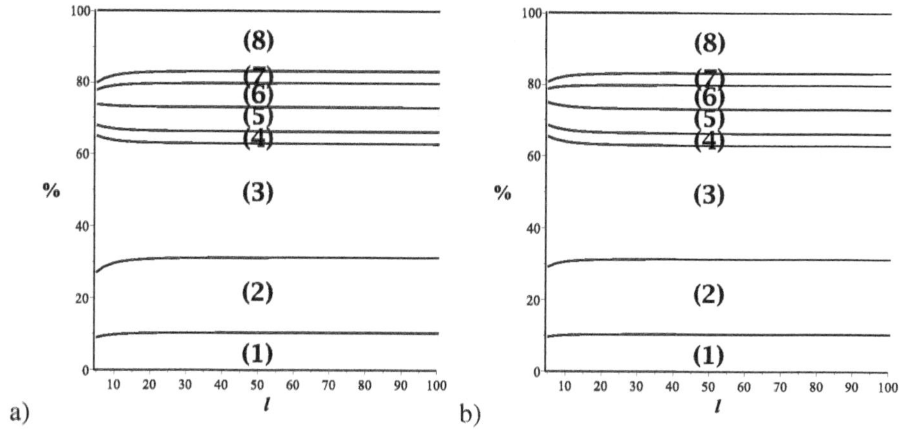

Figure 5.12. *The L^2 norm percentage of the various nonlinear terms in a) the part proportional to $\sin(t)$; and b) the part proportional to $\sin(3t)$ of the transverse equation, for varying l. $n = 1$, $\kappa_h = 10^{20}$*

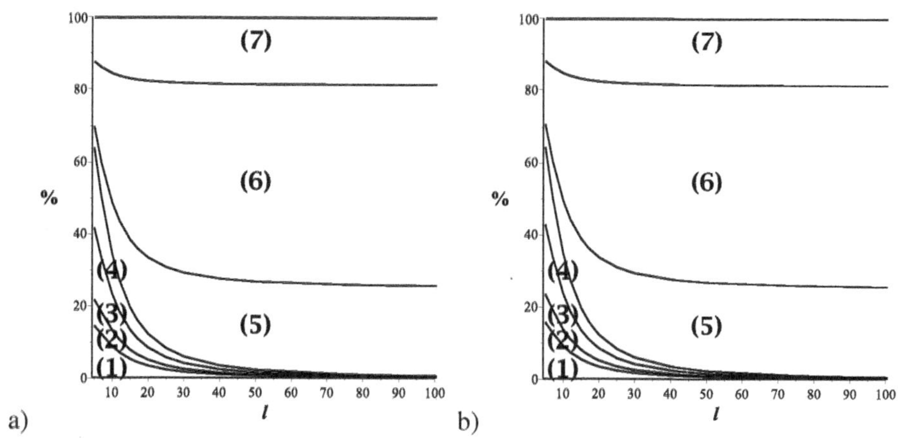

Figure 5.13. *The L^2 norm percentage of the various nonlinear terms in a) the part proportional to $\sin(t)$; and b) the part proportional to $\sin(3t)$ of the rotational equation, for varying l. $n = 1$, $\kappa_h = 10^{20}$*

5.4.4. *The effect of the resonance order*

To study the effect of the resonance order, we consider $n = 2$.

The results for the axial equation are shown in Figure 5.14 for $\kappa_h = 0$, and in Figure 5.15 for $\kappa_h = 10^{20}$. They further confirm that no nonlinear terms can be neglected in this equation. The strange behavior at $l \simeq 11.64$ and $l \simeq 48.64$ for $\kappa_h = 0$ is again due to the singularities of ω_2 and needs further study.

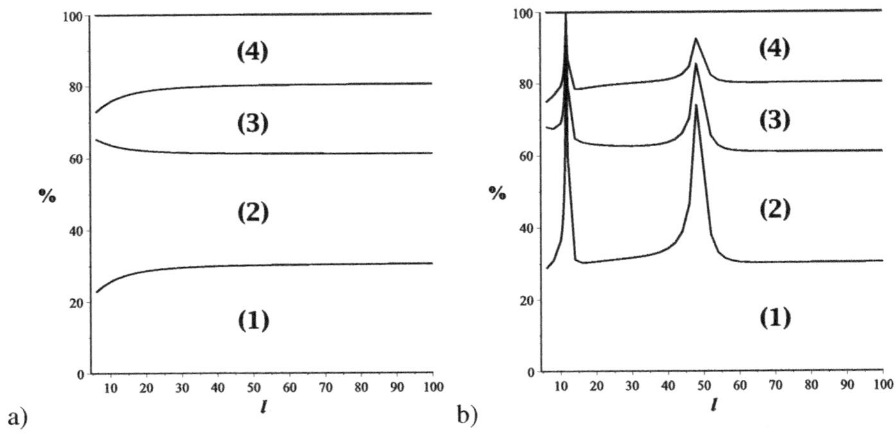

Figure 5.14. *The L^2 norm percentage of the various nonlinear terms in a) the constant part; and b) the part proportional to $\cos(2t)$ of the axial equation, for varying l. $n = 2$, $\kappa_h = 0$*

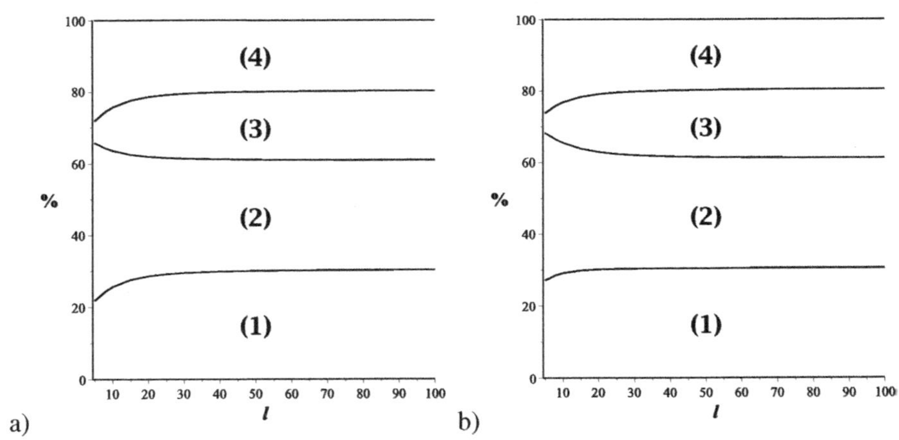

Figure 5.15. *The L^2 norm percentage of the various nonlinear terms in a) the constant part; and b) the part proportional to $\cos(2t)$ of the axial equation, for varying l. $n = 2$, $\kappa_h = 10^{20}$*

The results for the transversal equation are shown in Figure 5.16 for $\kappa_h = 0$, and in Figure 5.17 for $\kappa_h = 10^{20}$. They support the previous findings that the nonlinear terms **(4)** and **(7)** can be neglected.

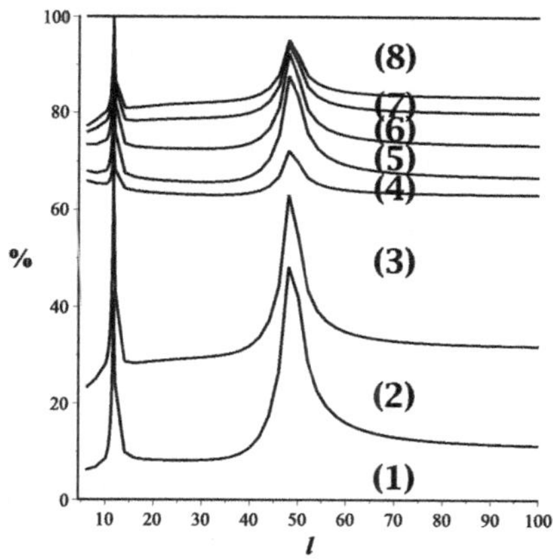

a)

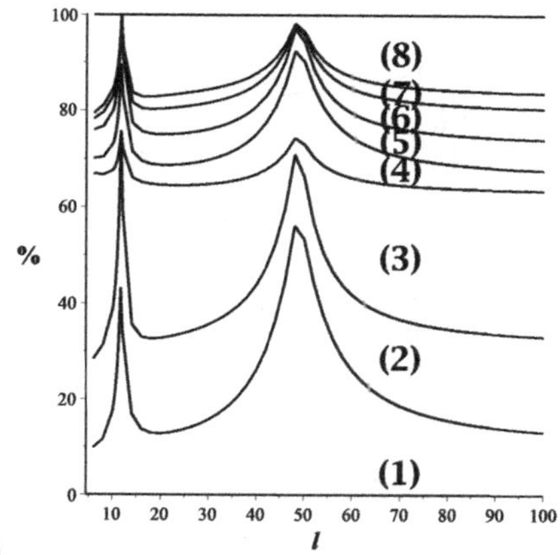

b)

Figure 5.16. *The L^2 norm percentage of the various nonlinear terms in a) the part proportional to $\sin(t)$; and b) the part proportional to $\sin(3t)$ of the transverse equation, for varying l. $n = 2$, $\kappa_h = 0$*

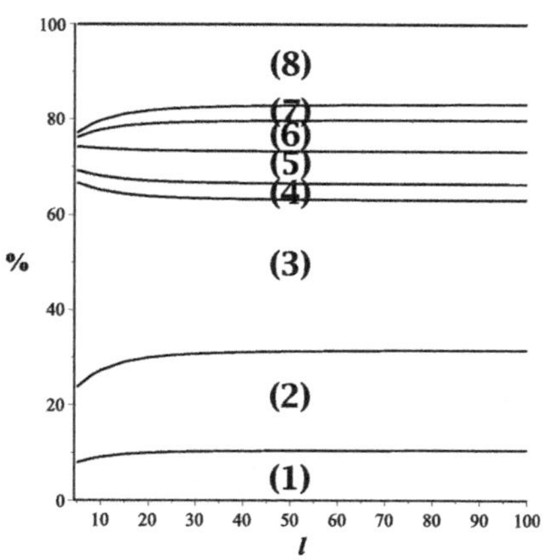

a)

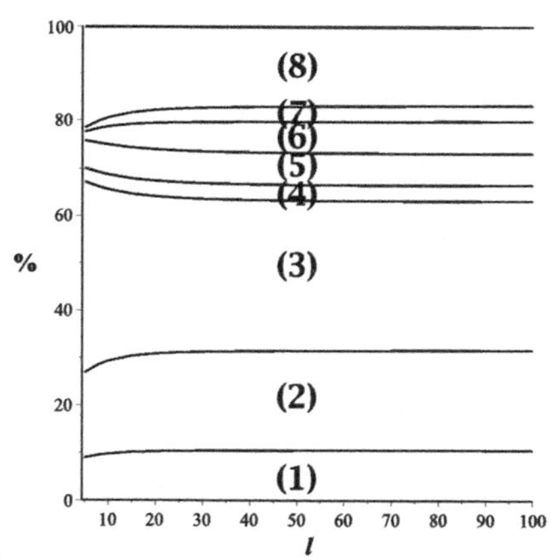

b)

Figure 5.17. *The L^2 norm percentage of the various nonlinear terms in a) the part proportional to $\sin(t)$; and b) the part proportional to $\sin(3t)$ of the transverse equation, for varying l. $n = 2$, $\kappa_h = 10^{20}$*

Finally, the results for the rotational equation are shown in Figure 5.18 for $\kappa_h = 0$, and in Figure 5.19 for $\kappa_h = 10^{20}$. Although the qualitative behavior is the same for the case $n = 1$, the curves are shifted towards higher values of l. This means that, for higher-order modes, the threshold of slenderness for negligibility of the nonlinear terms **(1)**, **(2)**, **(3)** and **(4)** is larger, a fact that aligns with common sense.

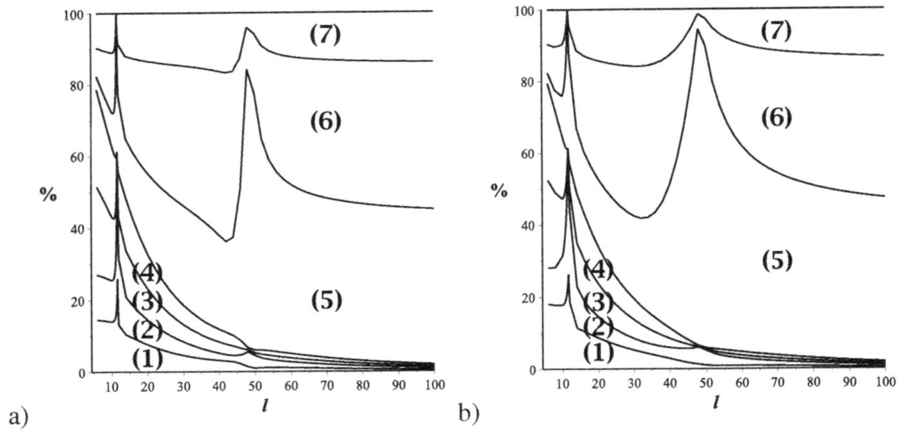

Figure 5.18. *The L^2 norm percentage of the various nonlinear terms in a) the part proportional to $\sin(t)$; and b) the part proportional to $\sin(3t)$ of the rotational equation, for varying l. $n = 2$, $\kappa_h = 0$*

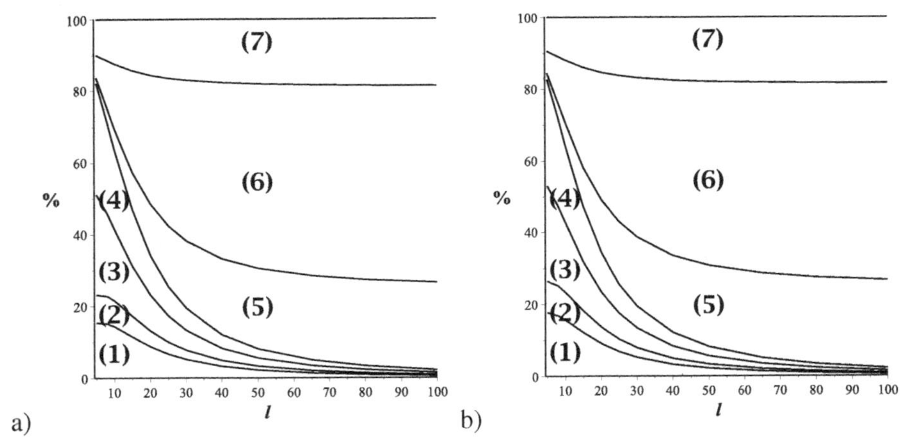

Figure 5.19. *The L^2 norm percentage of the various nonlinear terms in a) the part proportional to $\sin(t)$; and b) the part proportional to $\sin(3t)$ of the rotational equation, for varying l. $n = 2$, $\kappa_h = 10^{20}$*

5.5. Simplified models

Summarizing the previous results, we conclude that a simplified, but reliable, model is:

$$EA\,(W'' + U'U'') + GA\,(U''\theta + U'\theta' - 2U'U'') = \rho A\ddot{W}, \qquad [5.84]$$

$$GA\,(U'' - \theta') + EA\,(U''W' + U'W'' + 3/2\,U'^2U'')$$
$$+\,GA\,(-\mathbf{U''W'} - U'W'' + U'U''\theta + \mathbf{1/2\,U'^2}\theta' - 5/2\,U'^2U'') = \rho A\ddot{U}, \quad [5.85]$$

$$EJ\,\theta'' + GA\,(U' - \theta) + GA\,(-W'\theta - 1/2\,U'^2\theta + 1/6\,U'^3) = \rho J\,\ddot{\theta}. \quad [5.86]$$

Interestingly, no nonlinear terms proportional to the bending stiffness EJ are present. The terms reported in bold can possibly be neglected, with a small, but not infinitesimal, error (of the order of 3%).

We have not checked whether this model can be obtained by a Lagrangian approach, since it has been derived directly by eliminating negligible terms in the third-order approximation of equations [5.4]–[5.6], obtained by using a Newtonian approach. Actually, since the elimination of the nonlinear terms has been done without reference to the stationarity of any functional, we believe that it cannot be derived from a Lagrangian. The detailed check of this guess requires a large number of computations, which are out of the scope of this chapter, and are left for future developments.

5.5.1. *Neglecting axial inertia*

If in the previous model we neglect the axial inertia, $\rho A\ddot{W} = 0$, a hypothesis that *sometimes* can be acceptable (e.g. for axially immovable boundary conditions, see (Lenci *et al.* 2016)), from [5.84], we have

$$W' = -\frac{1}{2}U'^2 + \frac{GA}{EA}\,U'(U' - \theta) + c_3, \quad [5.87]$$

where c_3 is given by

$$c_3 = \frac{W(L) - W(0)}{L} + \frac{1}{2L}\int_0^L \left[U'^2 - \frac{GA}{EA}\,U'(U' - \theta)\right] dZ. \quad [5.88]$$

Note that c_3 is the second order.

Equation [5.88] is quite different from what is commonly used (see, for example, [5.1], where the term proportional to GA/EA is missing). $W(L)$ and $W(0)$ (equal to zero in the present case) can be determined by the boundary conditions in the axial direction.

It is not necessary to integrate [5.87] since only the derivatives of W appear in [5.85] and [5.86]. Indeed, the expression of W' can be introduced in [5.85] and [5.86] to have a further simplified, now two-field, model:

$$\Big[GA\,(U'-\theta)+GA\,U'^2\left(\frac{2}{3}U'-\frac{1}{2}\theta\right)$$
$$-\frac{(GA)^2}{EA}\,U'^2(U'-\theta)+(EA-GA)U'c_3\Big]'=\rho A\,\ddot{U}, \quad [5.89]$$

$$EJ\,\theta''+GA\,(U'-\theta)+GA\,\left(-\theta c_3+\frac{1}{6}U'^3\right)-\frac{(GA)^2}{EA}\,U'\theta(U'-\theta)=\rho J\,\ddot{\theta}. \quad [5.90]$$

5.5.2. *One-field equation*

Let us define F such that $U=F'$, i.e.

$$F(Z,t)=\int_0^Z U(\zeta,t)d\zeta. \quad [5.91]$$

Note that, by definition, $F(0,t)=0$, so that $\ddot{F}(0,t)=0$. Then from [5.89] and [5.90], we have

$$\Big[GA\,(F''-\theta)+GA\,F''^2\left(\frac{2}{3}F''-\frac{1}{2}\theta\right)$$
$$-\frac{(GA)^2}{EA}\,F''^2(F''-\theta)+(EA-GA)F''c_3\Big]'=\rho A\,\ddot{F}', \quad [5.92]$$

$$EJ\,\theta''+GA\,(F''-\theta)+GA\,\left(-\theta c_3+\frac{1}{6}F''^3\right)$$
$$-\frac{(GA)^2}{EA}\,F''\theta(F''-\theta)=\rho J\,\ddot{\theta}. \quad [5.93]$$

Equation [5.92] gives (where we can assume the constant of integration as null)

$$GA\,(F''-\theta)+GA\,F''^2\left(\frac{2}{3}F''-\frac{1}{2}\theta\right)$$
$$-\frac{(GA)^2}{EA}\,F''^2(F''-\theta)+(EA-GA)F''c_3=\rho A\,\ddot{F}, \quad [5.94]$$

which in turn gives

$$\theta = \frac{-\frac{\rho A}{GA}\ddot{F} + F'' + \left(\frac{2}{3} - \frac{GA}{EA}\right)F''^3 - \left(1 - \frac{EA}{GA}\right)F''c_3}{1 + \left(\frac{1}{2} - \frac{GA}{EA}\right)F''^2}. \qquad [5.95]$$

Coherently with the framework of this work, where only nonlinearities up to the third order are retained, we develop [5.95] up to the third order:

$$\theta = -\frac{\rho A}{GA}\ddot{F} + F'' + \frac{1}{6}F''^3 - \left(1 - \frac{EA}{GA}\right)F''c_3 + \left(\frac{\rho A}{2GA} - \frac{\rho A}{EA}\right)\ddot{F}F''^2. \qquad [5.96]$$

Note that this introduces nonlinear inertial terms.

From expression [5.96], we have

$$c_3 = \frac{W(L) - W(0)}{L} + \frac{1}{2L}\int_0^L \left[F''^2 - \frac{\rho A}{EA}F''\ddot{F}\right] dZ. \qquad [5.97]$$

It is worth noting that c_3 no longer depends on the shear stiffness GA.

Using the expression [5.96] of θ in equation [5.93] we obtain, after retaining terms up to the third order only:

$$\begin{aligned}
&EJ\,F^{VI} + \rho A\,\ddot{F} - \left(\rho J + \frac{EJ\,\rho A}{GA}\right)\ddot{F}'' + \frac{\rho J\,\rho A}{GA}\ddddot{F} \\
&+ \left[\left(1 - \frac{EA}{GA}\right)(\rho J\,\ddot{F}'' - EJ\,F^{VI}) + \rho A\,\ddot{F} - EA\,F''\right] c_3 \\
&+ \left(\frac{\rho A}{2GA} - \frac{\rho A}{EA}\right)\left[-\rho J(2\ddot{F}''\ddot{F}F'' + 2\ddot{F}\dot{F}''^2 + \ddddot{F}F''^2 + 4F''\dot{F}''\dddot{F})\right. \\
&\left.+EJ(\ddot{F}''F''^2 + 2F^{VI}\ddot{F}F'' + 2\ddot{F}F'''^2 + 4F''F'''\ddot{F}')\right] \\
&- \rho J\left(\frac{1}{2}F''^2\ddot{F}'' + F''\dot{F}''^2\right) + EJ\left(\frac{1}{2}F''^2F^{VI} + F''F'''^2\right) \\
&- \frac{\rho A}{2}\ddot{F}F''^2 + \frac{(\rho A)^2}{EA}F''\ddot{F}^2 = 0, \qquad [5.98]
\end{aligned}$$

which is the unique nonlinear equation in the unique unknown $F(Z, t)$. Only negligible axial inertia and third-order approximation have been assumed so far, and c_3 is given by [5.97].

In equation [5.98], the fourth-order time derivative of F also appears in the linear terms. It is a consequence of the rotational inertia (which in fact disappears for

$\rho J = 0$), and calls for initial (in time) conditions of F, $\dot{F}$, $\dddot{F}$ and $\dddot{F}$. From [5.91], the initial condition on F is linked with the initial condition on the displacement U; consequently, the initial condition on $\dot{F}$ follows from the initial translational velocity $\dot{U}$. Solving [5.96] with respect to $\ddot{F}$, we find that it is a function of θ and F''; thus, the initial condition on $\ddot{F}$ is linked with the initial rotation and (the second derivative of) the initial displacement. Consequently, the initial condition on $\dddot{F}$ follows from the initial rotational velocity, the initial displacement and the initial translational velocity.

Because of the presence of $\ddot{F}$, $\dddot{F}$ and $\ddddot{F}$ in the nonlinear terms, we cannot "easily" (i.e. without using the *inde nite* integral [5.91] and introducing integro-differential equations) come back to the physical displacement $U = F'$, not even deriving one time with respect to Z. In fact, this works only for the linear part, which becomes

$$EJ\,F^{VII} + \rho A\,\ddot{F}' - \left(\rho J + \frac{EJ\,\rho A}{GA}\right)\ddot{F}''' + \frac{\rho J\,\rho A}{GA}\ddddot{F}' \qquad [5.99]$$

and thus

$$EJ\,U^{VI} + \rho A\,\ddot{U} - \left(\rho J + \frac{EJ\,\rho A}{GA}\right)\ddot{U}'' + \frac{\rho J\,\rho A}{GA}\ddddot{U}. \qquad [5.100]$$

In the nonlinear part, this "problem" can be "solved" only in the expression [5.97] of c_3. In fact, integrating by parts and using [5.91], we obtain

$$\begin{aligned} c_3 &= \frac{W(L) - W(0)}{L} + \frac{1}{2L}\int_0^L F''^2 dZ - \frac{\rho A}{2L\,EA}\int_0^L F''\ddot{F} dZ \\ &= \frac{W(L) - W(0)}{L} + \frac{1}{2L}\int_0^L F''^2 dZ \\ &\quad - \frac{\rho A}{2L\,EA}\left[F'(L)\ddot{F}(L) - F'(0)\ddot{F}(0) - \int_0^L F'\ddot{F}' dZ\right] \\ &= \frac{W(L) - W(0)}{L} + \frac{1}{2L}\int_0^L U'^2 dZ \\ &\quad - \frac{\rho A}{2L\,EA}\left[U(L)\int_0^L \ddot{U} dZ - \int_0^L U\,\ddot{U} dZ\right]. \end{aligned} \qquad [5.101]$$

5.5.3. *The Euler–Bernoulli nonlinear beam*

Further simplified models in which rotational inertia is neglected, $\rho J = 0$, and/or shear deformation is disregarded, $GA \to \infty$, can be easily obtained from [5.98].

Using these two assumptions, we obtain the nonlinear Euler–Bernoulli beam equation (c_3 is still given by [5.97])

$$\begin{aligned} & EJ\,F^{VI} + \rho A\,\ddot{F} \\ & + (-EJ\,F^{VI} + \rho A\,\ddot{F} - EA\,F'')c_3 \\ & - \frac{\rho A\,EJ}{EA}(\ddot{F}''F''^2 + 2F^{VI}\ddot{F}F'' + 2\ddot{F}F'''^2 + 4F''F'''\ddot{F}') \\ & + EJ\left(\frac{1}{2}F''^2F^{VI} + F''F'''^2\right) - \frac{\rho A}{2}\ddot{F}F''^2 + \frac{(\rho A)^2}{EA}F''\ddot{F}^2 = 0, \end{aligned} \qquad [5.102]$$

which is the final model we propose for a slender beam. In this case

$$\theta = F'' + \frac{1}{6}F''^3 - F''c_3 - \frac{\rho A}{EA}\ddot{F}F''^2. \qquad [5.103]$$

Note that, because of the axial displacement, it is different from its linearized version $\theta = F'' = U'$, which is commonly assumed for Euler–Bernoulli beams.

Again, because of the presence of $\ddot{F}$ in the nonlinear terms, it is not possible to easily obtain a differential equation containing the transversal displacement $U = F'$.

5.6. Conclusion

Planar nonlinear dynamics of a Timoshenko beam have been considered with the aim of detecting the most important nonlinear terms.

Starting from a kinematically exact, and linear elastic, framework, the third-order approximation has been considered.

It has been first assumed that the axial displacement is one order of magnitude larger than the transversal and rotational ones. This allows us to initially neglect some nonlinear terms in the three equations of motion.

Then, using an asymptotic solution of the problem obtained elsewhere, all the nonlinear terms appearing in each equation have been considered and compared between each other. The L^2 norm has been used for a quantitative comparison.

A parametric analysis has been performed by varying the slenderness of the beam, and by considering movable and immovable axial boundary conditions. First- and second-order resonances have also been considered.

It has been shown that only four terms in the rotational equation can certainly be neglected for slender beams (and the slenderness threshold for negligibility increases with the order of the resonance), while two terms can possibly be neglected in the transversal equation with a small, but not infinitesimal, error.

Finally, ensuing simplified models have been obtained and briefly discussed.

5.7. References

Antman, S.S. (2005). *Problems of Nonlinear Elasticity*. Springer, New York.

Ballarini, R. (2003). The Da Vinci-Euler-Bernoulli beam theory? *Mechanical Engineering Magazine Online* [Online]. Available at: https://web.archive.org/web/20060623063248 and http://www.memagazine.org/contents/current/webonly/webex418.html.

Bolotin, V.V. (1956). *Dynamic Stability of Elastic Systems*, translated by NASA [in Russian]. Gostekhizdat, Moscow.

Clebsch, A. (1862). *Theorie der elasticität fester körper*. Druck und Verlag, Leipzig.

Clementi, F., Lenci, S., Rega, G. (2017). Cross-checking asymptotics and numerics in the hardening/softening behavior of Timoshenko beams with axial end spring and variable slenderness. *Arch. Appl. Mech.*, 87(5), 865–880. DOI: 10.1007/s00419-016-1159-z.

Elishakoff, I. (2019). Who developed the so-called Timoshenko beam theory? *Math. Mech. Solids*, 25, 97–116. DOI: 10.1177/1081286519856931.

Elishakoff, I., Kaplunov, J., Nolde, E. (2015). Celebrating the centenary of Timoshenko's study of effects of shear deformation and rotary inertia. *Appl. Mech. Rev.*, 67, 060802-1-11. DOI: 10.1115/1.4031965.

Euler, L. (1744). Additamentum I (de curvis elasticis), methodus invieniendi lineas curvas maximi minimive proprietate gaudentes. *Opera Omnia I*, 24, 231–297.

Hamdan, M.N. and Shabneh, N.H. (1997). On the large amplitude free vibrations of a restrained uniform beam carrying an intermediate lumped mass. *J. Sound Vib.*, 199, 711–736. DOI: 10.1006/jsvi.1996.0672.

Hodges, D.H., Ormiston, R.A., Peters, D.A. (1980). On the nonlinear deformation geometry of Euler-Bernoulli beams, NASA Technical paper 1566, AVRADCOM Technical report 80-A-1.

Kirchhoff, G. (1859). Ueber das gleichgewicht und die bewegung eines unendlich dünnen elastischen stabes. *J. Reine Angew. Math.*, 56, 285–343.

Kloda, L., Lenci, S., Warminski, J. (2018). Nonlinear dynamics of a planar beam-spring system: Analytical and numerical approaches. *Nonlin. Dyn.*, 94, 1721–1738. DOI: 10.1007/s11071-018-4452-2.

Kloda, L., Lenci, S., Warminski, J. (2019). Nonlinear dynamics of a planar hinged-simply supported beam with one end spring: Higher order resonances. In *IUTAM Symposium on Exploiting Nonlinear Dynamics for Engineering Systems*, Kovacic, I. and Lenci, S. (eds). IUTAM Bookseries 37, Springer, Cham. ISBN: 978-3-030-23691-5, e-ISBN: 978-3-030-23692-2, DOI: 10.1007/978-3-030-23692-2_14.

Kloda, L., Lenci, S., Warminski, J. (2020). Hardening vs softening dichotomy of a hinged-simply supported beam with one end axial linear spring: Experimental and numerical studies. *Int. J. Mech. Sci.*, 178, 105588 [Online]. Available at: https://doi.org/10.1016/j.ijmecsci.2020.105588.

Labuschagne, A., van Rensburg, N.F.J., van der Merwe, A.J. (2009). Comparison of linear beam theories. *Math. Comput. Modell.*, 49, 20–30. DOI: 10.1016/j.mcm.2008.06.006.

Lacarbonara, W. (2013). *Nonlinear Structural Mechanics: Theory, Dynamical Phenomena and Modeling*. Springer, New York.

Lenci, S. and Rega, G. (2016a). Nonlinear free vibrations of planar elastic beams: A unified treatment of geometrical and mechanical effects. *Procedia IUTAM, IUTAM Symposium Analytical Methods in Nonlinear Dynamics*, 19, 35–42. DOI: 10.1016/j.piutam.2016.03.007.

Lenci, S. and Rega, G. (2016b). Axial-transversal coupling in the free nonlinear vibrations of Timoshenko beams with arbitrary slenderness and axial boundary conditions. *Proc. Royal Soc. A*, 472, 20160057. DOI: 10.1098/rspa.2016.0057.

Lenci, S., Clementi, F., Rega, G. (2016). A comprehensive analysis of hardening/softening behaviour of shearable planar beams with whatever axial boundary constraint. *Meccanica*, 51(11), 2589–2606. DOI: 10.1007/s11012-016-0374-6.

Lenci, S., Clementi, F., Rega, G. (2017a). Comparing nonlinear free vibrations of Timoshenko beams with mechanical or geometric curvature definition. *Procedia IUTAM, 24th International Congress of Theoretical and Applied Mechanics*, 20, 34–41. DOI: 10.1016/j.piutam.2017.03.006.

Lenci, S., Clementi, F., Rega, G. (2017b). Reply to the discussion on 'A comprehensive analysis of hardening/softening behavior of shearable planar beams with whatever axial boundary constraint', by D. Genovese. *Meccanica*, 52, 3005–3008. DOI: 10.1007/s11012-016-0613-x.

Love, A.E.H. (1944). *A Treatise on the Mathematical Theory of Elasticity*. Dover Publications, New York.

Luongo, A., Pignataro, M., Rega, G., Vestroni, F. (1987). Nonlinear vibrations of beams including shear deformations and rotatory inertia. In *Re ned Dynamical Theories of Beams, Plates and Shells and Their Applications*, Elishakoff, I. and Irretier, G. (eds). Springer-Verlag, Berlin. DOI: 10.1007/978-3-642-83040-2_16.

Marur, S.R. (2001). Advances in nonlinear vibration analysis of structures. Part I: Beams. *Sadhana*, 26, 243–249. DOI: 10.1007/BF02703386.

Mettler, E. (1962). Dynamic buckling. In *Handbook of Engineering Mechanics*, Flugge, W. (ed). McGraw-Hill, New York.

Morris, K.A. and Vidyasagar, M. (1990). A comparison of different models for beam vibrations from the standpoint of control design. *ASME J. Dyn. Sys., Meas., Control.*, 112, 349–356. DOI: 10.1115/1.2896151.

Nayfeh, A.H. and Pai, P.F. (2004). *Linear and Nonlinear Structural Mechanics*. Wiley, New York.

Rao, S.S. (2007). *Vibrations of Continuous Systems*. John Wiley & Sons, Hoboken, New Jersey.

Rao, G.V., Raju, I.S., Kanaka Raju, K. (1976). Nonlinear vibrations of beams considering shear-deformations and rotary inertia. *AIAA J.*, 14, 685–687. DOI: 10.2514/3.7138.

Sathyamoorthy, M. (1982). Nonlinear analysis of beams. Part I: A survey of recent advances. *Shock Vib. Digest*, 14, 19–35.

Singh, G., Sharma, A.K., Venkateswara Rao, G. (1990). Large-amplitude free vibrations of beams – A discussion on various formulations and assumptions. *J. Sound Vib.*, 142, 77–85. DOI: 10.1016/0022-460X(90)90583-L.

Takahashi, K. (1979). Nonlinear free vibrations of inextensible beams. *J. Sound Vib.*, 64, 31–34. DOI: 10.1016/0022-460X(79)90568-6.

Timoshenko, S. (1983). *History of Strength of Materials*. Dover Publications, New York.

Villaggio, P. (1997). *Mathematical Models for Elastic Structures*. Cambridge University Press, Pisa.

Wagner, H. (1965). Large-amplitude free vibrations of a beam. *ASME J. Appl. Mech.*, 32, 997–892. DOI: 10.1115/1.3627331.

6

Statics, Dynamics, Buckling and Aeroelastic Stability of Planar Cellular Beams

6.1. Introduction

Homogenization techniques are very useful in modeling periodic structures as equivalent continua (Noor 1988; Pshenichnov 1993; Tollenaere and Caillerie 1998; Dos Reis and Ganghoffer 2012). Reticular and pantographic structures are studied in this context (Cioranescu and Paulin 1999; Boutin *et al.* 2017; dell'Isola *et al.* 2018). When one of the dimensions of the micro-structured body prevails over the other two, the structure is called a *cellular beam*. Multi-story frame buildings, composed of a sufficiently large number of floors, are examples of such structures, (Chajes *et al.* 1993, 1996). Recently, a series of papers by a research group from L'Aquila and Genoa Universities have been devoted to buildings and towers immersed in the 3D space, macroscopically modeled as (i) shear–shear–torsional beams (Piccardo *et al.* 2015, 2016; D'Annibale *et al.* 2019; Luongo and Zulli 2020) or (ii) Timoshenko beams (Ferretti 2018; Piccardo *et al.* 2019; Ferretti *et al.* 2020a). In these models, the floors represent the cross-sections and the columns represent the longitudinal fibers of the homogenized beam. Several mechanical problems have been addressed there, namely: statics (D'Annibale *et al.* 2019; Piccardo *et al.* 2019), dynamics (Piccardo *et al.* 2019; Luongo and Zulli 2020), buckling (Ferretti 2018; Ferretti *et al.* 2020a) and aeroelasticity (Piccardo *et al.* 2015, 2016), in the linear, as well as in the nonlinear field. More recently, the study has been extended to micro-structured beams, which are of interest, for example, in the automotive industry and in plastic furniture and shelf design. These cellular beams are made up of three orders of orthogonal micro-beams that are periodically arranged, and potentially have a very

Chapter written by Angelo LUONGO.

large number of cells and a multitude of boundary conditions. Some preliminary results of the research are sketched out in this chapter.

In order to model a cellular beam as a solid beam, some strong hypotheses are usually introduced, singularly or in combination, namely: (H1) the cross-section is rigid in its own plane (membrane rigidity); (H2) the cross-section is rigid out of its own plane (flexural rigidity) and (H3) the longitudinal fibers are inextensible (axial rigidity). If all the H1, H2 and H3 hypotheses are invoked, the structure behaves as a shear–shear–torsional beam in 3D, or a shear beam in 2D; if H3 is relaxed, the structure behaves as a 3D or 2D Timoshenko beam. In all cases, the cross-section is rigid, both in its plane and out of its plane. Relaxing rigidity (i.e. removing both H1 and H2) is quite a difficult task. Indeed, to account for the deformation of the cross-section, it would be necessary to resort to the generalized beam theory (GBT, see, for example, Silvestre and Camotim 2002; Ranzi and Luongo 2011; Piccardo *et al.* 2014a, b; Ferrarotti *et al.* 2017; Piccardo *et al.* 2017). This theory has been developed in the context of the semi-variational (or Kantorovich) method, in which in-plane and out-of-plane deflections of the cross-section are described by a set of properly selected trial functions, defined in the cross-sectional domain, which are longitudinally modulated by unknown amplitude functions. Thus, the displacement field of the generalized beam is made up of several additional components with respect to the classical translations and rotations of the cross-section. A reinterpretation of GBT, in the context of direct 1D models, has been discussed in Luongo and Zulli (2013), and some applications have been presented in Luongo and Zulli (2014), Luongo *et al.* (2018) and Zulli (2019). Such an approach deeply differs from the usual homogenization techniques, in which the continuous model is derived via asymptotic techniques applied to the micro-structure (see, for example, Auriault *et al.* (2009), Challamel *et al.* (2016) and Hache *et al.* (2017, 2018)), since it *a priori* selects the macro-model on the ground of qualitative mechanical interpretations. Accordingly, kinematics and dynamics are defined in a 1D environment, while the constitutive law calls for micro–macro identification. This approach, although partially heuristic, has the advantage of bringing back a cellular structure to a known mathematical model, while also allowing the identification of generalized stress (as the bimoment of the Vlasov theory).

In this chapter, cellular beams, made up of longitudinal and transverse fibers, equipped with flexural and shear stiffnesses, are studied. Attention is confined, for brevity, to planar beams in the linear field. The study is developed in the framework of the "classical" theories of beams, in which the hypothesis of rigid cross-sections holds. However, it is proved here that, by resorting to the concept of "shear factor" (as suggested by the procedure adopted to include the de Saint-Venant results in the Timoshenko beam theory), it is possible to account for the warping of the cross-sections (i.e. for flexibility of the transverse fibers) in a rigid cross-section model (thus removing H2). The shear factor reduces the macroscopic shear stiffness of the longitudinal fibers, by approximately accounting, on an energy balance ground, for the neglected bending of the transverse fibers. Hypothesis H1, concerning in-plane deformability, is instead kept valid, and no energy corrections are introduced. Accordingly, and similarly to the Vlasov theory of thin-walled beams, the cellular structure should be equipped with stiff transverse diaphragms or bracings.

This chapter is organized as follows. In section 6.2, continuous models for planar cellular structures are derived, namely: (i) the shear beam and (ii) the Timoshenko beam. Derivation is performed in the framework of the direct 1D approach, while the constitutive law is determined by a homogenization procedure based on an energy equivalence between a cell of the periodic model and a segment of the solid beam. In section 6.3, the static problem is addressed for a grid beam, with regular mesh, for which the elastic constants are identified, under different hypotheses concerning the stiffness of the transverse fibers. The existence of a *shear factor*, which accounts for the warping of the cross-section, via artificial correction of the microscopic characteristics of the cellular beam, is proved. A comparison between results provided by micro finite element models and macro homogenized models is performed. In section 6.4, the linear model is updated to include geometrical effects related to uniform compression. The existence of local and global contributions of prestress to the stiffness of the cellular beam is discussed. The linear bifurcation problem is solved, and the critical loads and modes are evaluated. In section 6.5, the free oscillations of the macro-beam are investigated. The natural frequencies and modes are compared with those of fine models, obtained by finite element analyses or by analytically solving difference equations. In section 6.6, aeroelastic phenomena are studied. The cellular structure is immersed in a steady wind flow producing velocity-dependent aerodynamic forces. The critical wind velocity triggering galloping (Hopf bifurcation) is determined for the beam. The efficiency of a base-isolation control device, used to passively control the phenomenon, is also discussed.

6.2. Continuous models of planar cellular structures

Two linear one-dimensional models of beam are derived, namely: (i) the Timoshenko beam (TB) and (ii) the shear beam (SB). The body is constrained to belong to the $(\mathbf{i}, \mathbf{j})$-plane of the $(\mathbf{i}, \mathbf{j}, \mathbf{k})$-space.

6.2.1. *Timoshenko beam*

An internally unconstrained model of the beam is formulated. The beam is made up of material points, equipped with orientation (Timoshenko beam or, equivalently, a one-dimensional Cosserat continuum). For a comprehensive review about the historical aspects of the Timoshenko beam model, the reader is referred to Elishakoff (2020a, b). In the reference configuration, the material points lay on the segment $\overline{AB}$

of the $\mathbf{i}$-axis, spanned by the material abscissa $x \in (0, \ell)$. Each point undergoes a translation $\mathbf{u} = u(x)\,\mathbf{i} + v(x)\,\mathbf{j}$ and a rotation $\theta(x)$ around $\mathbf{k}$ (dependence on time is understood). The following strain–displacement relationships hold:

$$\begin{aligned} \varepsilon &= u' \\ \gamma &= v' - \theta \\ \kappa &= \theta' \end{aligned} \qquad [6.1]$$

where $\varepsilon, \gamma, \kappa$ are the unit extension, the shear strain and the curvature, respectively. If the beam is clamped at A, displacements vanish there: $u_A = v_A = \theta_A = 0$.

Equilibrium is enforced via the virtual work principle:

$$\begin{aligned} \int_0^\ell (N\delta\varepsilon + T\delta\gamma + M\delta\kappa)\,dx &= \int_0^\ell (p_x\delta u + p_y\delta v + c_z\delta\theta)\,dx \\ &\quad + P_x\delta u_B + P_y\delta v_B + C_z\delta\theta_B \qquad \forall(\delta u, \delta v, \delta\theta) \end{aligned} \qquad [6.2]$$

where: N, T, M are the normal and shear internal forces and the bending moment, respectively; p_x, p_y, c_z are distributed external forces and couples per unit length, respectively; and P_x, P_y, C_z are lumped forces and couple applied at the free end B, respectively. By using equations [6.1], the variational principle [6.2] supplies the indefinite equilibrium equations:

$$\begin{aligned} N' + p_x &= 0 \\ T' + p_y &= 0 \\ M' + T + c_z &= 0 \end{aligned} \qquad [6.3]$$

and the boundary conditions $(N_B, T_B, M_B) = (P_x, P_y, C_z)$.

It is assumed that the beam is made of hyperelastic material so that the existence of a strain energy function, quadratic in the strains, is postulated:

$$\phi = \frac{1}{2}\left(c_{11}\varepsilon^2 + c_{22}\gamma^2 + c_{33}\kappa^2 + 2c_{12}\varepsilon\gamma + 2c_{13}\varepsilon\kappa + 2c_{23}\gamma\kappa\right) \qquad [6.4]$$

depending on six independent elastic constants c_{ij}, to be identified according to the micro-structure of the beam. After that, Green's law supplies:

$$\begin{pmatrix} N \\ T \\ M \end{pmatrix} = \begin{pmatrix} c_{11} & c_{12} & c_{13} \\ & c_{22} & c_{23} \\ sym & & c_{33} \end{pmatrix} \begin{pmatrix} \varepsilon \\ \gamma \\ \kappa \end{pmatrix} \qquad [6.5]$$

By expressing equilibrium in terms of displacements, the following field equations are derived:

$$\mathbf{K}_2\mathbf{u}'' + \mathbf{K}_1\mathbf{u}' + \mathbf{K}_0\mathbf{u} + \mathbf{p} = \mathbf{0} \qquad [6.6]$$

with the boundary conditions:

$$\begin{aligned} \mathbf{u}_A &= \mathbf{0} \\ \mathbf{B}_1\mathbf{u}'_B + \mathbf{B}_0\mathbf{u}_B &= \mathbf{P} \end{aligned} \qquad [6.7]$$

Here, $\mathbf{K}_i \ \ (i = 0, 1, 2)$ are *symmetric or skew-symmetric stiffness matrices*, and $\mathbf{u}, \mathbf{p}$ are displacement and load vectors, respectively:

$$\mathbf{K}_2 := \begin{bmatrix} c_{11} & c_{12} & c_{13} \\ & c_{22} & c_{23} \\ sym & & c_{33} \end{bmatrix}, \quad \mathbf{K}_1 := \begin{bmatrix} 0 & 0 & -c_{12} \\ & 0 & -c_{22} \\ skw & & 0 \end{bmatrix}$$
$$\mathbf{K}_0 := \begin{bmatrix} 0 & 0 & 0 \\ & 0 & 0 \\ sym & & -c_{22} \end{bmatrix}, \quad \mathbf{u} := \begin{pmatrix} u \\ v \\ \theta \end{pmatrix}, \quad \mathbf{p} := \begin{pmatrix} p_x \\ p_y \\ c_z \end{pmatrix} \qquad [6.8]$$

Moreover:

$$\mathbf{B}_1 := \begin{bmatrix} c_{11} & c_{12} & c_{13} \\ & c_{22} & c_{23} \\ sym & & c_{33} \end{bmatrix}, \qquad \mathbf{B}_0 := \begin{bmatrix} 0 & 0 & -c_{12} \\ 0 & 0 & -c_{22} \\ 0 & 0 & -c_{23} \end{bmatrix} \qquad [6.9]$$

are boundary operators. Equation [6.6] extends the classic Timoshenko beam because of the fully coupled constitutive law [6.5].

If dynamic effects must be taken into account, the load $\mathbf{p}$ must include inertial forces, i.e. $\mathbf{p} = \mathbf{p}_{ext} + \mathbf{p}_{in}$, with:

$$\mathbf{p}_{in} = -\mathbf{M}\ddot{\mathbf{u}} \qquad [6.10]$$

where $\mathbf{M} := \mathrm{diag}\,(m_{11}, m_{11}, m_{33})$ is the mass matrix, assumed diagonal, and $\ddot{\mathbf{u}}$ is the acceleration.

6.2.2. *Shear beam*

A simplified model of *shear beam* can be derived by the Timoshenko model by introducing the internal constraint $\kappa\,(x) = 0 \quad \forall x$. Accordingly, all the cross-sections remain parallel in the current configuration. If $\theta_A = 0$, then $\theta\,(x) = 0 \quad \forall x$, so that $\mathbf{u} = (u, v)^T$. Since the bending moment M becomes a reactive internal stress, the Lagrangian equations only express the translation equilibrium. They are obtained by using equation [6.6], by dropping the third row and column in the algebraic operators, thus obtaining:

$$\begin{aligned} & c_{11}u'' + c_{12}v'' + p_x = 0 \\ & c_{12}u'' + c_{22}v'' + p_y = 0. \end{aligned} \qquad [6.11]$$

together with:

$$\begin{aligned} & u_A = 0 \\ & v_A = 0 \\ & c_{11}u'_B + c_{12}v'_B = P_x \\ & c_{12}u'_B + c_{22}v'_B = P_y \end{aligned} \qquad [6.12]$$

6.2.3. *Elastic constant identification*

To identify the six elastic constants of the TB, or the three constants of the SB, it needs to account for the micro-structure of the underlying 3D cellular system, here assumed to be *periodic*, of period h. The identification procedure, first calls for identifying the displacements of the two bodies. Due to the non-homogeneous nature of the 3D structure, identification is carried out at selected sampled abscissas $x_i := i\,h \quad (i = 0, 1, \ldots, n)$. At these abscissas, the 3D structure possesses equal cross-sections $\mathcal{D}_i$ occupying a domain of the (y, z)-plane (generally not simply connected, or even constituted by isolated points). Two successive cross-sections, located at the abscissas $x_i,\ x_{i+1}$, bound a *cell*. A kinematic constraint is introduced: *the sampled cross-sections $\mathcal{D}_i$ remain plane*, while displacements inside the cells are unconstrained. The rigid motion of $\mathcal{D}_i$, $\tilde{u}(x_i, y, z)$, $\tilde{v}(x_i, y, z)$ (the tilde denoting, from now on, a quantity relevant to the 3D body), can then be described via the configuration variables of the 1D beam, *sampled at the abscissas x_i*, i.e.:

$$\begin{aligned} & \tilde{u}(x_i, y, z) = u(x_i) - \theta(x_i)\,y \\ & \tilde{v}(x_i, y, z) = v(x_i) \end{aligned} \qquad \text{in } \mathcal{D}_i,\ i = 0, 1, \ldots, n \qquad [6.13]$$

These equations constitute a *discrete map*, which relates displacements in the two models at the selected abscissas $x_i \ (i = 0, 1, \ldots, n)$.

The identification of the elastic constants is carried out by equating the elastic energies stored by a 3D cell and a segment of equal length of the 1D beam, when the two models undergo suitably chosen "equal" (in the meaning of equation [6.13]) displacements at the ends. Due to the map [6.13], the elastic energy of the cell is a function of the configuration variables of the 1D beam, namely $\tilde{U} = \tilde{U}(u(x_i), v(x_i), \theta(x_i), u(x_{i+1}), v(x_{i+1}), \theta(x_{i+1}))$; the *averaged* energy density of the cell is evaluated as $\tilde{\phi} := \tilde{U}/h$. However, to compare it to equation [6.4], the displacements must be expressed in terms of strains. The task is accomplished by integrating the strain–displacement relationships [6.1] in the (x_i, x_{i+1}) interval, given a suitable *strain test field*. The simplest one is a constant field $(\varepsilon, \gamma, \kappa) = (\varepsilon_0, \gamma_0, \kappa_0)$, which is compatible with the following end displacements (rigid motions removed):

$$\begin{aligned} u(x_i) &= 0, \qquad u(x_{i+1}) = \varepsilon h \\ v(x_i) &= 0, \qquad v(x_{i+1}) = \gamma h + \frac{1}{2}\kappa h^2 \\ \theta(x_i) &= 0, \qquad \theta(x_{i+1}) = \kappa h \end{aligned} \tag{6.14}$$

These displacements can be read as the superposition of three independent *deformation modes* of the cell: (i) an extensional mode, in which the right cross-section translates axially; (ii) a shear mode, in which the same cross-section translates transversely; and (iii) a flexural mode, in which the same cross-section rotates and translates transversely (h being finite).

In conclusion, $\tilde{\phi} = \frac{1}{h}\tilde{U} = \frac{1}{h}\tilde{U}\left(0, 0, 0, \varepsilon h, \ \gamma h + \frac{1}{2}\kappa h^2, \ \kappa h\right)$ (index 0 dropped). By the equality $\phi = \tilde{\phi} \quad \forall(\varepsilon, \gamma, \kappa)$, the elastic constants $\mathbf{C} := [c_{ij}]$ follow. They, of course, depend on h, as a memory of the underlying micro-structure.

From the previous discussion, it emerges that the main difficulty of the identification consists of writing the elastic energy of the cell, consequent to prescribed displacements assigned at the end cross-sections. This operation calls for a *cell analysis*, which, with a few exceptions, cannot be carried out in closed form. In these cases, a finite element analysis can be performed, consisting of (a) assigning nodal displacements at the boundary joints, which are compatible with the deformation modes; (b) evaluating the reactive forces at the same joints; and (c) computing the elastic energy via the Clapeyron theorem, as half the product between reactions and displacements. Three distinct analyses, one for each deformation mode, and their linear combinations, are sufficient to evaluate all the constants (see Ferretti *et al.* (2020b) for details).

6.3. The grid beam

As an illustrative example, a planar frame is considered, made up of two orders of micro-beams (hereafter referred to as *fibers*), forming equal rectangular $b \times h$ meshes (Figure 6.1). The longitudinal x-fibers are modeled as Timoshenko micro-beams, having axial, shear and bending stiffness EA_x, GA_x^*, EI_x, respectively; the transverse y-fibers, EA_y, GA_y^*, EI_y, with $EA_y \to \infty$ (inextensible). The frame is considered as a periodic cellular structure, in which the y-fibers, or only its nodal points, constitute the cross-sections $\mathcal{D}_i$.

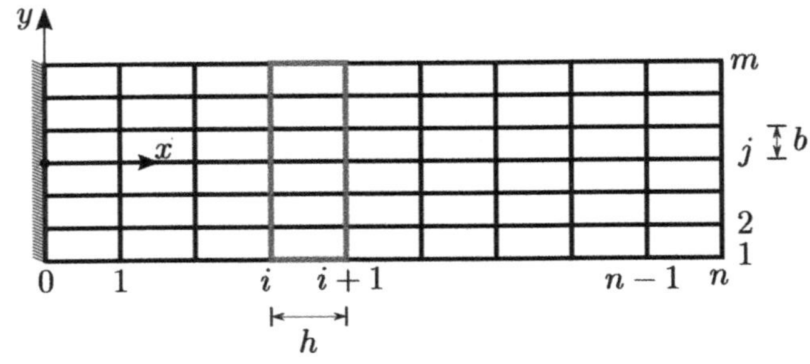

Figure 6.1. *Grid beam. For a color version of this figure, see www.iste.co.uk/challamel/mechanics2.zip*

A cell made up of m x-fibers and two y-fibers of half stiffnesses each, is considered. Two models of cell are distinguished, according to hypothesized mechanical behavior:

– rigid transverse (RT) (Figure 6.2a): the y-fiber does not bend so that all of its material points remain aligned (i.e. $\mathcal{D}_i$ coincides with the fiber itself);

– flexible transverse (FT) (Figure 6.2b): only the nodal points on the y-fiber remain aligned (i.e. $\mathcal{D}_i$ coincides with the set of the isolated joints), but the same fiber is free to warp around the line.

Warping described by the FT will be called *micro-warping*. A more general macro-warping, describing the misalignment of the nodal points, would call for formulating a richer 1D model of beam, equipped with a deformable local structure, along the lines described in Luongo and Zulli (2013, 2014), Luongo *et al.* (2018) and Zulli (2019); therefore, it will not be dealt with here.

6.3.1. *Rigid transverse model*

By assigning the deformation modes in Figure 6.2a and adding up the relevant elastic energies of the m x-fibers, the elastic energy per unit length of the cell is found as:

$$\tilde{\phi} = \frac{1}{2}\left(c_{11}\varepsilon^2 + c_{22}\gamma^2 + c_{33}\kappa^2\right) \qquad [6.15]$$

where:

$$c_{11} := mEA_x, \quad c_{22} := m\frac{12EI_x}{(1+\alpha_x)\,h^2}, \quad c_{33} := mEI_x + EA_x\sum_{i=1}^{m} y_i^2 \qquad [6.16]$$

and where a non-dimensional stiffness ratio α_x appears (while α_y will be used later):

$$\alpha_x := \frac{12EI_x}{GA_x^* h^2}, \qquad \alpha_y := \frac{12EI_y}{GA_y^* b^2} \tag{6.17}$$

Due to the geometric and mechanic regularity of the lattice, no coupling terms among strains appear in the energy (i.e. $c_{12} = c_{23} = c_{13} = 0$), so that $\mathbf{C} = \text{diag}(c_{11}, c_{22}, c_{33})$. More general constitutive laws occur for different topologies, for example, when diagonals, braking the symmetry with respect to the x-axis, exist in the cell (see Ferretti *et al.* (2020b)).

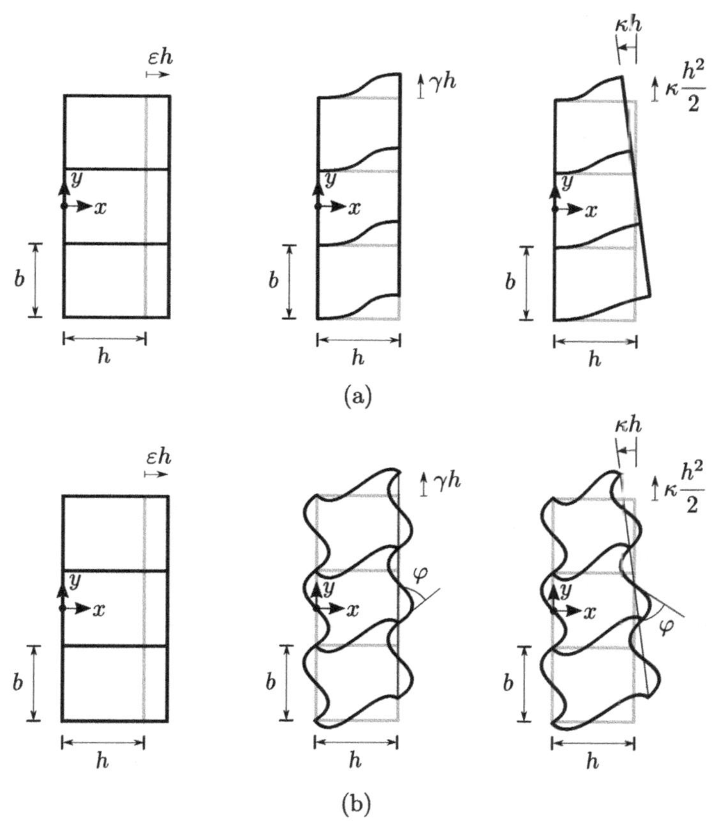

Figure 6.2. *Single cell of the grid beam and its deformation modes for two different models: (a) rigid transverse (RT) and (b) flexible transverse (FT)*

6.3.2. *Flexible transverse model*

The analysis is repeated by assigning the deformation modes in Figure 6.2b, identical to the previous ones, but in which the transverse fibers are allowed to warp around the line joining the nodes. To simplify the analysis, it is assumed that all of the joints undergo the same unknown rotation φ. Referring to the shear mode, the energy of the x- and y-fibers is given by:

$$\tilde{U}_x = m\frac{6EI_x}{h\left(1+\alpha_x\right)}\left(\gamma-\varphi\right)^2, \qquad \tilde{U}_y = \left(m-1\right)\frac{6EI_y}{b\left(1+\alpha_y\right)}\varphi^2 \qquad [6.18]$$

so that $\tilde{U} := \tilde{U}_x + \tilde{U}_y$. By requiring $\frac{\partial \tilde{U}}{\partial \varphi} = 0$, the unknown φ is condensed, and the energy per unit length is written as:

$$\tilde{\phi} = \frac{1}{2}\chi_s m\frac{12EI_x}{\left(1+\alpha_x\right)h^2}\gamma^2 \qquad [6.19]$$

where:

$$\chi_s := \frac{\frac{m-1}{m}\varrho\frac{h}{b}}{1+\frac{m-1}{m}\varrho\frac{h}{b}} \qquad \text{with} \quad \varrho := \frac{EI_y}{EI_x}\frac{1+\alpha_x}{1+\alpha_y} \qquad [6.20]$$

is a non-dimensional quantity named the *shear factor*, with ϱ as a stiffness ratio. By comparing equations [6.19] and [6.15], it appears that the shear constant supplied by the flexible model is proportional, via the χ_s factor, to that furnished by the rigid model. Thus, the shear factor accounts for the flexibility of the transverse fibers in the same way it accounts, in the Timoshenko beam, for the warping of the solid cross-section. In other words, we can consider the cross-section as not warpable, provided we artificially modify the elastic properties of the longitudinal fibers (namely $\chi_s EI_x$, instead of the true EI_x) in the same way we consider, for example, for a rectangular cross-section, $A^* = \frac{5}{6}A$.

It is worth noting that, since $\chi_s \in (0,1)$, it reduces the macroscopic shear stiffness of the grid beam. As an example, if the two orders of fibers are equal, and $h = b$, then $\chi_s \simeq 0.5$ for large m. If, however, $EI_y/\left(1+\alpha_y\right) = 10\,EI_x/\left(1+\alpha_x\right)$, with the same squared mesh, then $\chi_s \simeq 0.91$, i.e. the factor quickly tends to 1 with the stiffness ratio.

The same analysis, carried out in Luongo *et al.* (2020) for the flexural deformation mode, has shown that an analogous corrective factor χ_f can be defined. However, in real cases, $\chi_f \simeq 1$, denoting that the bending of the transverse fibers is energetically negligible with respect to the extension of the longitudinal fibers. Therefore, the constant c_{33} previously determined is sufficiently accurate. By summarizing, the TB with flexible transverse fibers has elastic matrix $\mathbf{C} = \text{diag}\left(c_{11}, \chi_s c_{22}, c_{33}\right)$.

6.3.3. *Comparison among models*

From the previous discussion, it appears that four different models of increasing refinement, accounting, or not, for the extension of the longitudinal fibers and/or the bending of the transverse fibers, can be adopted, namely: (i) the shear beam with rigid transverses (SB-RT), (ii) the Timoshenko beam with rigid transverses (TB-RT), (iii) the shear beam with flexible transverses (SB-FT) and (iv) the Timoshenko beam with flexible transverses (TB-FT). To compare their accuracy against exact solutions, a numerical analysis is carried out on a cantilever grid beam, made up of $n = 10, 25$ cells, constituted by fibers of comparable stiffnesses ($\varrho = 1.71$), subjected to equal lateral forces applied to any transverses (see Luongo *et al.* (2020) for details). Results are compared to internally unconstrained finite element (FE) analyses. Figure 6.3 reports the deflections $v(x)$ of the beam for the different models.

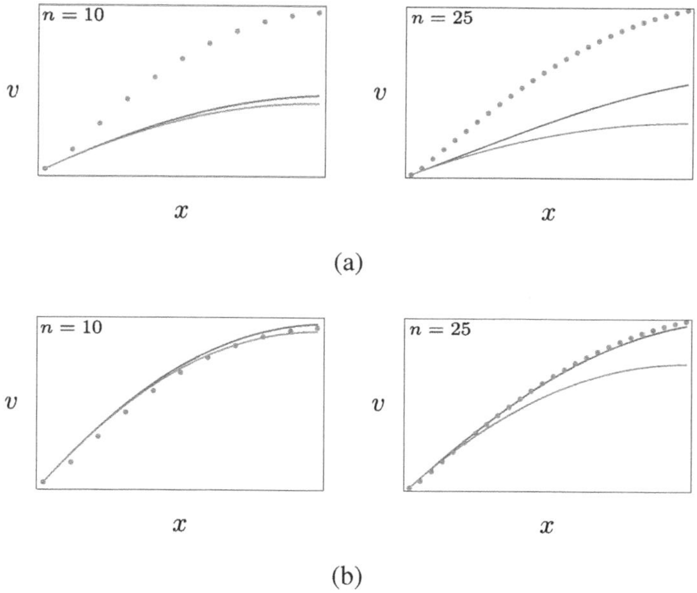

Figure 6.3. *Static response of a grid beam to lateral forces: (a) rigid transverse fibers and (b) flexible transverse fibers; red line: TB, green line: SB, bullets: FE. For a color version of this figure, see www.iste.co.uk/challamel/mechanics2.zip*

When the rigid transverse model is adopted (Figure 6.3a), significant errors occur, for both TB-RT and SB-RT, with the former being more accurate than the latter and the error being larger for shorter cellular beams. However, the error reduces when the stiffness of the transverse fibers increases. For example, when $\varrho = 7.85$ (the results are not shown here, see Luongo *et al.* (2020)), the error at the tip of the TB-RT is about 25%. The approximation remarkably improves when the flexibility of the transverse is accounted for (Figure 6.3b). An excellent agreement is found between the TB-FT and the numerical model. The SB-FT also gives very good results for short cellular beams and not as good results for longer beams.

6.4. Buckling

The elastic buckling of a planar cellular beam, uniformly compressed by an axial force $-P\mathbf{i}$ applied at the end B, is studied (see Ferretti (2018) and Ferretti *et al.* (2020a) for more general cases). A 1D model is formulated, and local and global buckling is discussed.

6.4.1. *Formulation*

To analyze the buckling of a cellular beam, it needs to extend the linear 1D model, to account for geometric effects. As it will be clear below, this task calls for two different steps: (i) to enforce equilibrium in the adjacent (slightly varied) configuration, to describe *global* geometrical effects, and (ii) to incorporate the prestress in the constitutive law, to describe *local* geometrical effects. This latter item, which could not appear obvious at first glance, accounts for geometrical effects *inside the cell*, which, if analyzed uniquely on the ground of the equilibrium, would disappear when the periodic system has been homogenized. The circumstance clearly emerges when we consider, for example, the purely torsional buckling of an open thin-walled beam (TWB). There, the (local) geometrical effect of the inclination of the longitudinal fibers, cannot be captured by a 1D model; indeed, it is described by a constrained 3D model, according to the Vlasov theory.

The equilibrium equations of the Timoshenko beam in the adjacent configuration can be conveniently derived by the virtual work principle, equation [6.2], in which the virtual work spent by the prestress $\mathring{N} = -P$ over the second-order part of the unit extension $\varepsilon^{(2)}$, is added. An exact measure of the unit strain is offered by (see, for example, Luongo and Zulli (2013)):

$$\begin{aligned} \varepsilon &= (1+u')\cos\theta + v'\sin\theta - 1 \\ &= u' - \frac{1}{2}\theta^2 + v'\theta + O\left(\|\mathbf{u}\|^3\right) \end{aligned} \qquad [6.21]$$

which represents the projection of the stretched initially unitary segment, $(1+u')\,\mathbf{i} + v'\mathbf{j}$, onto the current direction of the rotated i-vector, $\cos\theta\,\mathbf{i} + \sin\theta\,\mathbf{j}$. After expansion at the second order, the VWP is given by (no incremental loads applied):

$$\int_0^\ell (N\delta u' + T\delta(v'-\theta) + M\delta\theta')\,dx - \int_0^\ell P\delta\left(v'\theta - \frac{1}{2}\theta^2\right)dx = 0 \qquad [6.22]$$

$$\forall(\delta u, \delta v, \delta\theta)$$

From this, the indefinite equilibrium equations follow:

$$\begin{aligned} N' &= 0 \\ T' - P\theta' &= 0 \\ M' + T + P(v' - \theta) &= 0 \end{aligned} \qquad [6.23]$$

together with the natural boundary conditions:

$$N_B = 0, \qquad T_B = P\theta_B, \qquad M_B = 0 \qquad [6.24]$$

in which $u_A = v_A = \theta_A = 0$ have been accounted.

To evaluate the elastic constants, the procedure illustrated in section 6.2.3 is applied. It calls for evaluating the elastic energy $\tilde{U}$ of the cell, when the deformation modes are assigned at the boundaries of the cell. Now, since the body is prestressed, $\tilde{U} := \tilde{U}^0 + \tilde{U}^*$, with $\tilde{U}^0$ as the elastic energy of the unprestressed cell and $\tilde{U}^*$ as the geometric contribution proportional to the prestress in the cell. For example, for the grid beam studied in section 6.3, the energy is given by:

$$\tilde{U} = \frac{1}{2}\sum_{j=1}^{m} \mathbf{u}_j^T \left(\mathbf{K}_j^0 + \overset{\circ}{N}_j \mathbf{K}_j^*\right) \mathbf{u}_j \qquad [6.25]$$

where $\mathbf{u}_j$ are the displacements at the ends of the x-fibers, $\mathbf{K}_j^0$, $\mathbf{K}_j^*$ are the elastic and geometric stiffness matrices of a micro-beam, respectively, and $\overset{\circ}{N}_j$ is the quota of the prestress relevant to the jth element; for a regular mesh, $\overset{\circ}{N}_j = -\frac{P}{m}$. The identification procedure, therefore, leads to an elastic matrix $\mathbf{C} := \mathbf{C}^0 - P\mathbf{C}^*$, in which $\mathbf{C}^0$ is the elastic matrix of the unprestressed body and $\mathbf{C}^*$ is its geometric correction. For rigid transverses:

$$\mathbf{C} := \begin{bmatrix} c_{11} & 0 & 0 \\ & c_{22} & 0 \\ sym & & c_{33} \end{bmatrix} - P \begin{bmatrix} 0 & 0 & 0 \\ & c_{22}^* & c_{23}^* \\ sym & & c_{33}^* \end{bmatrix} \qquad [6.26]$$

where (c_{11}, c_{22}, c_{33}) are given by equation [6.16], and:

$$c_{22}^* := \frac{6 + 10\alpha_x}{5(1 + \alpha_x)^2}, \qquad c_{23}^* = \frac{h}{2}, \qquad c_{33}^* := \frac{h^2}{3} \qquad [6.27]$$

with α_x defined in equation [6.17a] (see Ferretti *et al.* (2020a) for details). Consequently, equations [6.23] become:

$$\begin{aligned} \mathbf{K}_2\mathbf{v}'' + \mathbf{K}_1\mathbf{v}' + \mathbf{K}_0\mathbf{v} &= \mathbf{0} \\ \mathbf{v}_A &= \mathbf{0} \\ \mathbf{B}_1\mathbf{v}_B' + \mathbf{B}_0\mathbf{v}_B &= \mathbf{0} \end{aligned} \qquad [6.28]$$

where $\mathbf{v} = (v, \theta)^T$, and:

$$\mathbf{K}_2 := \begin{bmatrix} c_{22} - Pc^*_{22} & -Pc^*_{23} \\ -Pc^*_{23} & c_{33} - Pc^*_{33} \end{bmatrix},$$

$$\mathbf{K}_1 := \begin{bmatrix} 0 & -c_{22} + P\left(c^*_{22} - 1\right) \\ c_{22} - P\left(c^*_{22} - 1\right) & 0 \end{bmatrix},$$

$$\mathbf{K}_0 := \begin{bmatrix} 0 & 0 \\ 0 & -c_{22} + P\left(c^*_{22} - 1\right) \end{bmatrix}, \qquad [6.29]$$

$$\mathbf{B}_1 := \begin{bmatrix} c_{22} - Pc^*_{22} & -Pc^*_{23} \\ -Pc^*_{23} & c^*_{33} - Pc^*_{33} \end{bmatrix},$$

$$\mathbf{B}_0 := \begin{bmatrix} 0 & -c_{22} + P\left(c^*_{22} - 1\right) \\ 0 & Pc^*_{23} \end{bmatrix}$$

and $u = 0 \; \forall x$ was accounted for.

6.4.2. *Critical loads*

Since the elastic matrix [6.26] depends on prestress, there exist (two) critical values of P which render it singular, namely det $\left(\mathbf{C}^0 - P\mathbf{C}^*\right) = 0$. The smallest root P is the cell critical load P_{cell}. At this load, a local buckling manifests, in which the cell loses its linear carrying capacity. This, however occurs when the cell is clamped at one end, consistently with the deformation modes (Figure 6.2) used to build up $\mathbf{C}$. When the cells are assembled in a cellular beam, they interact elastically, so it is expected that bifurcation occurs at a lower critical load, $P_c < P_{cell}$.

To determine the beam critical load, equations [6.28] must be solved. The task can be accomplished in closed form, since the field equation admits simple characteristic exponents, i.e. $\lambda_{1,2} = 0$, $\lambda_{3,4} = \pm\sqrt{f\left(P\right)}$, with f a function of P. Boundary conditions provide a system of linear algebraic equations in the arbitrary constants; by zeroing the determinant of the matrix of the coefficients, a transcendent equation in P is found, whose smallest root is the critical load of the beam, P_c.

As a case study, Figure 6.4a shows the critical mode of a grid beam made up of 10 cells. Analytical results are compared with finite element analyses. The relevant critical load is found with an error of 1%, although the number of cells is low. Figure 6.4b reports the first three critical loads P_i $(i = 1, 2, 3)$, non-dimensionalized with respect to P_{cell}, versus the number of cells, in the range $n \in (1, 30)$. It is seen that all the P_i's tend to P_{cell} when the number of cell is low. In this case, a large number of close eigenvalues exist (i.e. the beam has a large modal density), possibly leading to interaction in the nonlinear field, or localization phenomena (see, for example, Luongo (1991, 1992)). In contrast, when n is sufficiently large, the first eigenvalues separate among them, and the lowest takes values sensibly lower than P_{cell}. For example, for $n = 30$, $P_c \simeq 0.6 P_{cell}$.

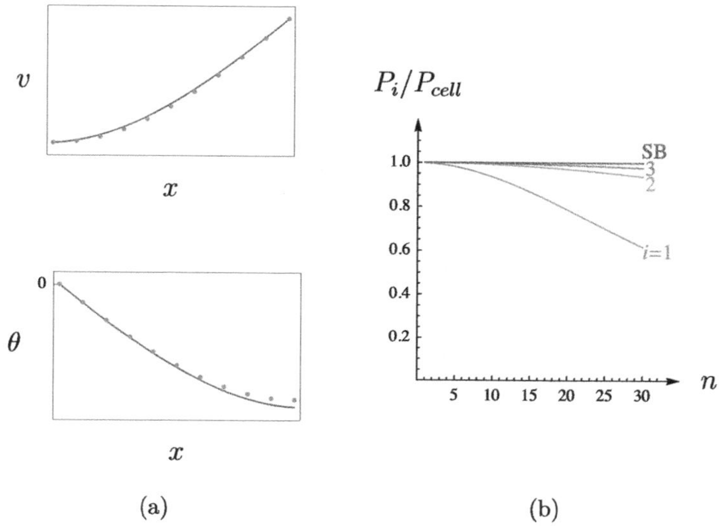

Figure 6.4. *Buckling of a uniformly compressed grid beam (Ferretti (2018)): (a) critical mode ($n = 10$), continuous line TB-RT, bullets FE; (b) critical load versus the number of cells, SB shear beam. For a color version of this figure, see www.iste.co.uk/challamel/mechanics2.zip*

It is worth discussing the behavior of the purely shear beam, which, in principle, may be able to describe the behavior of cellular beams made up of longitudinal fibers of high axial stiffness. For this beam, the equilibrium equations [6.28] reduce to:

$$\begin{aligned} (c_{22} - P c_{22}^*)\, v'' &= 0 \\ v_A &= 0 \\ (c_{22} - P c_{22}^*)\, v'_B &= 0 \end{aligned} \qquad [6.30]$$

which admits *infinite coincident critical loads* $P_c = P_{cell} = \frac{c_{22}}{c_{22}^*}$, which are independent of the number of cells (see Figure 6.4b). The corresponding critical mode is arbitrary. Therefore, *this model is inadequate* to capture the behavior of the beam at the bifurcation. Indeed, if the cells are not allowed to rotate, but only to deform in shear, they are in the constraint conditions in which P_{cell} has been evaluated. Since the cells are all equal, the assembly has equal eigenvalues. In contrast, even a small extensibility of the longitudinal fibers, accounted in the

Timoshenko model, reduces the critical load, as well as splitting the coalescence. It is interesting to note that the same phenomenon occurs in the purely torsional buckling of TWB, whose equation has the same structure of equation [6.30a], and where the critical load is found to be independent of length.

6.5. Dynamics

To analyze dynamical problems, the inertial properties of the cellular beam must be identified. For simplicity, a reference is made here to a regular grid beam, having masses lumped at joints. The hypothesis, from a kinetic point of view, makes the difference between the RT and RF models irrelevant, which would instead exist for distributed masses.

To perform identification, the kinetic energy of the cell, $\tilde{\mathcal{T}}$, is equated to the kinetic energy $\mathcal{T}$ of a segment of a beam of the same length h, subjected to a rigid field of velocity:

$$\tilde{\mathcal{T}} = \frac{1}{2} M \sum_{j=1}^{m} \left(\dot{u}_j^2 + \dot{v}_j^2 \right), \qquad \mathcal{T} = \frac{h}{2} \left[m_{11} \left(\dot{u}^2 + \dot{v}^2 \right) + m_{33} \dot{\theta}^2 \right] \qquad [6.31]$$

where M are the lumped masses, $\dot{u}_j, \dot{v}_j$ are the translational velocities at joints, m_{11} is the mass per unit length and m_{33} is the mass inertia moment of the homogenized beam. By using the displacement map [6.13], it follows:

$$m_{11} := \frac{mM}{h}, \qquad m_{33} = \frac{M}{h} \sum_{j=1}^{m} y_j^2 \qquad [6.32]$$

6.5.1. *Timoshenko beam*

By adopting the TB-FT model, the transverse vibrations are governed by:

$$\begin{aligned} -m_{11} \ddot{v} + \chi_s c_{22} \left(v' - \theta \right)' &= 0 \\ -m_{33} \ddot{\theta} + c_{33} \theta'' + c_{22} \left(v' - \theta \right) &= 0 \end{aligned} \qquad [6.33]$$

with the boundary conditions $v_A = \theta_A = 0$, $v'_B - \theta_B = 0$, $\theta'_B = 0$. By letting $\left(v(x,t), \theta(x,t) \right) = \left(\hat{v}(x), \hat{\theta}(x) \right) \exp(i\omega t)$, with ω as the unknown natural frequency, a boundary value problem follows:

$$\begin{aligned} \chi_s c_{22} \left(\hat{v}' - \hat{\theta} \right)' + \omega^2 m_{11} \hat{v} &= 0 \\ c_{33} \hat{\theta}'' + c_{22} \left(\hat{v}' - \hat{\theta} \right) + \omega^2 m_{33} \hat{\theta} &= 0 \end{aligned} \qquad [6.34]$$

with the boundary conditions $\hat{v}_A = \hat{\theta}_A = 0, \hat{v}'_B - \hat{\theta}_B = 0, \hat{\theta}'_B = 0$. It can be solved, for example, by a finite-difference approach (see Ferretti *et al.* (2020b)). Results relevant to a sample grid beam of $n = 25$ cells are displayed in Figure 6.5. They concern the first three natural modes, compared with FE analyses. Errors on the frequencies are found to be of the order of 1%; the natural modes are almost exact, too.

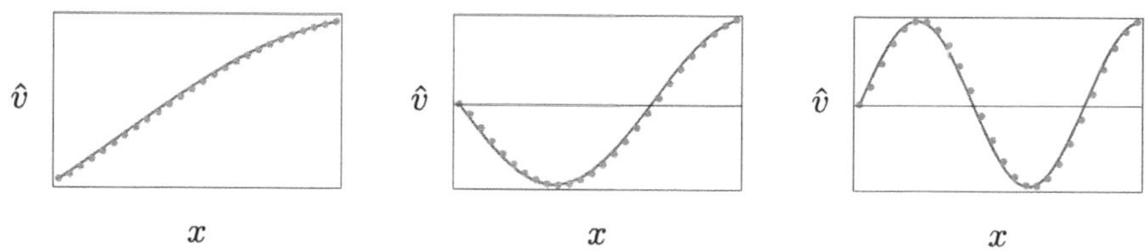

Figure 6.5. *Lateral displacements $\hat{v}(x)$ for the first three modes ($n = 25$), continuous line TB-FT, bullets FE. For a color version of this figure, see www.iste.co.uk/challamel/mechanics2.zip*

6.5.2. *Shear beam and discrete spring–mass model*

The shear beam model has been adopted in the literature, to study complex nonlinear dynamical problems (see Luongo and Zulli (2011), Zulli and Luongo (2012) and Zulli and Di Egidio (2015)). When the model is linearized, the boundary value problem [6.34] reduces to:

$$\chi_s c_{22} \hat{v}'' + \omega^2 m_{11} \hat{v} = 0 \qquad [6.35]$$

with the boundary conditions $\hat{v}_A = \hat{v}'_B = 0$. The solution is:

$$\hat{v}_k = \sin\left(\frac{k\pi x}{2\ell}\right), \qquad \omega_k = \frac{k\pi}{2\ell}\sqrt{\frac{\chi_s c_{22}}{m_{11}}}, \qquad k = 1, 3, 5, \ldots \qquad [6.36]$$

On the other hand, since only transverse displacements are allowed, the cellular beam behaves as a chain of equal masses $mM = m_{11}h$, and equal springs of stiffness $\chi_s \frac{m12EI_x}{(1+\alpha_x)h^3} = \chi_s \frac{c_{22}}{h}$, whose discrete (non-homogenized) exact equations of motion are given by:

$$\begin{aligned} & m_{11} h \ddot{v}_j + \chi_s \frac{c_{22}}{h}\left(-v_{j-1} + 2v_j - v_{j+1}\right) = 0 \qquad j = 1, \ldots n-1 \\ & v_0 = 0 \\ & m_{11} h \ddot{v}_j + \chi_s \frac{c_{22}}{h}\left(-v_{n-1} + v_n\right) = 0 \end{aligned} \qquad [6.37]$$

with $v_j(t)$ as the displacement of the jth transverse. By separating the variables, via $v_j(t) = \hat{v}_j \exp(i\omega t)$, and solving the relevant difference equations, it follows (Luongo and Zulli 2011):

$$\hat{v}_j = \sin(\beta j), \qquad \cos\beta := 1 - \frac{1}{2}\omega^2 \frac{m_{11}}{\chi_s c_{22}} h^2 \tag{6.38}$$

with the characteristic equation:

$$-\sin[\beta(n-1)] + [2\cos(\beta) - 1]\sin(\beta n) = 0 \tag{6.39}$$

The asymptotic solution $\beta_k = \frac{k\pi}{2n}$ holds for large n and small k; it entails:

$$\begin{aligned} \hat{v}_{kj} &= \sin\left(\frac{k\pi}{2n} j\right), \\ \omega_k &= \frac{k\pi}{2n}\frac{1}{h}\sqrt{\frac{\chi_s c_{22}}{m_{11}}}, \end{aligned} \qquad k = 1, 3, 5, \ldots; \; j = 0, 1, \ldots n \tag{6.40}$$

This is fully consistent with the solution [6.36] of the homogenized model, in which $x = j\,h$ and $\ell = n\,h$ are taken. Therefore, the continuous shear beam accurately captures the low-frequency dynamics of long cellular beams (when rotations of the transverse fibers are negligible).

6.6. Aeroelastic stability

It is well known that wind can trigger aerodynamic instability (galloping, or Hopf bifurcation) of prismatic structures such as tall buildings, towers and chimneys (see, for example, Novak (1972)). This phenomenon is due to the cross-wind motion of the structure immersed in a steady wind flow, which makes the *attack angle* (i.e. the angle formed by the wind relative velocity and a material axis of the cylinder cross-section) dependent on velocity. The linear part of the aerodynamic forces, being proportional to the structural velocity, acts as an apparent *aerodynamic damping*, which, if negative, can trigger self-excited vibrations of a large amplitude.

6.6.1. *Modeling a base-isolated tower*

A possible control of galloping consists of artificially increasing the antagonist structural damping. An idea was proposed in Di Nino and Luongo (2019), to isolate the structure from the ground, by interposing an elastically sliding base, similar to what is done against seismic actions. By placing a dashpot in parallel with the spring and exploiting the relative motion between the structure and the ground, energy is dissipated (Figure 6.6a). By referring to the simpler case in which the structure behaves as a shear-type frame, the system is modeled as a continuous *viscoelastic* shear beam, equipped with a viscoelastic device at the bottom end (Figure 6.6b).

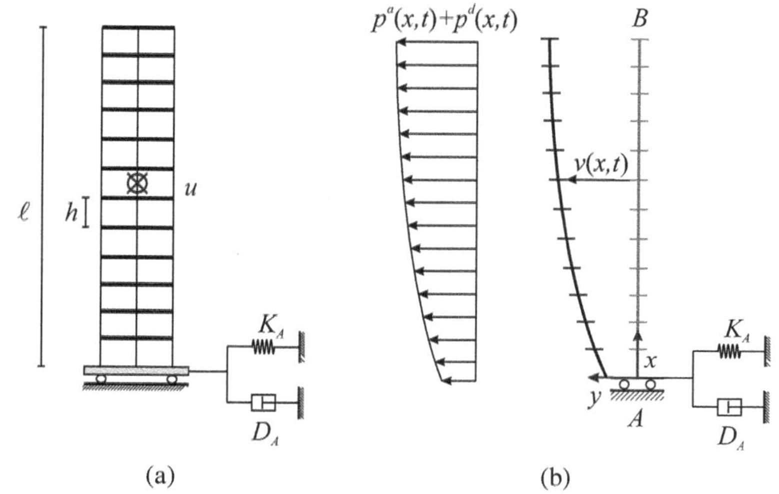

Figure 6.6. *Base-isolated tower building under wind flow: (a) cellular beam, (b) continuous shear beam*

The equations of motion of the system are:

$$\begin{aligned}&m_{11}\ddot{v} - \chi_s c_{22}\left(1+\bar{\eta}\partial_t\right)v'' = p^a + p^d\\&-\chi_s c_{22}\left(1+\bar{\eta}\partial_t\right)v'_A + K_A v_A + D_A \dot{v}_A = 0\\&\chi_s c_{22}\left(1+\bar{\eta}\partial_t\right)v'_B = 0\end{aligned} \qquad [6.41]$$

ruling the motion in the domain and at two boundaries, respectively. Here, K_A, D_A are the stiffness and damping of the device, respectively; $\bar{\eta}$ is a viscosity coefficient, according to the Kelvin–Voigt rheological model, describing *internal damping*; moreover, $p^a := -\bar{\xi}_a \bar{u}\dot{v}$ is the linear part of the aerodynamic force, proportional to the wind velocity $\bar{u}$ and the structural velocity $\dot{v}$, via the aerodynamic coefficient $\bar{\xi}_a$; $p^d := -\bar{\xi}_e \dot{v}$ is the *external damping* force, with $\bar{\xi}_e$ as a coefficient.

In the non-dimensional form, the previous equations can be written as:

$$\begin{aligned}&\ddot{v} - \left(1+\eta\partial_t\right)v'' + \left(\xi_e + \xi_a u\right)\dot{v} = 0\\&-\left(1+\eta\partial_t\right)v'_A + \chi v_A + \zeta\dot{v}_A = 0\\&\left(1+\eta\partial_t\right)v'_B = 0\end{aligned} \qquad [6.42]$$

in which unbarred symbols keep their meaning, and, moreover, the stiffness ratio $\chi := \frac{K_A \ell}{\chi_s c_{22}}$ and the damping ratio $\zeta := \frac{D_A}{\sqrt{\chi_s c_{22} m_{11}}}$, characterize the device.

6.6.2. *Critical wind velocity*

To evaluate the critical velocity of wind, $u = u_c$, at which the beam loses stability, harmonic solutions to equations [6.42] are sought for denoting the branching of a family of limit cycles from the trivial equilibrium path. By letting $v(x,t) = \hat{v}(x) \exp(i\omega_c t)$, with ω_c as the unknown frequency at the bifurcation, the following eigenvalue problem is found:

$$\begin{aligned} \hat{v}'' + \beta^2 \hat{v} &= 0 \\ \hat{v}'_A - \alpha \hat{v}_A &= 0 \\ \hat{v}'_B &= 0 \end{aligned} \qquad [6.43]$$

where:

$$\beta := \sqrt{\frac{\omega_c^2 - i\omega_c (\xi_e + \xi_a u_c)}{1 + i\omega_c \eta}}, \qquad \alpha := \frac{\chi + i\omega_c \zeta}{1 + i\omega_c \eta} \qquad [6.44]$$

By solving the field equations and enforcing boundary conditions, the transcendental characteristic equation follows:

$$f(\omega_c, u_c) := -\alpha + \beta \tan \beta = 0 \qquad [6.45]$$

By splitting it into its real and imaginary parts, two real equations in the two unknowns ω_c, u_c are found. The smallest root u_c is the critical velocity.

Numerical results are reported in Figure 6.7 for a case study (see Di Nino and Luongo (2019) for details). First (Figure 6.7a), a purely elastic base ($\zeta = 0$) is considered. When $\chi \to \infty$, u_c tends to the value $u_{c\infty}$ of the clamped beam. For finite values of χ, $u_c < u_{c\infty}$, i.e. the elastic basis has a detrimental effect on the system behavior. This is due to the fact that, since frequency is reduced, the internal damping also reduces. If, instead, an external damping ζ is added, the critical velocity can exceed $u_{c\infty}$. However, in order for this beneficial effect to take place, damping must be larger than a threshold value $\zeta^* = \zeta^*(\chi)$: the larger the χ, the larger the ζ^*. Therefore, soft springs work better, since they allow large translations of the base, entailing larger dissipation. On the other hand, too large translations could be unwanted, and therefore, parameters must be optimized.

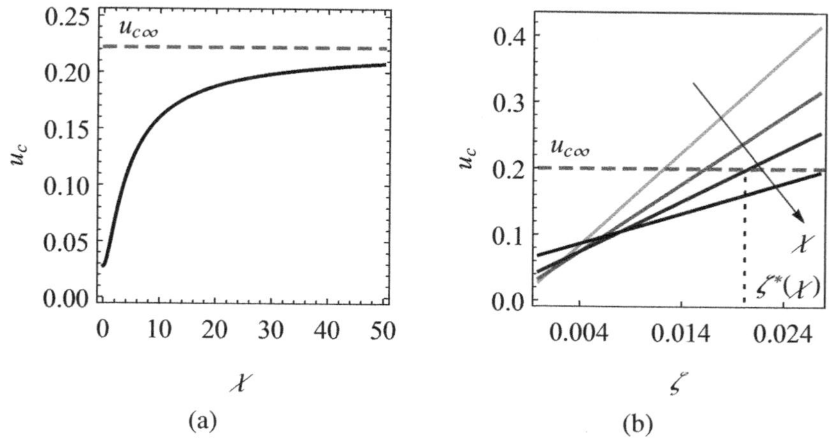

Figure 6.7. *Critical wind velocity of the base-isolated beam: (a) versus the stiffness ratio, for purely elastic device* ($\zeta = 0$)*; (b) versus the damping ratio, for different stiffnesses ratios.* $u_{c\infty}$ *is the critical velocity of the clamped beam. For a color version of this figure, see www.iste.co.uk/challamel/mechanics2.zip*

6.7. References

Auriault, J.-L., Boutin, C., Geindreau, C. (2009). *Homogenization of Coupled Phenomena in Heterogenous Media*, ISTE Ltd, London, and John Wiley & Sons, New York.

Boutin, C., dell'Isola, F., Giorgio, I., Placidi, L. (2017). Linear pantographic sheets: Asymptotic micro-macro models identification. *Mathematics and Mechanics of Complex Systems*, 5, 127–162.

Chajes, M., Romstad, K., McCallen, D. (1993). Analysis of multiple-bay frames using continuum model. *Journal of Structural Engineering*, 119(2), 522–546.

Chajes, M., Zhang, L., Kirby, J. (1996). Dynamic analysis of tall building using reduced-order continuum model. *Journal of Structural Engineering*, 122(11), 1284–1291.

Challamel, N., Hache, F., Elishakoff, I., Wang, C. (2016). Buckling and vibrations of microstructured rectangular plates considering phenomenological and lattice-based nonlocal continuum models. *Composite Structures*, 149, 145–156.

Cioranescu, D. and Paulin, J.S.J. (1999). *Homogenization of Reticulated Structures*. Springer-Verlag, New York.

D'Annibale, F., Ferretti, M., Luongo, A. (2019). Shear-shear-torsional homogeneous beam models for nonlinear periodic beam-like structures. *Engineering Structures*, 184, 115–133.

dell'Isola, F., Eremeyev, V.A., Porubov, A. (eds) (2018). *Advances in Mechanics of Microstructured Media and Structures*. Springer, Cham.

Di Nino, S. and Luongo, A. (2019). Nonlinear aeroelastic behavior of a base-isolated beam under steady wind flow. *International Journal of Non-Linear Mechanics*, 119, 103340.

Dos Reis, F. and Ganghoffer, J.F. (2012). Equivalent mechanical properties of auxetic lattices from discrete homogenization. *Computational Materials Science*, 51(1), 314–321.

Elishakoff, I. (2020a). *Handbook on Timoshenko-Ehrenfest Beam and Uflyand-Mindlin Plate Theories*. World Scientific, Singapore.

Elishakoff, I. (2020b). Who developed the so-called Timoshenko beam theory? *Mathematics and Mechanics of Solids*, 25(1), 97–116.

Ferretti, M. (2018). Flexural torsional buckling of uniformly compressed beam-like structures. *Continuum Mechanics and Thermodynamics*, 30(5), 977–993.

Ferrarotti, A., Piccardo, G., Luongo, A. (2017). A novel straightforward dynamic approach for the evaluation of extensional modes within GBT 'cross-section analysis'. *Thin-Walled Structures*, 114, 52–69.

Ferretti, M., D'Annibale, F., Luongo, A. (2020a). Buckling of tower buildings on elastic foundation under compressive tip forces and self-weight. *Continuum Mechanics and Thermodynamics*, 1–21. doi: 10.1007/s00161-020-00911-2.

Ferretti, M., D'Annibale, F., Luongo, A. (2020b). Modeling beam-like planar structures by a one-dimensional continuum: An analytical-numerical method. *Journal of Applied and Computational Mechanics*. doi: 10.22055/jacm.2020.33100.2150.

Hache, F., Challamel, N., Elishakoff, I., Wang, C. (2017). Comparison of nonlocal continualization schemes for lattice beams and plates. *Archive of Applied Mechanics*, 87(7), 1105–1138.

Hache, F., Challamel, N., Elishakoff, I. (2018). Lattice and continualized models for the buckling study of nonlocal rectangular thick plates including shear effects. *International Journal of Mechanical Sciences*, 145, 221–230.

Luongo, A. (1991). On the amplitude modulation and localization phenomena in interactive buckling problems. *International Journal of Solids and Structures*, 27(15), 1943–1954.

Luongo, A. (1992). Mode localization by structural imperfections in one-dimensional continuous systems. *Journal of Sound and Vibration*, 155(2), 249–271.

Luongo, A. and Zulli, D. (2011). Parametric, external and self-excitation of a tower under turbulent wind flow. *Journal of Sound and Vibration*, 330(13), 3057–3069.

Luongo, A. and Zulli, D. (2013). *Mathematical Models of Beams and Cables*. ISTE Ltd, London, and John Wiley & Sons, New York.

Luongo, A. and Zulli, D. (2014). A non-linear one-dimensional model of cross-deformable tubular beam. *International Journal of Non-Linear Mechanics*, 66, 33–42.

Luongo, A. and Zulli, D. (2020). Free and forced linear dynamics of a homogeneous model for beam-like structures. *Meccanica*, 55, 907–925.

Luongo, A. and Zulli, D., Scognamiglio, I. (2018). The Brazier effect for elastic pipe beams with foam cores. *Thin-Walled Structures*, 124, 72–80.

Luongo, A., D'Annibale, F., Ferretti, M. (2021). Shear and flexural factors for static analysis of homogenized beam models of planar frames. *Engineering Structures*, 228, 111440. doi: 10.1016/j.engstruct.2020.111440.

Noor, A. (1988). Continuum modeling for repetitive lattice structures. *Applied Mechanics Reviews*, 41(7), 285–296.

Novak, M. (1972). Galloping oscillations of prismatic structures. *Journal of the Engineering Mechanics Division*, 98(1), 27–46.

Piccardo, G., Ranzi, G., Luongo, A. (2014a). A complete dynamic approach to the Generalized Beam Theory cross-section analysis including extension and shear modes. *Mathematics and Mechanics of Solids*, 19(8), 900–924.

Piccardo, G., Ranzi, G., Luongo, A. (2014b). A direct approach for the evaluation of the conventional modes within the GBT formulation. *Thin-Walled Structures*, 74, 133–145.

Piccardo, G., Tubino, F., Luongo, A. (2015). A shear–shear torsional beam model for nonlinear aeroelastic analysis of tower buildings. *Zeitschrift für Angewandte Mathematik und Physik*, 66(4), 1895–1913.

Piccardo, G., Tubino, F., Luongo, A. (2016). Equivalent nonlinear beam model for the 3-d analysis of shear-type buildings: Application to aeroelastic instability. *International Journal of Non-Linear Mechanics*, 80, 52–65.

Piccardo, G., Ferrarotti, A., Luongo, A. (2017). Nonlinear generalized beam theory for open thin-walled members. *Mathematics and Mechanics of Solids*, 22(10), 1907–1921.

Piccardo, G., Tubino, F., Luongo, A. (2019). Equivalent Timoshenko linear beam model for the static and dynamic analysis of tower buildings. *Applied Mathematical Modelling*, 71, 77–95.

Pshenichnov, G.I. (1993). *A Theory of Latticed Plates and Shells*, vol. 5. World Scientific, Singapore.

Ranzi, G. and Luongo, A. (2011). A new approach for thin-walled member analysis in the framework of GBT. *Thin-Walled Structures*, 49(11), 1404–1414.

Silvestre, N. and Camotim, D. (2002). First-order generalised beam theory for arbitrary orthotropic materials. *Thin-Walled Structures*, 40, 755–789.

Tollenaere, H. and Caillerie, D. (1998). Continuous modeling of lattice structures by homogenization. *Advances in Engineering Software*, 29(7), 699–705.

Zulli, D. (2019). A one-dimensional beam-like model for double-layered pipes. *International Journal of Non-Linear Mechanics*, 109, 50–62.

Zulli, D. and Di Egidio, A. (2015). Galloping of internally resonant towers subjected to turbulent wind. *Continuum Mechanics and Thermodynamics*, 27(4–5), 835–849.

Zulli, D. and Luongo, A. (2012). Bifurcation and stability of a two-tower system under wind-induced parametric, external and self-excitation. *Journal of Sound and Vibration*, 331(2), 365–383.

7

Collapse Limit of Structures under Impulsive Loading via Double Impulse Input Transformation

7.1. Introduction

Around 1960, many structural and mechanical engineering researchers focused on two topics, i.e. the finite element method and the elastic–plastic response analysis. Both research topics were accelerated by the enhanced development of computer science and related devices. Although the invention of the finite element method enabled the computation of complicated elastic–plastic structures, the theoretical development of limit states of structures under extreme loadings needs sophisticated novel treatment leading to a breakthrough in mechanical science.

The resonance curve of an elastic–plastic single-degree-of-freedom (SDOF) system is an important fundamental tool for clarifying the intrinsic property of such a system. Historically the elastic–plastic earthquake responses of structures were derived originally for the steady-state response to sinusoidal input, or the transient response to an extremely simple sinusoidal input in the 1960–1970s (Caughey 1960a, 1960b; Iwan 1961, 1965a, 1965b) and those methods were applied to more complex problems. On the contrary, Kojima and Takewaki (2015a, b, c) introduced a completely different innovative approach and demonstrated that the peak elastic–plastic response (in terms of continuation of free vibrations) can be derived by an energy approach, without directly solving the equations of motion. Their approach led to a breakthrough in the field of resonant nonlinear vibration. Comparison between the conventional approach and the innovative approach is shown in Figure 7.1. The critical response derived by the innovative approach is

Chapter written by Izuru TAKEWAKI, Kotaro KOJIMA and Sae HOMMA.

positioned in the resonance curves of an elastic–plastic SDOF system, which are plotted for constant acceleration and constant velocity in Figure 7.2. While the conventional approach deals with the resonance curve for constant acceleration, the present critical response corresponds to the resonance curve for a constant velocity.

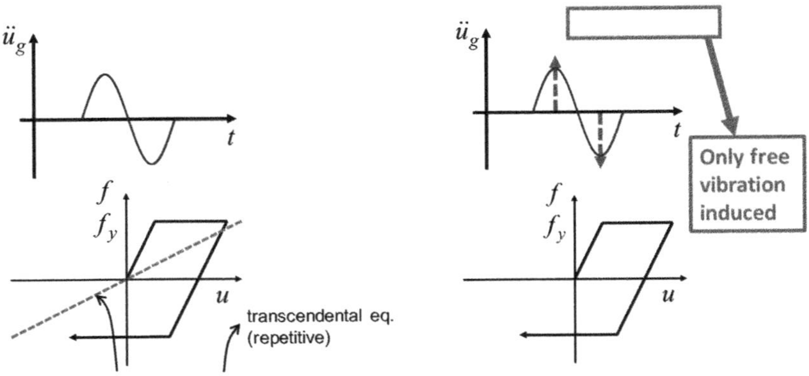

Figure 7.1. *Comparison between the conventional approach and the innovative approach. For a color version of this figure, see www.iste.co.uk/challamel/mechanics2.zip*

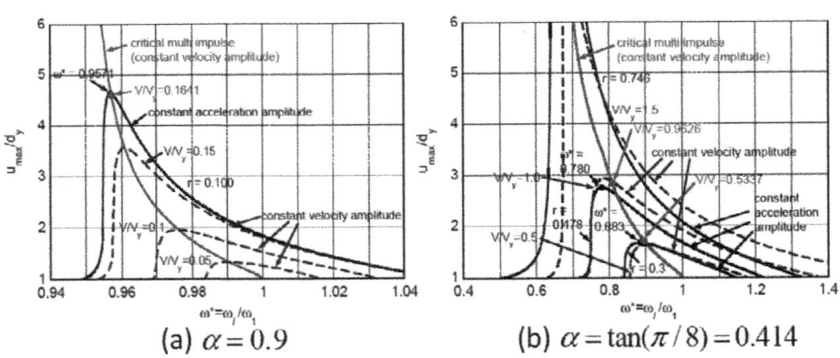

Figure 7.2. *Critical response, derived by the innovative approach, positioned in the resonance curves of an elastic–plastic SDOF system for constant acceleration and constant velocity (Kojima and Takewaki 2017). For a color version of this figure, see www.iste.co.uk/challamel/mechanics2.zip*

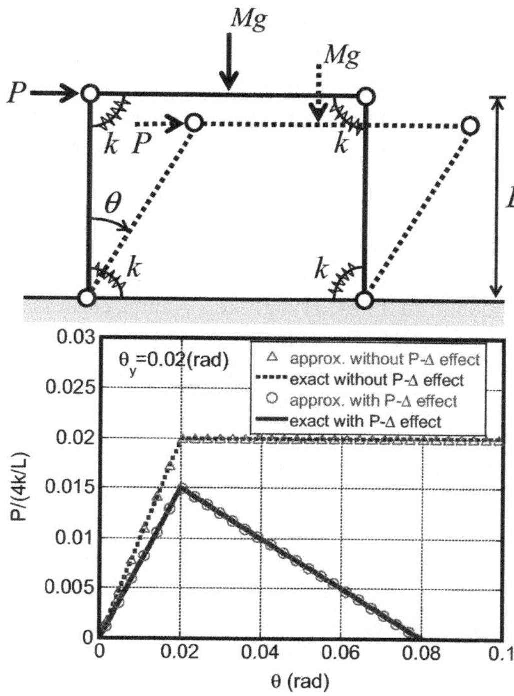

Figure 7.3. *SDOF elastic–plastic system with negative post-yield stiffness. For a color version of this figure, see www.iste.co.uk/challamel/mechanics2.zip*

The limit state is also one of the most important concepts in the earthquake-resistant design. The P-delta effect is apt to cause the limit state, i.e. the collapse of building structures under earthquake ground motions. The negative post-yield stiffness in the elastic–plastic restoring-force characteristic represents the P-delta effect of elastic–plastic structures, and numerous theoretical and numerical investigations have been conducted to derive the limit for the dynamic collapse problem. Jennings and Husid (1968) started a theoretical study on the dynamic collapse of an SDOF system, consisting of an elastic–plastic rotational spring with negative post-yield stiffness. They defined the limit state, which was called the critical rotation in their literature, as the zero restoring moment in the plastic range where the moment due to gravity force coincides with the rotational resistance. Sun *et al.* (1973) derived a similar collapse condition for a braced SDOF system in the free vibration with initial velocity and displacement. These static or dynamic collapse conditions for the SDOF system were expanded to multi-degree-of-freedom (MDOF) systems (Takizawa and Jennings 1980). Araki and Hjelmstad (2000) treated a model consisting of a rigid bar and a rotational spring with the effect of gravity load and investigated the tangent stiffness in view of instability. The exponential growth of rotation in one direction was defined as the dynamic collapse as in the work by Ishida and Morisako (1985). However, it seems that there is no theoretical approach including explicit mathematical expressions on the limit of dynamic collapse.

Long-period and pulse-like ground motions with short duration have been observed near earthquake source faults during large-scale earthquake events, such as the Northridge earthquake in 1994, Hyogo-ken Nanbu earthquake in 1995 and the Kumamoto earthquake in 2016. Recently, it was found that the main part of these pulse-like ground motions can be characterized by a one-cycle sinusoidal wave (Mavroeidis and Papageorgiou 2003; Makris and Black 2004; Kalkan and Kunnath 2006; Rupakhety and Sigbjörnsson 2011; Kojima and Takewaki 2015a). In a critical case where the pulse frequency coincides with the equivalent natural frequency of the elastic–plastic structure, the elastic–plastic response is amplified, and the structure may collapse. Although the critical resonant case plays a key role in the earthquake-resistant design, a repetitive procedure of changing the input pulse frequency for a specified acceleration or velocity amplitude is necessary to compute the resonant frequency for the elastic–plastic systems (Caughey 1960a, 1960b; Iwan 1961, 1965a, 1965b). In contrast, Kojima and Takewaki (2015a) proposed a method to approximately calculate the critical frequency and elastic–plastic response via a double impulse input without iteration. In this theory, the critical frequency can be characterized as the time interval of the critical double impulse, where the second impulse acts at the zero restoring-force timing in the unloading process. This advantageous feature can be expanded to the system with negative post-yield stiffness (Kojima and Takewaki 2016). The maximum elastic–plastic response can be computed by an energy approach, where the kinetic energy computed by the mass velocity, just after each impulse input is transformed into the sum of hysteretic and elastic strain energies. Kojima and Takewaki (2016) derived the collapse-limit input level of the critical double impulse for an elastic–plastic structure in closed form by using this simple approach. It was demonstrated that this closed-form solution enables a simple evaluation of collapse input level of the pulse-like ground motion without the repetitive procedure. Recently, it was shown that this approach can be applied to the model with viscous damping and the collapse-limit input velocity level has been derived (Saotome *et al.* 2019).

In this chapter, the double impulse is treated again as an input and a method is explained which enables the judgment of collapse or non-collapse of an elastic–plastic SDOF system with negative post-yield stiffness, under the double impulse with a fixed set of the velocity amplitude and the interval t_0 of two impulses. In the first part, the resonant case is treated by introducing the work of Kojima and Takewaki (2016). In the second part, the non-resonant case is investigated (Homma *et al.* 2020).

7.2. Collapse limit corresponding to the critical timing of second impulse

In this section, the work of Kojima and Takewaki (2016) is introduced, in which it was assumed that the timing of the second impulse in the double impulse input at the time of zero restoring force in the unloading path provides a critical timing. The critical timing means that the double impulse with such timing of the second impulse gives the minimum collapse limit velocity for varied timing of the second impulse. It will be clarified in this chapter that this assumption is valid for some velocity levels, but not for others.

Consider an SDOF elastic–plastic system with negative post-yield stiffness, as shown in Figure 7.3. The negative post-yield stiffness is induced by the P-delta effect. Let us introduce the following notations: m; mass, k; initial stiffness, α; ratio of post-yield stiffness to initial stiffness, u; displacement, f; restoring force, d_y; yield deformation, f_y; yield force, $u_{\max 1}$; maximum displacement in the negative direction, $u_{\max 2}$; maximum displacement in the positive direction, u_{p1}; plastic deformation in the negative direction, u_{p2}; plastic deformation in the positive direction, V; velocity amplitude of double impulse, V_y; velocity amplitude of double impulse under which the SDOF system just attains the yield deformation after the first impulse, v_c; mass velocity at the time of zero restoring force after the first impulse. The collapse of the SDOF system is characterized by the zero restoring force in the post-yield stiffness range.

Figure 7.4 presents an elastic–plastic collapse response of the SDOF system subjected to the critical double impulse, which is defined by the action of the second impulse at the zero restoring-force point in the unloading stage. This second impulse timing corresponds to the resonant case in which the plastic deformation u_{p2} exhibits the maximum value for varied second impulse timing.

To facilitate the evaluation of the maximum response of the SDOF system under the critical double impulse in a compact way, the energy balance law is exploited as shown in Figure 7.5. With this energy balance law, the maximum response can be obtained without complicated computation.

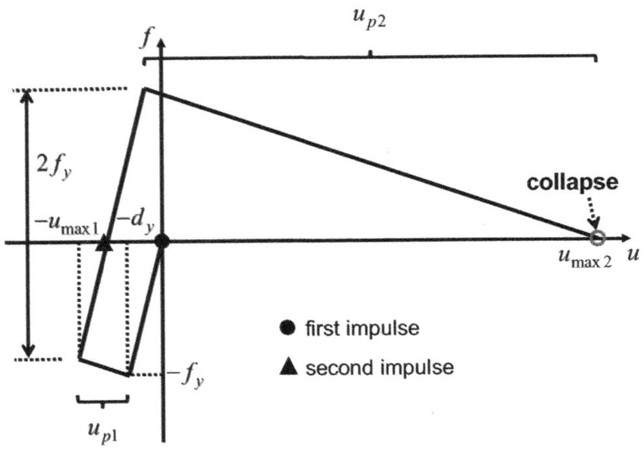

Figure 7.4. *Elastic–plastic collapse response of the SDOF system subjected to critical double impulse (Kojima and Takewaki 2016). For a color version of this figure, see www.iste.co.uk/challamel/mechanics2.zip*

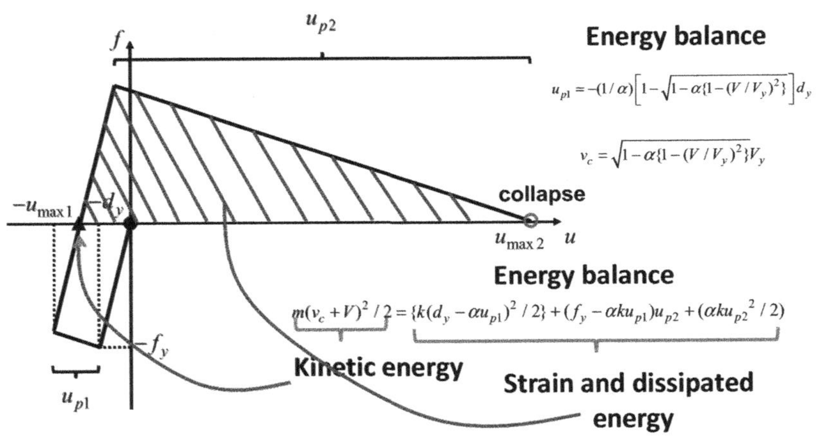

Figure 7.5. *Energy balance law for the elastic–plastic collapse response of the SDOF system subjected to critical double impulse. For a color version of this figure, see www.iste.co.uk/challamel/mechanics2.zip*

Figure 7.6 shows the collapse limit input velocity for various post-yield stiffness ratios (resonant case). The input velocity level is classified into three cases. In CASE 1, the SDOF system remains elastic even after the second impulse. CASE 2 means that the SDOF system goes into the plastic range for the first time after the second impulse. In CASE 3, the SDOF system goes into the plastic range after the first impulse. After some investigation, it was made clear that five collapse patterns exist, i.e. Patterns 1, 2, 3, Additional Patterns 1 and 2. Pattern 1 indicates that the SDOF system attains the collapse limit after the second impulse with the remaining elastic after the first impulse. In Pattern 2, the SDOF system attains the collapse limit after the second impulse experiencing the plastic response after the first impulse. In Pattern 3, the SDOF system attains the collapse limit after the second impulse experiencing the plastic response after the first impulse and a loop after the second impulse. Additional Pattern 1 means that the SDOF system attains the collapse limit after the first impulse. Additional Pattern 2 means that the SDOF system attains the collapse limit after the second impulse without plastic response after the second impulse. It can be observed that an isolated collapse region exists for $\alpha < -1/3$. In the range of $\alpha > -1/3$, Collapse Pattern 3 gives the minimum collapse limit velocity. Through the response analysis to the corresponding one-cycle sine wave, it was found that Collapse Pattern 1 gives the minimum collapse limit velocity in the range of $\alpha < -1/3$.

To investigate the accuracy and reliability of the proposed method, the main part of a recorded ground motion (Rinaldi Station, 1994 Northridge Earthquake) was focused (modeling into one-cycle sine wave as shown in Figure 7.7) and it was transformed into a double impulse.

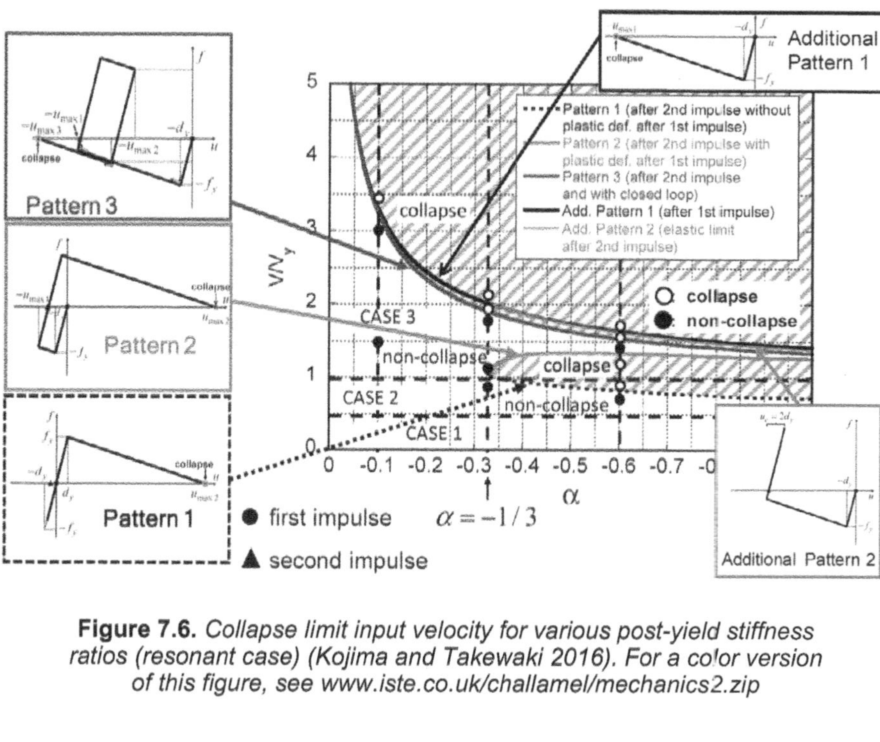

Figure 7.6. *Collapse limit input velocity for various post-yield stiffness ratios (resonant case) (Kojima and Takewaki 2016). For a color version of this figure, see www.iste.co.uk/challamel/mechanics2.zip*

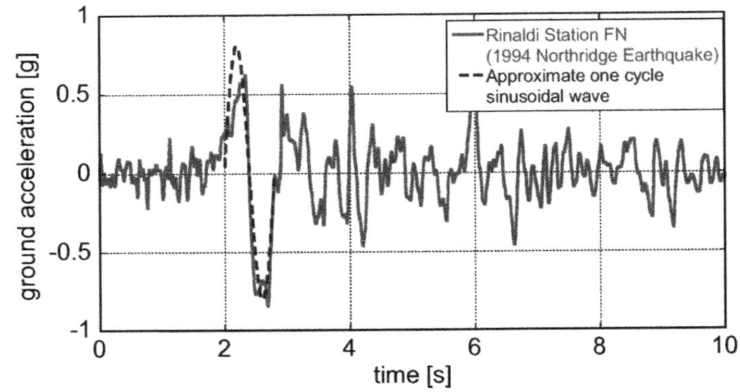

Figure 7.7. *Modeling of the main part of a near-fault ground motion into one-cycle sine wave (Kojima and Takewaki 2016). For a color version of this figure, see www.iste.co.uk/challamel/mechanics2.zip*

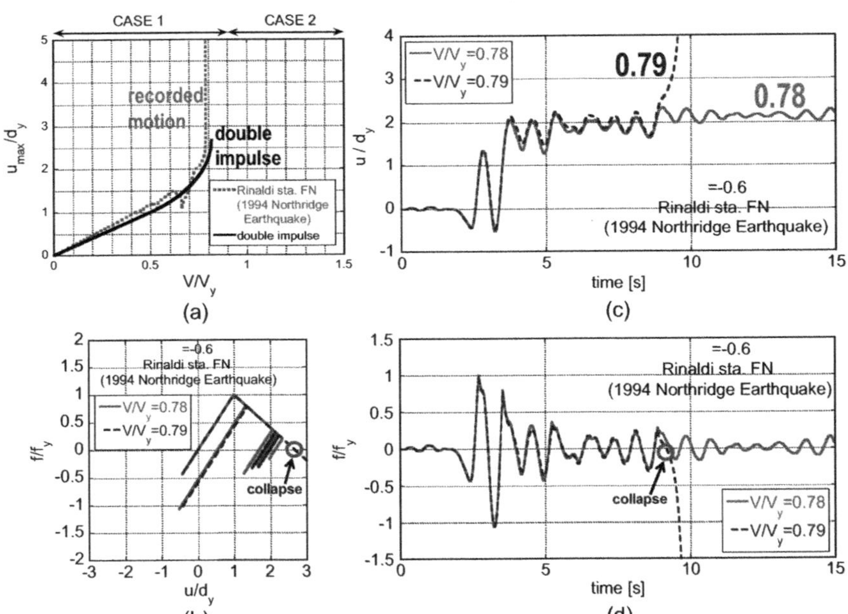

Figure 7.8. *Displacement and restoring-force responses under Rinaldi Sta. FN with two input velocities close to the collapse limit input velocity (Kojima and Takewaki 2016). For a color version of this figure, see www.iste.co.uk/challamel/mechanics2.zip*

Figure 7.8 shows the displacement and restoring-force responses under Rinaldi Sta. FN with two input velocities close to the collapse limit input velocity. It can be observed that, while the SDOF system does not collapse under Rinaldi Sta. FN with $V / V_y = 0.78$, it collapses under Rinaldi Sta. FN with $V / V_y = 0.79$.

7.3. Classification of collapse patterns in non-critical case

By using energy balance laws similar to those introduced by Kojima and Takewaki (2016), the input velocity corresponding to the collapse limit can be derived by specifying the period ratio t_0 / T_1 (T_1: the natural period of the SDOF system). The velocity amplitude and time interval of the double impulse correspond to the maximum velocity and input period of the one-cycle sinusoidal wave. It should be kept in mind that non-resonant cases are dealt with differently here in comparison to previous research (Kojima and Takewaki 2016) and the resonant case gives the smallest input velocity level of collapse limit in some cases. With this criterion, researchers can judge, without time-history response, analysis whether the given set of input velocity and input period provides a safe or unsafe state for a given structure.

In the case where the input timing of the second impulse is not the critical case, the influence of the second impulse may decrease the response of the structure. This is because the direction of the mass velocity may be against the input direction of the second impulse. To investigate this influence in detail, some patterns to the collapse state are classified and the combination between the input velocity and the second impulse timing for the collapse is clarified using the energy balance law and the solutions of free vibration. It is made clear that the resonant case does not necessarily give the smallest input velocity level of collapse limit depending on the collapse patterns.

Figure 7.9 shows the trajectories in restoring-force characteristics after the first impulse for three cases (CASEs-A, B, C) with three input velocity levels:

CASE-A: Elastic just after the first impulse;

CASE-B: Yield just after the first impulse;

CASE-C: Just attain the collapse under only the first impulse.

Figure 7.10 illustrates the domain decomposition in the input velocity–impulse timing relation. The left figure shows three cases (CASEs-A, B, C) in the input velocity–impulse interval relation. On the other hand, the right figure indicates four cases (CASEs-I, II, III, IV) for four timings of the second impulse under various input velocity levels.

CASE-I: The structure does not yield after the first impulse and the second impulse acts;

CASE-II: The structure goes into the plastic range after the first impulse and the second impulse acts before the structure attains the maximum displacement or before the structure collapses under only the first impulse;

CASE-III: The structure goes into the plastic range after the first impulse and the second impulse acts while the structure exhibits a harmonic free vibration after the attainment of the maximum displacement;

CASE-IV: The structure collapses before the action of the second impulse.

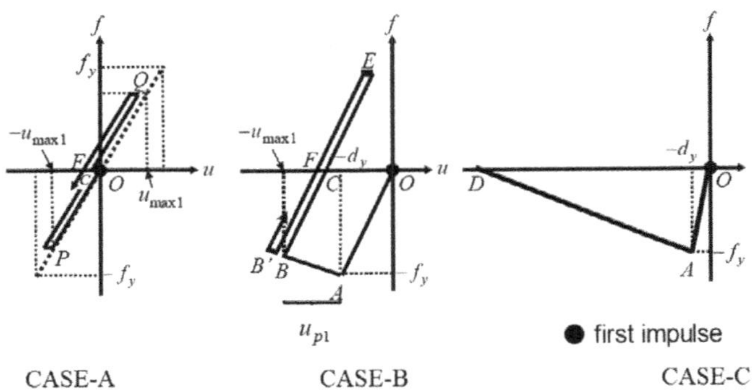

Figure 7.9. *Trajectories in restoring-force characteristics for CASEs-A, B, C (Homma* et al. *2020)*

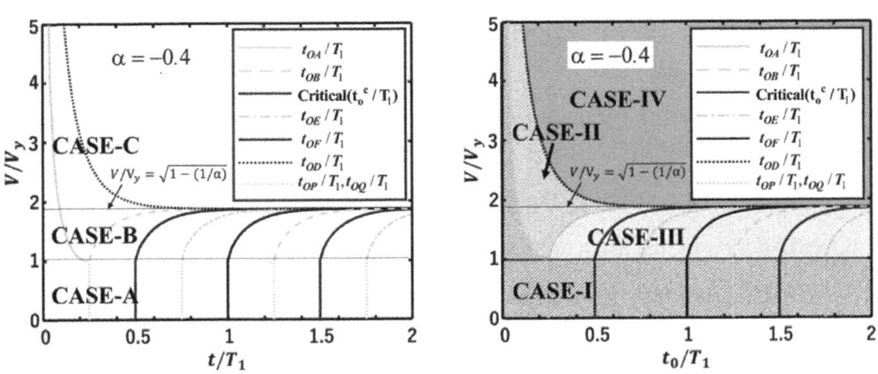

Figure 7.10. *Three cases (A, B, C) for three input velocity levels and four cases (I, II, III, IV) for four timings of the second impulse under various input velocity levels (Homma* et al. *2020). For a color version of this figure, see www.iste.co.uk/challamel/mechanics2.zip*

7.4. Analysis of collapse limit using energy balance law

It is important to consider the combination of CASEs-A, B, C and CASEs-I, II, III, IV shown in Figure 7.10 to derive the combination of the collapse limit velocity and impulse interval. The collapse limit velocity level can be obtained in closed form in some cases by using the energy balance law. However, in some other cases, the collapse velocity level can be obtained by solving a transcendental equation derived from the energy balance law.

Consider here several collapse patterns, i.e. Collapse Patterns 1'–4', as shown in Figure 7.11. The naming of these patterns comes from the similarity to the formulation for the nonlinear resonant case shown above (Kojima and Takewaki 2016). Collapse Pattern 4' represents a new type. Each collapse pattern is explained briefly in the following. More detailed derivation and explanation can be found in the reference (Homma *et al.* 2020).

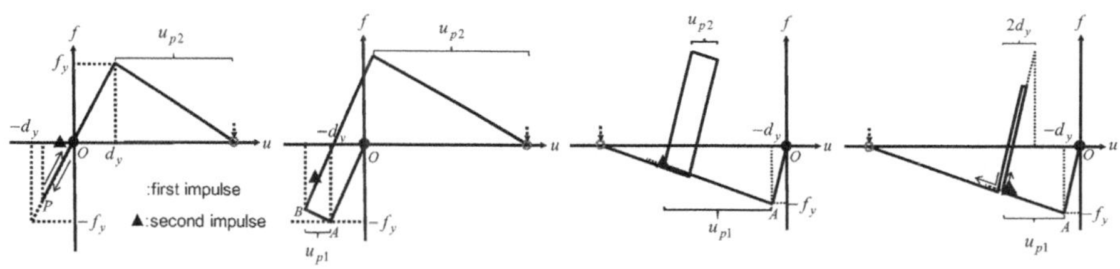

Collapse Pattern 1' Collapse Pattern 2' Collapse Pattern 3' Collapse Pattern 4'

Figure 7.11. *Restoring-force characteristics for Collapse Patterns 1'–4'. For a color version of this figure, see www.iste.co.uk/challamel/mechanics2.zip*

7.4.1. *Collapse Pattern 1'*

In the first collapse pattern, the structure remains elastic after the first impulse and attains the collapse limit after the second impulse with arbitrary timing, as shown in Figure 7.12(a) (see Figure 7.11).

Let O and A denote the point of the first impulse (origin of the restoring-force characteristic) and the point of initial yielding in the negative direction. The interval of two impulses is denoted by t_0 and the required time between Points O and A is indicated by t_{OA}. Since the structure does not yield after the first impulse in this case, the following two cases exist.

$$\text{CASE-I}= \begin{matrix} 0 \le V/V_y \le 1\ (\text{CASE-A}) \\ 1.0 < V/V_y \quad (\text{CASEs-B, C}) \text{ and } 0 < t_0 \le t_{OA} \end{matrix} \qquad [7.1]$$

$[0 \le V / V_y \le 1$ **(CASE-A)]**

The energy balance law between the state just after the second impulse and the collapse point H in Figure 7.12(a) provides the collapse limit input velocity for CASE-A in Collapse Pattern 1'.

$$V / V_y = \sqrt{\frac{1-(1/\alpha)}{2-2\cos(2\pi t_0 / T_1)}} \qquad [7.2]$$

Since $0.5 < V / V_y \le 1$ is necessary, the following condition for α and t_0 must be satisfied.

$$1/\alpha \ge 2\cos(2\pi t_0 / T_1) - 1 \qquad [7.3]$$

Figure 7.12(b) shows the collapse limit input velocity for $\alpha = -0.4$ for CASE-A in Collapse Pattern 1' (red line).

It can be observed that the critical case for $t_0 / T_1 = 0.5$ (Kojima and Takewaki 2016) gives the minimum collapse limit input velocity in this case.

$[1 < V / V_y$ **(CASEs-B, C) and** $0 < t_0 \le t_{OA}]$ This case does not appear in the ordinary model with $\alpha > -1$.

7.4.2. *Collapse Pattern 2'*

The second collapse pattern is the case where the structure exhibits plastic deformation after the first impulse and attains the collapse limit after the second impulse (see Figure 7.11). Since the structure exhibits plastic deformation after the first impulse in this case, $V / V_y > 1$ must be satisfied.

Because the second impulse acts after the structure goes into a plastic region under the first impulse, the case is divided into the following two cases, CASE-II and CASE-III.

$$\begin{cases} \text{CASE-II}=\begin{cases} 1 < V / V_y < \sqrt{1-(1/\alpha)} \ \text{(CASE-B) and } t_{OA} < t_0 \le t_{OB} \\ \sqrt{1-(1/\alpha)} \le V / V_y \quad \text{(CASE-C) and } t_{OA} < t_0 \le t_{OD} \end{cases} \\ \text{CASE-III}= 1 < V / V_y < \sqrt{1-(1/\alpha)} \ \text{(CASE-B) and } t_{OB} \le t_0 \end{cases} \qquad [7.4]$$

[$1 < V/V_y < \sqrt{1-(1/\alpha)}$ **(CASE-B) and (CASE-II)]** This case does not appear in the ordinary model with $\alpha > -1$.

[$\sqrt{1-(1/\alpha)} \leq V/V_y$ **(CASE-C) and (CASE-II)]** This case does not appear in the ordinary model with $\alpha > -1$.

[$1 < V/V_y < \sqrt{1-(1/\alpha)}$ **(CASE-B) and (CASE-III)]** The energy balance law provides the collapse limit input velocity.

7.4.3. *Collapse Pattern 3'*

In the third collapse pattern, the structure exhibits plastic deformation after the first impulse and attains the collapse limit with a closed loop after the second impulse (see Figure 7.11).

[$1 < V/V_y < \sqrt{1-(1/\alpha)}$ **(CASE-B) and (CASE-II)]** The energy balance law provides the collapse limit input velocity.

[$\sqrt{1-(1/\alpha)} \leq V/V_y$ **(CASE-C) and (CASE-II)]** Since the input velocity level in this case is too large, the solution to satisfy the collapse condition does not exist in this case.

[$1 < V/V_y < \sqrt{1-(1/\alpha)}$ **(CASE-B) and (CASE-III)]** The energy balance law provides the collapse limit input velocity. The critical state (Kojima and Takewaki 2016) corresponding to the nonlinear resonance does not necessarily provide the minimum input velocity level with respect to arbitrary impulse timing.

7.4.4. *Collapse Pattern 4'*

The fourth collapse pattern is the case where the structure exhibits plastic deformation after the first impulse and attains the collapse limit after experiencing unloading (positive direction) and reloading–reyielding (negative direction) for the second impulse (see Figure 7.11).

[$1 < V/V_y < \sqrt{1-(1/\alpha)}$ **(CASE-B) and (CASE-II)]** The energy balance law can be used.

[$\sqrt{1-(1/\alpha)} \leq V/V_y$ **(CASE-C) and (CASE-II)]** The energy balance law can be used.

[$1 < V/V_y < \sqrt{1-(1/\alpha)}$ (CASE-B) and (CASE-III)] The energy balance law provides the collapse limit input velocity. The critical state (Kojima and Takewaki 2016) corresponding to the nonlinear resonance does not necessarily provide the minimum input velocity level with respect to arbitrary impulse timing.

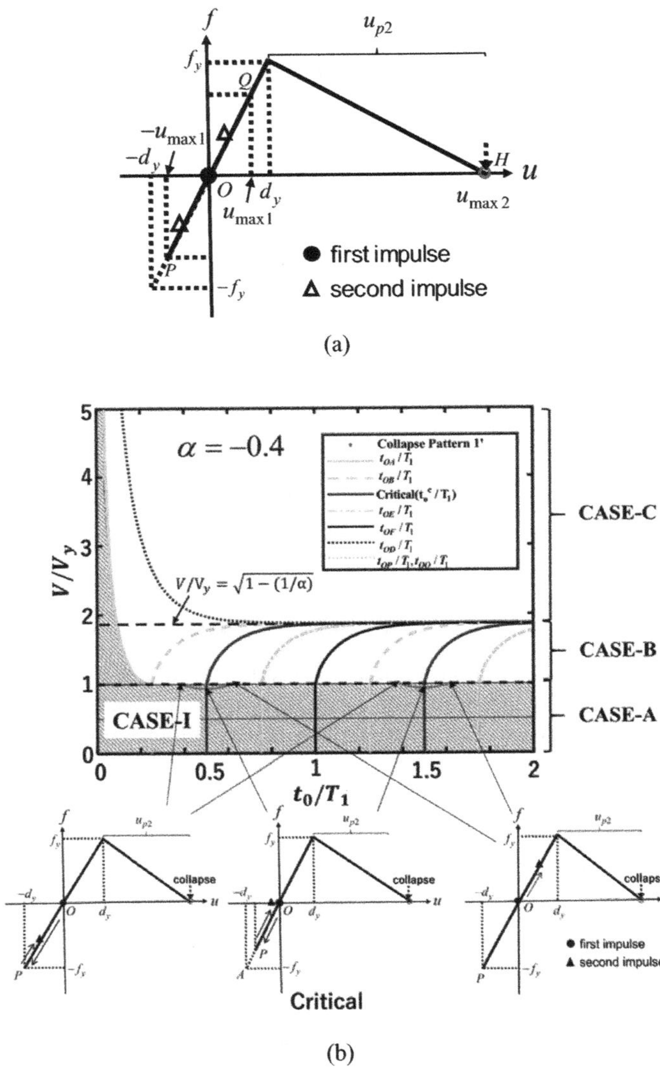

Figure 7.12. *Collapse Pattern 1' (CASE-A): (a) Restoring-force characteristic and (b) second impulse timing t_0/T_1 –input velocity relation (Homma* et al. *2020). For a color version of this figure, see www.iste.co.uk/challamel/mechanics2.zip*

Figure 7.13 presents the collapse limit input velocity for various intervals of two impulses. The line passing the point t_0 / T_1=0.5 in the horizontal axis indicates the resonant case (Kojima and Takewaki 2016). It should be remarked that the present SDOF model is an undamped model, the states of t_0 / T_1=0.5 and t_0 / T_1=1.5 provide the same collapse limit. It can be observed that an isolated region of the collapse state exists around the level of t_0 / T_1=0.5 (also 1.5) and the level of V / V_y=1. The most important point to note is that the critical state (Kojima and Takewaki 2016) corresponding to the nonlinear resonance does not necessarily provide the minimum input velocity level with respect to arbitrary impulse timing in the case of $\alpha > -1/3$ where the Collapse Patterns 1' and 2' do not exist.

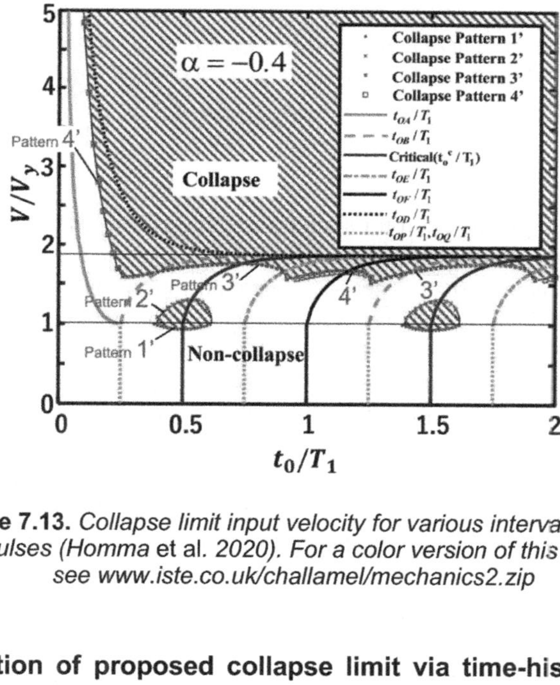

Figure 7.13. *Collapse limit input velocity for various intervals of two impulses (Homma* et al. *2020). For a color version of this figure, see www.iste.co.uk/challamel/mechanics2.zip*

7.5. Verification of proposed collapse limit via time-history response analysis

To investigate the accuracy and reliability of the proposed collapse limit velocity, response analysis under the double impulse with arbitrary velocity amplitude and impulse interval has been conducted.

Figure 7.14 shows the results on response analysis simulation of the SDOF models ($\alpha = -0.2, -0.4$) for various combinations of input velocity and interval of two impulses. It is found that strict and accurate classification into the collapse state and the non-collapse state has been made with the proposed limit curve.

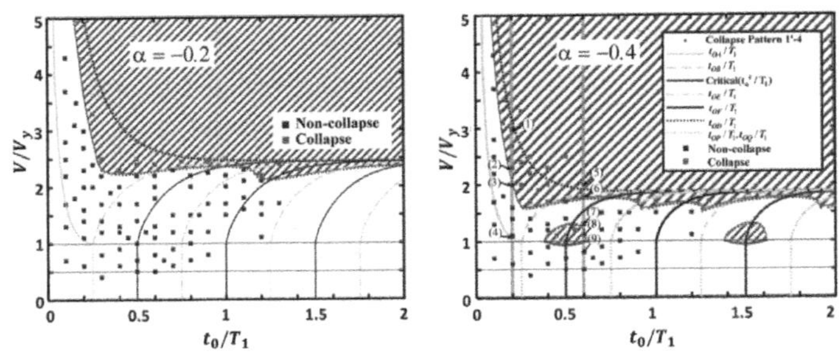

Figure 7.14. *Response analysis simulation for various combinations of input velocity and interval of two impulses (Homma* et al. *2020). For a color version of this figure, see www.iste.co.uk/challamel/mechanics2.zip*

7.6. Conclusion

A dynamic collapse criterion for simple softening elastic–plastic structures under peculiar near-fault ground motions was derived analytically by approximately transforming the main part of a near-fault ground motion into the double impulse and using an energy balance law. A softening single-degree-of-freedom (SDOF) model with negative post-yield stiffness modeling the P-delta effect was introduced to enable a simple formulation of the dynamic collapse criterion. The double impulse with impact only at two stages induces only free vibration and enables the efficient use of the energy approach in the derivation of compact expressions of complicated elastic–plastic responses of structures with the negative post-yield stiffness. In contrast to the previous work (Kojima and Takewaki 2016) for the resonant critical case, a general collapse criterion was provided for the velocity amplitude and the frequency of the double impulse. It should be remarked that no iteration is needed in the derivation of the dynamic collapse criterion except the solution of transcendental equations. It is shown that discussions on several patterns of dynamic collapse behaviors introduced in the previous critical case are useful for deriving a boundary between the collapse and the non-collapse in the plane of the input velocity and the input frequency. Note that the critical state (Kojima and Takewaki 2016) corresponding to the nonlinear resonance does not necessarily provide the minimum input velocity level with respect to arbitrary impulse timing. The validity of the proposed dynamic collapse criterion was examined by the numerical response analysis for SDOF systems under double impulses with collapse or non-collapse parameters.

7.7. References

Araki, Y. and Hjelmstad, K.D. (2000). Criteria for assessing dynamic collapse of elastoplastic structural systems. *Earthquake Engng. Struct. Dyn.*, 29, 1177–1198.

Caughey, T.K. (1960a). Sinusoidal excitation of a system with bilinear hysteresis. *J. Appl. Mech.*, 27(4), 640–643.

Caughey, T.K. (1960b). Random excitation of a system with bilinear hysteresis. *J. Appl. Mech.*, 27(4), 649–652.

Homma, S., Kojima, K., Takewaki, I. (2020). General dynamic collapse criterion for elastic-plastic structures under double impulse as substitute of near-fault ground motion. *Front. Built Environ.*, 6, 84.

Ishida, S. and Morisako, K. (1985). Collapse of SDOF system to harmonic excitation. *J. Eng. Mech.*, ASCE, 111(3), 431–448.

Iwan, W.D. (1961). The dynamic response of bilinear hysteretic systems. PhD Thesis, California Institute of Technology, Pasadena.

Iwan, W.D. (1965a). The dynamic response of the one-degree-of-freedom bilinear hysteretic system. *Proc. of the Third World Conf. on Earthq. Eng.*, New Zealand.

Iwan, W.D. (1965b). The steady-state response of a two-degree-of-freedom bilinear hysteretic system. *J. Appl. Mech.*, 32(1), 151–156.

Jennings, P.C. and Husid, R. (1968). Collapse of yielding structures during earthquakes. *J. Eng. Mech.*, ASCE, 94(EM5), 1045–1065.

Kalkan, E. and Kunnath, S.K. (2006). Effects of fling step and forward directivity on seismic response of buildings. *Earthquake Spectra*, 22(2), 367–390.

Kojima, K. and Takewaki, I. (2015a). Critical earthquake response of elastic-plastic structures under near-fault ground motions (Part 1: Fling-step input). *Front. Built Environ.*, 1, 12.

Kojima, K. and Takewaki, I. (2015b). Critical earthquake response of elastic-plastic structures under near-fault ground motions (Part 2: Forward-directivity input). *Front. Built Environ.*, 1, 13.

Kojima, K. and Takewaki, I. (2015c). Critical input and response of elastic-plastic structures under long-duration earthquake ground motions. *Front. Built Environ.*, 1, 15.

Kojima, K. and Takewaki, I. (2016). Closed-form dynamic stability criterion for elastic-plastic structures under near-fault ground motions. *Front. Built Environ.*, 2, 6.

Kojima, K. and Takewaki, I. (2017). Critical steady-state response of SDOF bilinear hysteretic system under multi impulse as substitute of long-duration ground motions. *Front. Built Environ.*, 3, 41.

Makris, N. and Black, C.J. (2004). Dimensional analysis of rigid-plastic and elastoplastic structures under pulse-type excitations. *J. Eng. Mech.*, ASCE, 130, 1006–1018.

Mavroeidis, G.P. and Papageorgiou, A.S. (2003). A mathematical representation of near-fault ground motions. *Bull. Seism. Soc. Am.*, 93(3), 1099–1131.

Rupakhety, R. and Sigbjörnsson, R. (2011). Can simple pulses adequately represent near-fault ground motions? *J. Earthq. Eng.*, 15, 1260–1272.

Saotome, Y., Kojima, K., Takewaki, I. (2019). Collapse-limit input level of critical double impulse for damped bilinear hysteretic SDOF system with negative post-yield stiffness. *Front. Built Environ.*, 5, 106.

Sun, C.-K., Berg, G.V., Hanson, R.D. (1973). Gravity effect on single-degree inelastic system. *J. Eng. Mech. Div.*, ASCE, 99(EM1), 183–200.

Takizawa, H. and Jennings, P.C. (1980). Collapse of a model for ductile reinforced concrete frames under extreme earthquake motions. *Earthquake Engng. Struct. Dyn.*, 8, 117–144.

8

Nonlinear Dynamics and Phenomena in Oscillators with Hysteresis

Many mechanical systems are characterized by hysteretic behaviors with a restoring force dependent on their deformation history. Several materials and elements base their capacity to dissipate vibratory energy on hysteresis. The wide variety of engineering applications explains the high number of studies devoted to the dynamic hysteretic response of structural systems and the hysteretic models proposed, with different levels of complexity. The Bouc–Wen model has been adopted here, because it is simple, yet, at the same time, is able to represent diverse types of hysteretic behaviors. Hysteresis can be classified among material nonlinearities and is recognized as a strong nonlinearity due to the high variation of stiffness and damping with deformation. The main characteristics of the dynamic response are first illustrated by means of frequency response curves of a hysteretic oscillator, highlighting the dependence of the response on the oscillation amplitude. The aim of this chapter is to investigate nonlinear modal interactions in the dynamic response of a two degree-of-freedom system (2DOF). These phenomena are notably important in internal resonance conditions; since when increasing excitation intensity frequencies of hysteretic system change and in turn their ratio changes, several internal resonance conditions occur, where the interaction phenomena between the two modes produce strong modifications of the response with possible beneficial effects. Two configurations are investigated: the hysteretic element at the top and the hysteretic element at the base. Qualitative similar results are obtained, characterized by a transfer of energy between the two modes, which greatly influences the evolution of the response amplitude with the excitation intensity and can be exploited in the vibration mitigation of the forced response around the first mode.

Chapter written by Fabrizio VESTRONI and Paolo CASINI.

8.1. Introduction

Over the last decades, several theoretical and applied contributions have been made in the field of dynamics of nonlinear structural systems. Due to the fact that the most involved researchers worked in aeronautical and mechanical engineering, attention is mainly devoted to geometrical/polynomial nonlinearities, with contributions dealing with piecewise/non-smooth constitutive relationships, in all the cases single-valued laws (Nayfeh and Mook 1979; Nayfeh and Balachandran 1995; Leine and Nijmeijer 2007).

Among nonlinear dynamic phenomena, the simplest are represented by the modification of the modal quantities with oscillation amplitude, which, like the energy involved in motion, governs the role of nonlinearity for a given system. Therefore, the natural frequencies of periodic motions depend on the oscillation amplitude, as well as the modes which now define, not planes in the configuration space, but curved manifolds (Rosenberg 1962; Shaw and Pierre 1991; Rand *et al.* 1992; Vakakis 1992, 1997; Vestroni *et al.* 2008; Haller and Ponsioen 2016). In a cycle of a periodic motion, modes do not maintain their shape and lose their fundamental orthogonality property.

Moreover, response amplifications are observed at driven frequencies close to an integer fraction or multiple of the natural frequency, the well-known super- and sub-harmonics, as well as aperiodic motions and chaos can be easily encountered (Nayfeh and Balachandran 1995; Wiercigroch and de Kraker 2000; Awrejcewicz *et al.* 2017). When the nonlinearities become significant, or for some systems with particular characteristics, such as non-linearizable systems (Anand 1972; Pak 1989; Vakakis 1992; Gendelman 2004; Casini and Vestroni 2011), a peculiar phenomenon arises, the onset of novel periodic motions, in addition to those expected in a linear response, as a result of bifurcations.

In civil engineering, as well as in aeronautical, mechanical and electrical engineering, one frequently deals with structural elements characterized by hysteresis, such as materials with a limited elastic field, micro-sliding friction, shape-memory alloys, magnetostrictive materials and elastomeric absorbers, to mention just a few. Hysteretic nonlinearities produce much more clear effects than geometrical ones (Vestroni and Noori 2002; Lacarbonara and Vestroni 2003; Al-Bender *et al.* 2004; Muravskii 2005; Awrejcewicz 2007; Hassani *et al.* 2014) and can be classified among strong nonlinearities with a wide variety of dynamical phenomena, such as significant modal coupling, bifurcations and super-abundant modes (Capecchi and Vestroni 1995; Masiani *et al.* 2002; Casini and Vestroni 2011).

The hysteresis means that the output depends on the history of the input, which does not make analytical development easy. Some mathematical models have been proposed to represent hysteretic behaviors; they can be roughly grouped into two families: (a) operator-based models, which introduce hysteresis operators; and (b) differential equations models, where hysteresis behavior is described by a differential equation. Among the models of the second family, the Bouc–Wen model (Bouc 1967; Wen 1976) is the most widely used in applications as it is simple and, at the same time, can describe various hysteretic behaviors. Here, reference is made to its basic formulation, characterized by only three parameters, because the goal is not to describe the behavior of specific materials but rather the modification of the dynamic response of a structure with a hysteretic element (Ni *et al.* 1998; Awrejcewicz *et al.* 2008; Ismail *et al.* 2009).

In this chapter, the response of a 2DOF chain system with one hysteretic element to harmonic excitation is considered. Two configurations are examined: in the first, the hysteretic element is between the two masses; in the second, it is between the fixed support and the first mass, already dealt with separately in previous contributions (Casini and Vestroni 2018; Vestroni and Casini 2020). The general nonlinear dynamic characteristics of the two systems are highlighted, extending and enlarging previous investigations, with the aim of pointing out the use of these phenomena in modifying and mitigating the dynamic resonant response of a structure.

8.2. Hysteresis model and SDOF response to harmonic excitation

Different models have been proposed to describe hysteresis; among the class of differential equation models, the Bouc–Wen model is used here due to its simple formulation, as well as because it can represent the main characteristics of real hysteretic restoring force by suitably adjusting its parameters (Ni *et al.* 1998; Ismail *et al.* 2009).

The hysteretic single-degree-of-freedom (SDOF) oscillator (Figure 8.1a) consists of a mass m_1 and an element that furnishes a restoring force resulting from an elastic component, characterized by k_1 stiffness, and a hysteretic component $z(x_1)$ described by the Bouc–Wen model:

$$f(x_1) = k_1 x_1 + z(x_1) \qquad [8.1]$$

where the hysteretic part $z(x_1)$ is obtained by the nonlinear differential equation:

$$\frac{dz}{dt} = \left\{k_d - \left[\gamma + \beta\, sgn\left(z\frac{dx_1}{dt}\right)\right] |z|^n\right\}\frac{dx_1}{dt} \qquad [8.2]$$

and the constitutive parameters of the Bouc–Wen law k_d, γ, β, n are introduced.

The element stiffness changes with the oscillation amplitude from an initial value $k_A = k_1 + k_d$ to a final value k_B, which coincides with the post-elastic stiffness, $k_B = k_1$ (Figure 8.1b). The yield restoring force f_y and the yield displacement $x_y = f_y/k_A$ are denoted in Figure 8.1b, while the hardening coefficient δ is defined as the ratio between the final and initial stiffness values: $\delta = \frac{k_B}{k_A}$.

Among the Bouc–Wen parameters, n governs the transition from the elastic and post-elastic branches, and here, it is assumed equal to 1; $(\beta+\gamma)$ influences the maximum value of the hysteretic force but, more importantly, the ratio γ/β strongly modifies the shape of the cycles and, in turn, the dissipation capacity of the element (Figure 8.1c). In particular, $\gamma = \beta$ means a fully hysteretic loop with the maximum energy dissipation, whereas $\gamma/\beta > 1$ means a reduced hysteretic loop with a decreasing energy dissipation compared to the case $\gamma = \beta$.

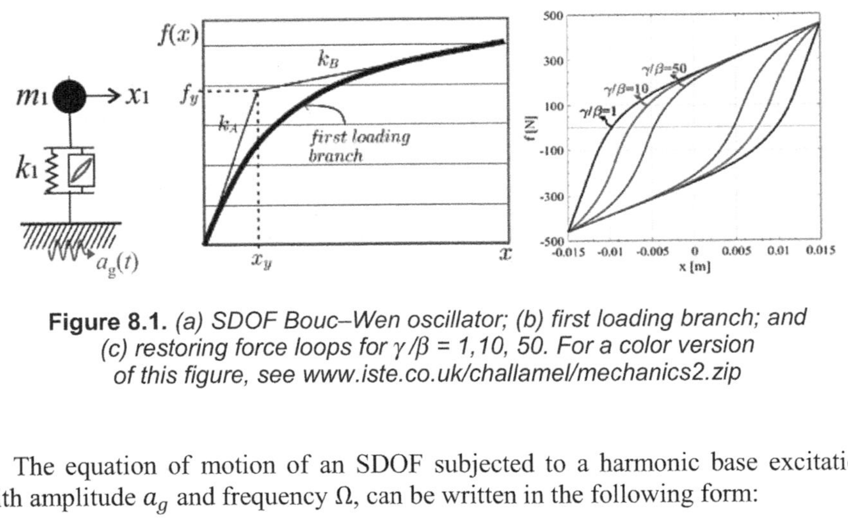

Figure 8.1. *(a) SDOF Bouc–Wen oscillator; (b) first loading branch; and (c) restoring force loops for γ/β = 1,10, 50. For a color version of this figure, see www.iste.co.uk/challamel/mechanics2.zip*

The equation of motion of an SDOF subjected to a harmonic base excitation, with amplitude a_g and frequency Ω, can be written in the following form:

$$m_1 \frac{d^2 x_1}{dt^2} + k_1 x_1 + z(x_1) = -m_1 a_g \sin(\Omega t) \quad [8.3]$$

where $z(x_1)$ is obtained from equation [8.2]. It is useful to denote some quantities related to the parameters of the Bouc–Wen hysteretic element, but with clear mechanical meaning: the hardening coefficient δ, the yield strength f_y, and the yield displacement x_y:

$$\delta = \frac{k_B}{k_A}, f_y = \frac{k_A}{k_d}\left(\frac{k_d}{\beta+\gamma}\right)^{\frac{1}{n}}, x_y = \frac{f_y}{k_A}, \quad [8.4]$$

as well as the initial and final frequencies, ω_A and ω_B, associated with small and large oscillation amplitudes, and the adimensional force intensity A and driven frequency η:

$$\omega_A = \sqrt{\frac{k_A}{m}},\ \omega_B = \sqrt{\frac{k_B}{m}},\ A = \frac{ma_g}{f_y},\ \eta = \frac{\Omega}{\omega_A} \qquad [8.5]$$

Figure 8.2 shows the frequency response curves (FRCs) of the hysteretic SDOF with $\gamma/\beta = 1, 10$. The case of full hysteresis is presented in Figure 8.2a. As expected, the resonance peaks follow a frequency–amplitude curve that is bent on the left, due to the softening nonlinearity of hysteresis. The peaks correspond to the nonlinear frequency of the oscillator which depends on the force intensity; for low levels, the frequency is close to ω_A, while for high levels, it moves towards the post-elastic frequency ω_B. As is known, all the points of the curve represent stable periodic solutions, the curve is said to be marginally stable and the passage between the resonant and non-resonant branches is nearly vertical (Capecchi and Vestroni 1990).

With the increase in intensity, the equivalent stiffness decreases and a greater response amplitude is to be expected; on the contrary, higher intensities correspond to higher dissipation and the relation between force and resonance amplitude is not far from linear in the range of intensity medium values, while the curve is steeper for small and large intensities when the damping is smaller.

Figure 8.2b shows the response with reduced hysteresis cyclic laws, $\gamma/\beta = 10$. For the same excitation intensity, larger response amplitudes are observed at resonance, due to the smaller dissipation capacity, but, more importantly, the resonant and non-resonant branches exist simultaneously in a limited range of driven frequency, which increases with γ/β. Thus, the curves are no longer single-valued and a typical jump phenomenon occurs. For a given value of γ/β, the range of multiple solutions will vanish with increasing amplitude, as already observed for different constitutive laws (Capecchi and Vestroni 1990), because at large response amplitudes, the difference in the dissipation capacity between full and reduced hysteresis tends towards zero.

The dependence of the oscillator frequency ω on the cycle amplitude is shown in Figure 8.3. Figure 8.3a refers to a full hysteresis laws ($\gamma/\beta = 1$) for three δ values. As already observed, ω changes from the initial frequency ω_A and final frequency ω_B, related to the stiffnesses k_A and k_B, and the ratio $\alpha = \omega_B/\omega_A$ between the frequencies increases with the parameter $\delta = k_B/k_A$. In the case of an elasto-perfectly plastic oscillator ($\delta = 0$), the nonlinear resonant response would be unbounded. In Figure 8.3b, for $\delta = 0.30$, the small influence of the dissipation parameter γ/β can be appreciated.

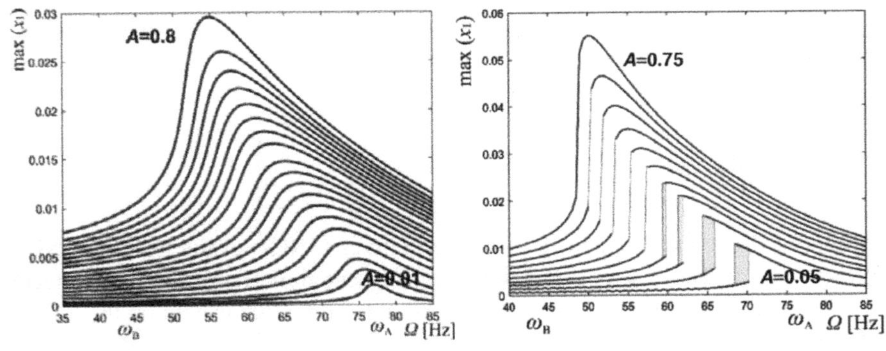

Figure 8.2. *SDOF oscillator: frequency response curves for varying forcing amplitude: (a) γ/β =1, A = [0.01-0.8] and (b) γ/β =10, A = [0.05-0.75] (ω_B = 43.1 Hz, ω_A = 78.3 Hz). For a color version of this figure, see www.iste.co.uk/challamel/mechanics2.zip*

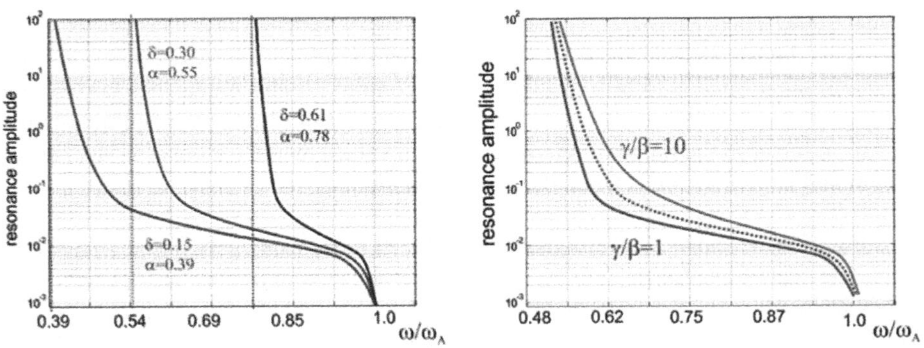

Figure 8.3. *Response amplitude versus resonance frequency: (a) $\frac{\gamma}{\beta} = 1$ and (b) $\frac{\gamma}{\beta} > 1$. For a color version of this figure, see www.iste.co.uk/challamel/mechanics2.zip*

8.3. 2DOF hysteretic systems

A 2DOF chain system is suitable to study the features of the main nonlinear coupling phenomena in a multi-degree-of-freedom system (Masiani *et al.* 2002; Casini and Vestroni 2018). Two different configurations are considered: in the first, the hysteretic element is located between the two masses (top configuration) when the added mass m_2 is devoted to mitigating the response of the principal mass m_1, while, in the second, the hysteretic element is between the constrained base and the mass m_1 (base configuration) when the aim is mainly to reduce the transmission of excitation from the base to the superstructure (Figure 8.4).

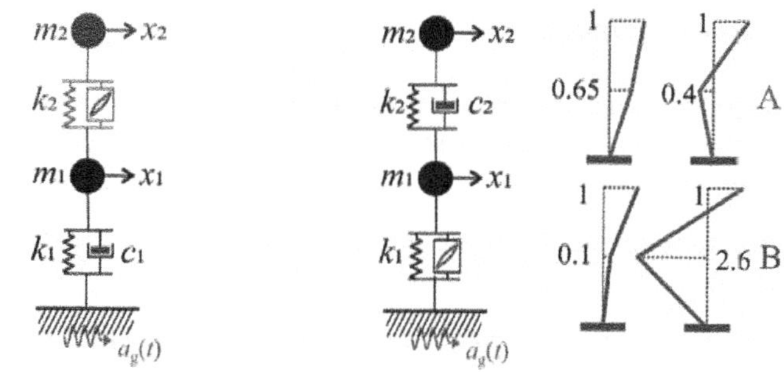

Figure 8.4. *2DOF system: (a) top-hysteresis and (b) base-hysteresis configurations and modes at small (A) and large (B) amplitudes for the BC1 system. For a color version of this figure, see www.iste.co.uk/challamel/mechanics2.zip*

8.3.1. *Equations of motion*

The equations of motion for the two systems consist of the dynamic balance equations of the two masses and the differential equation which governs the hysteretic component of the nonlinear element; the presence of one hysteretic element leads to a system of five first-order differential equations in the state space, only one equation more than the linear case.

For the top-hysteresis configuration, the equations are as follows:

$$m_1 \frac{d^2x_1}{dt^2} + k_1x_1 - k_2(x_2 - x_1) + c_1 \frac{dx_1}{dt} - z(x_2 - x_1) = -m_1 a_g \sin(\Omega t) \quad [8.6]$$

$$m_2 \frac{d^2x_2}{dt^2} + k_2(x_2 - x_1) + z(x_2 - x_1) = -m_2 a_g \sin(\Omega t) \quad [8.7]$$

$$\frac{dz}{dt} = \left\{k_d - \left[\gamma + \beta \operatorname{sgn}\left(z \frac{d(x_2 - x_1)}{dt}\right)\right] |z|^n\right\} \frac{d(x_2 - x_1)}{dt} \quad [8.8]$$

For the base-hysteresis configuration, the equations are as follows:

$$m_1 \frac{d^2x_1}{dt^2} + k_1x_1 - k_2(x_2 - x_1) + z(x_1) = -m_1 a_g \sin(\Omega t) \quad [8.9]$$

$$m_2 \frac{d^2x_2}{dt^2} + k_2(x_2 - x_1) = -m_2 a_g \sin(\Omega t) \quad [8.10]$$

$$\frac{dz}{dt} = \left\{k_d - \left[\gamma + \beta \operatorname{sgn}\left(z \frac{dx_1}{dt}\right)\right] |z|^n\right\} \frac{dx_1}{dt} \quad [8.11]$$

8.3.2. *Modal characteristics*

For both configurations, the presence of a strong nonlinear element causes a modification of the modal characteristics with oscillation amplitude and two limit cases can be observed:

A) small amplitudes, when the hysteretic element exhibits a linear behavior with stiffness k_A;

B) large amplitudes, when the hysteretic element exhibits a linear behavior with stiffness k_B.

In between, the system response resembles that of a system with an element characterized by an equivalent stiffness $k_A \geq k_{eq} \geq k_B$, where the range of interest is mainly that close to the small amplitudes. The ratio of the frequencies varies with oscillation amplitude as well. Thus, depending on the initial characteristics, the system can be involved in internal resonance conditions, where the nonlinear modal coupling produces a notable modification of the response to harmonic force.

The mechanical characteristics of the systems, dealt with below, are reported in the Appendix (section 8.7). For a sample 2DOF system, such as BC1, the modes at small and large amplitudes are reported in Figure 8.4b; they represent a reference for nonlinear modal shapes which, depending on the amplitude, start with shapes A and move towards shapes B.

8.4. Nonlinear modal interactions in 2DOF hysteretic systems

8.4.1. *Top-hysteresis configuration (TC)*

When the top mass of the system is a fraction of the first mass, the configuration is the typical configuration of a principal structure with an attachment which, if suitably designed, has the ability to split the one peak of the principal structure FRC into two much lower peaks. This is the well-known behavior of the tuned mass damper (Den Hartog 1934); here, the attachment is a hysteretic device. In this case, the design of the attachment requires a more refined tuning with respect to that of the Den Hartog viscoelastic tuned mass damper (VTMD), because the characteristics of the device depend on the oscillation amplitude, and thus, only in a well-defined range of amplitude, the ratio ω_2/ω_1 of the 2DOF frequencies reaches the optimal value. The use of nonlinear attachments is the subject of numerous papers after the pioneering studies (Vakakis *et al.* 2003) among which only a few have adopted hysteretic devices (Laxalde *et al.* 2006; Carpineto *et al.* 2013; Carboni and Lacarbonara 2015; Vestroni and Casini 2020).

To resemble the behavior of the classical tuned mass damper, the characteristics of the hysteretic attachment must be selected in order to obtain the condition where the frequency of the attachment is similar to the frequency of the principal structure, i.e. the system is close to the internal resonance condition 1:1. However, tuning is made harder due to frequency dependence on the oscillation amplitude. This aspect is highlighted by examining the response of a 2DOF system to harmonic excitation (Figure 8.5). You can find the system's TC1 characteristics in the Appendix (section 8.7) and its initial frequencies are $\omega_{1A} = 5.27$ Hz and $\omega_{2A} = 6.95$ Hz, which are not in optimal ratio, close to 1.25 according to Den Hartog (1934), but are candidates to reach this value with increasing excitation intensity.

In Figure 8.5a,b the FRCs of the principal mass displacement and of the attachment relative displacement are reported, respectively. If the optimal tuning is assumed as that which produces two equal peaks in the FRC of m_1, this is obtained for the intensity associated with the red curve (0.068 g). For lower intensities, k_{eq} of the attachment is lower than the VTMD optimal value k_0 and, as expected, the first peak is greater than the second. Contrarily, for higher intensities, k_{eq} of the attachment is greater than k_0 and the second peak prevails.

In Figure 8.5c, the variation of the ratio ω_2/ω_1 with the amplitude of the hysteretic vibration attachment (HVA) shows that in a small range of amplitudes, the ratio is close to the value of an optimal viscoelastic tuned mass damper; in this range, the effectiveness of HVA is practically the same as the VTMD, about 80% (Figure 8.5d), which means a reduction of 80% with respect to the non-controlled case.

Although many examples have proposed the use of nonlinear attachments, as with the present hysteretic vibration absorber, their mode of operation is only a variation of the classical VTMD, with the advantage of a very simple realization, with an element that combines elastic and dissipation capacity, and the drawback of a challenging effectiveness in a small range of excitation intensity. The nonlinearity is in the device behavior, but no typical phenomena of nonlinear dynamics are activated; the phenomenon illustrated above is similar to what already occurs in linear dynamics. Cases associated with internal resonance conditions n:1, with n integer greater than unity (Capecchi and Vestroni 1995; Jo and Yabuno 2009; Vestroni and Casini 2020), dealt with in the following, are more interesting.

Since the hysteretic restoring force is a rich constitutive law that collects different nonlinearities, even and odd, various internal resonance conditions can be encountered. Below the case of 3:1, internal resonance is illustrated; however, similar results can be shown for the other integer n, and also for more complex resonances of the kind $n\omega_1 = m\omega_2$.

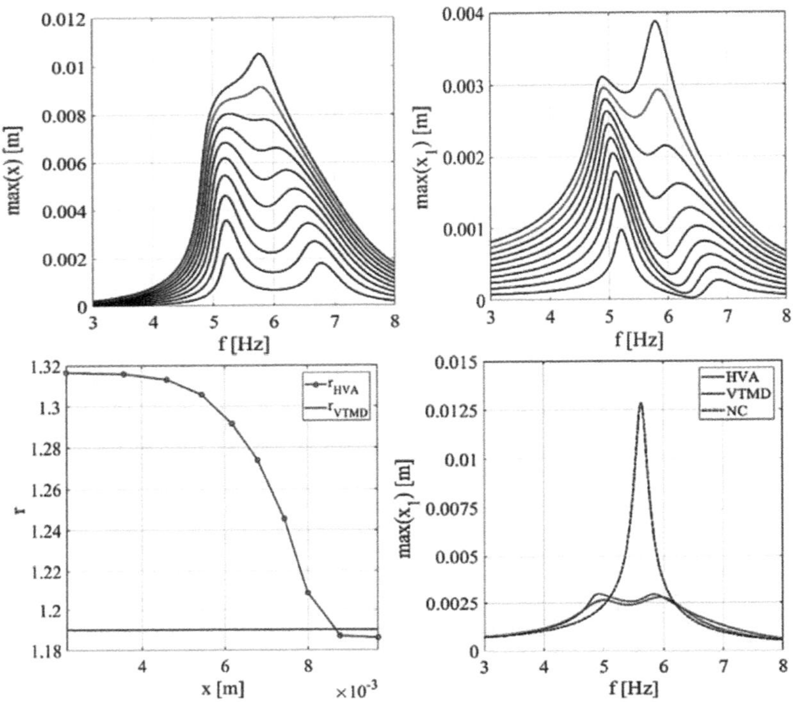

Figure 8.5. *2DOF top configuration close to a (1:1) internal resonance ($\omega_{2A}/\omega_{1A} \cong 1.32$): FRCs of oscillation amplitudes of mass m_1 (a) and of the relative displacement of mass m_2; (b) for increasing intensity (a_g = [0.005, 0.01-0.06, 0.065, 0.068, 0.08] g); (c) frequency ratio variation; and (d) non-controlled (NC), SDOF and VTMD responses. For a color version of this figure, see www.iste.co.uk/challamel/mechanics2.zip*

To obtain a system with a frequency ratio 3:1, maintaining the same principal structure of TC1, the mass and stiffness of the attachment should be varied according to TC2 (see Appendix, section 8.7). In this case, the FRCs, shown in Figure 8.6, exhibit two peaks, for low intensities at frequencies close to ω_{1A} = 5.10 Hz and ω_{2A} = 18 Hz (Figure 8.6a), with an initial value $r_A = \omega_{2A}/\omega_{1A} \cong 3.5$ and a final value for high intensities $r_B = \omega_{2B}/\omega_{1B} \cong 1.6$. With the increase in intensity, the ratio approaches the internal resonance condition 3:1, where the FRC around the first mode experiences a bifurcation (red curve) and instantaneously the curve of the 2DOFs system shows three peaks. The two peaks around the first resonance are very close in frequency near the bifurcation, while the nonlinear modal shapes, although maintaining the characteristic of the first mode (only one node, at the constraint), are distinctively different with a change in curvature (Figure 8.6b): the branch of the I_{A1}

peaks maintains the curvature of the original first mode, while the I_{A2} branch has an opposite curvature.

The first peak is split into two peaks with a behavior to a certain extent similar to what happens with a TMD. Actually, the situation is very different when a damper is added, as in the previous case TC1, the extra peak derives from a new DOF; here, the extra peak is related to a novel mode from the nonlinear coupling with the second mode close to a 3:1 internal resonance. An insight into this phenomenon is developed along with a discussion of other cases in base-hysteresis configurations.

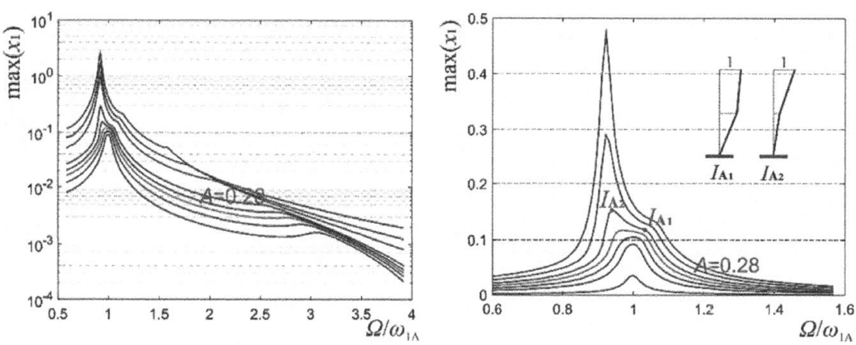

Figure 8.6. *2DOF top-hysteresis oscillator close to a (3:1) internal resonance $\omega_{2A}/\omega_{1A} \cong 3.5$, $\omega_{2B}/\omega_{1B} \cong 1.6$): (a) FRCs logarithmic scale for increasing forcing amplitude; and (b) detail of FRCs around the first resonance. For a color version of this figure, see www.iste.co.uk/challamel/mechanics2.zip*

8.4.2. *Base-hysteresis configuration (BC)*

The system with a hysteretic element at the base (Figure 8.4b) and the BC1 characteristics, reported in the Appendix (section 8.7), is considered. The small amplitude frequencies are $\omega_{1A} = 38.94$ Hz and $\omega_{2A} = 113.19$ Hz with a ratio $\omega_{2A}/\omega_{1A} = 2.90$ close to internal resonance condition 3:1, a case similar to the last case dealt with in the previous section.

Figure 8.7a shows the FRCs centered on the first resonance. Similar to the behavior illustrated in Figure 8.6b, for a critical intensity value, the fundamental branch $\omega - A$ of the first resonance with a single peak I_A has a bifurcation and then shows two peaks, I_{A1} corresponding to the pre-existence branch, destined to disappear, and I_{A2} which is the peak of the novel mode with a different shape, still belonging to the first mode shape, which is the only surviving one for large force intensity.

To better understand the role of nonlinear coupling and its exploitation in design (Vakakis 2017), the response of the BC1 system close to internal resonance is compared to the response of a system BC2 not in internal resonance conditions, obtained by only changing the stiffness, $k_2 = 2.54\ k_1$, leading to an initial frequency ratio $\omega_{2A}/\omega_{1A} = 3.40$. For both masses, Figure 8.7b,c shows the sizeable reduction of the maximum displacement in conditions of internal resonance, due to the transfer of most of the excitation energy to the second mode, with a smaller amplification of the first.

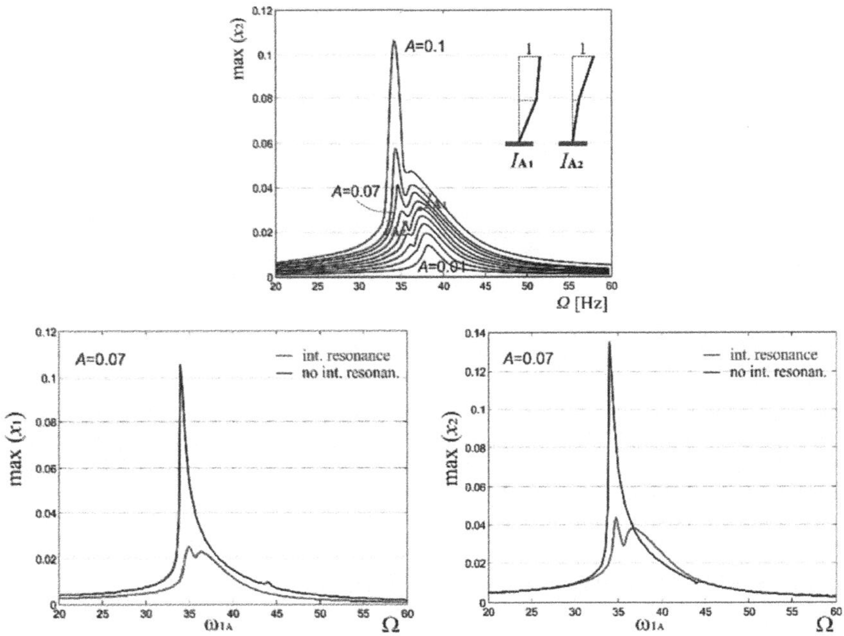

Figure 8.7. *2DOF base-hysteresis configuration close to a (3:1) resonance: (a) FRCs for different forcing amplitudes; comparison of FRCs of mass m_1 (b) and mass m_2 and (c) (A = 0.07) for the BC1 system, close to resonance, and BC2 system, far from resonance. For a color version of this figure, see www.iste.co.uk/challamel/mechanics2.zip*

In Figure 8.8, the analysis of the response in time and frequency domains can highlight some details of the modal coupling phenomenon; in particular, for each resonance peak, I_{A1} in Figure 8.8a and I_{A2} in Figure 8.8b, the time histories of a few periods, its frequency content and the modal trajectories in the configuration plane are drawn. When the frequency of first mode I_A coincides with $\omega_2/3$, bifurcation occurs, with the first peak I_{A1} remaining close to $\omega_2/3$ until it disappears, and the other I_{A2} moving away towards ω_{1B}. The time history is periodic and consists of the main frequency Ω and a small contibution of its super-harmonic w_2. The modal trajectories are, of course, nonlinear and, as anticipated, clearly bent in opposite directions.

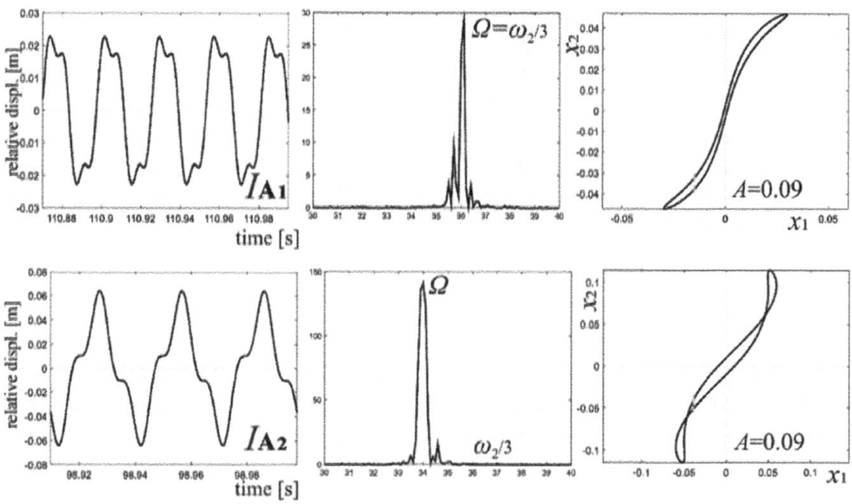

Figure 8.8. *2DOF base-hysteresis configuration close to (3:1) resonance for I_{A1} (upper) and I_{A2} (lower): time history of mass m_1, frequency content and modal trajectories in the coordinate plane x_1-x_2*

Due to the richness of the hysteretic nonlinearity, internal resonance conditions n:1 with n even can also occur; a BC3 system (see Appendix, section 8.7) close to internal resonance 2:1 is considered (Figure 8.9). At small intensities, the amplification of the second mode is much smaller than the first (Figure 8.9a). For increasing intensity, the zone of the first resonance widens and two peaks emerge, the first being more pronounced, the only one destined to survive at large intensity. The consequence is a flattening of the FRCs with a notable reduction of the first resonance peaks in a large range of force intensities.

The dependence of the frequency ratio r on the oscillation amplitude, drawn in Figure 8.9b, is not monotonic; it initially decreases crossing twice the exact resonance 2:1 before regularly increasing towards the limit value r = 2.41 at large amplitudes. The bifurcation happens at the force intensity A = 0.35 when the ratio r is close to 1.9.

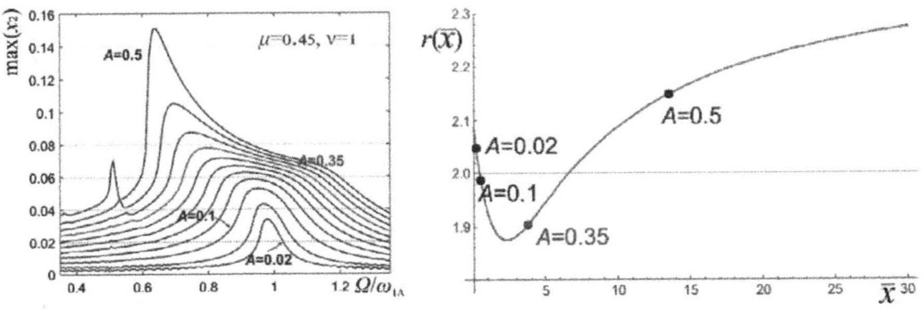

Figure 8.9. *2DOF base-hysteresis oscillator close to (2:1) internal resonance ($\omega_{2A}/\omega_{1A} \cong 2.09$): (a) FRCs for increasing force intensity and (b) the ratio ω_2/ω_1 versus the oscillation amplitude. For a color version of this figure, see www.iste.co.uk/challamel/mechanics2.zip*

8.5. Conclusion

The phenomenon of nonlinear modal interaction in a dynamical system with hysteresis has been investigated by means of a simple 2DOF chain structure that consists of two masses and two elements. In particular, two different configurations are considered: base-hysteresis configuration (BC) with the hysteretic element at the base and top-hysteresis configuration with the hysteretic element at the top (TC), between the two masses. The hysteretic behavior is described by the Bouc–Wen law, characterized by a full hysteretic loop; the response to a harmonic force of an SDOF system is first analyzed to pinpoint the main features of dynamic behavior. The strong hysteresis nonlinearity, due to the variation of stiffness and damping with the oscillation amplitude, produces a softening frequency–amplitude curve and an amplitude–force relationship which is almost linear, apart from the ranges of small and large amplitudes. The FRCs are always single-valued functions implying that the steady-state solutions are always stable. According to the variation of stiffness, the resonance frequency changes from that of the linear oscillator, with a stiffness equal to the initial stiffness of the hysteretic restoring force for small amplitudes, to that of the linear oscillator with stiffness equal to the post-elastic stiffness for high amplitudes. This characteristic is encountered again in 2DOF systems.

The response of the two configurations considered is examined by means of FRCs to harmonic excitation. For increasing intensity, the modal characteristics of system change, and in turn the frequency's ratio changes, and due to the richness of hysteretic nonlinearity, several conditions of internal resonance, where the modal interaction is stronger, are easy to meet.

The case of hysteretic attachment in the top configuration, which resembles the classical tuned mass damper, has been initially investigated. The hysteretic attachment requires a more refined tuning; only in a limited range of force intensity can the same effectiveness of a viscoelastic damper be guaranteed, the only advantage being that it is simpler to achieve. More interesting are the other cases dealt with, close to internal resonance conditions n:1 with n>1, the case 3:1 for the TC systems and the cases 3:1 and 2:1 for BC systems.

The nonlinear response phenomena are quite similar for both configurations: around the first resonance, at a critical oscillation amplitude, the branch ω_1-A experiences a bifurcation and two branches take origin: one is the continuation of the previous mode with a similar shape, destined to disappear, and the new one, with a clearly different shape, remains the only first resonance for large amplitude. Therefore, the FRCs exhibit two peaks, as expected for linear systems, in the small and large intensity range, while in the middle, three peaks appear in the FRCs, two of these close to the first resonance. Notwithstanding that the origin and the phenomenon are quite different, the beneficial effect on the amplification of the first mode is similar to some extent to the case of a tuned mass damper. The effectiveness of the nonlinear coupling is well explained by comparing two systems, close to and far from internal resonance conditions. What clearly emerges is the usefulness of exploiting these nonlinear phenomena in vibration mitigation for both configurations considered.

8.6. Acknowledgments

The support of Italian MIUR under the grant PRIN-2015, 2015TTJN95 P.I., Fabrizio Vestroni, “Identification and monitoring of complex structural systems” is gratefully acknowledged.

8.7. Appendix: Mechanical characteristics of SDOF and 2DOF systems

SDOF name	γ	β	m_1 kg	k_1 N/m	ω_A Hz	ω_B Hz	γ/β	δ
BW1	55	55	0.06	4400	78.3	43.1	1	0.3
BW2	10	100	0.06	14514	78.3	43.1	10	0.3

Table 8.1. *Mechanical characteristics of SDOF systems*

2DOF name	γ	β	m_1 kg	k_1 N/m	ω_{1A} Hz	ω_{2A} Hz	μ	ζ_1	ν	δ
TC1	82	82	1220.7	$1.5326\ 10^6$	5.27	6.95	0.05	0.02	0.0093	0.14
TC2	60	60	1220.7	$1.5326\ 10^6$	5.10	18.0	0.20	0.02	0.25	0.15
BC1	60	60	0.0256	4400	38.94	113.2	1.64	-	1	0.6
BC2	60	60	0.0256	4400	39.28	133.61	1.64	-	2.54	0.6
BC3	60	60	1220	$4.4\ 10^5$	4.1	8.57	0.45	-	1	0.15

Table 8.2. *Mechanical characteristics of 2DOF systems*

8.8. References

Al-Bender, F., Symens, W., Swevers, J., Van Brussel, H. (2004). Theoretical analysis of the dynamic behavior of hysteresis elements in mechanical systems. *Int. J. Non-Linear Mech.*, 39, 1721–1735.

Anand, G.V. (1972). Natural modes of coupled non-linear systems. *Int. J. Non-Linear Mech.*, 7, 81–91.

Awrejcewicz, J. (2007). Hysteresis modelling and chaos prediction in one and two-DOF hysteretic models. *Arch. Appl. Mech.*, 77, 261–279.

Awrejcewicz, J., Dzyubak, L., Lamarque, C.H. (2008). Modelling of hysteresis using Masing–Bouc-Wen's framework and search of conditions for the chaotic responses. *Commun. Nonlin. Sci. Numeri. Simul.*, 13, 939–958.

Awrejcewicz, J., Krysko, A.V., Papkova, I.V., Erofeev, N.P., Krysko, V.A. (2017). Chaotic dynamics of structural members under regular periodic and white noise excitations. *Lecture Notes in Computer Science*, April.

Bouc, R. (1967). Forced vibrations of mechanical systems with hysteresis. *Proceedings of the Fourth Conference on Non-Linear Oscillations*, Prague.

Capecchi, D. and Vestroni, F. (1990). Periodic response of a class of hysteretic oscillators. *Int. J. Non-Linear Mech.*, 25(2) 309–317.

Capecchi, D. and Vestroni, F. (1995). Asymptotic response of a two DOF elastoplastic system under harmonic excitation. Internal resonance case. *Nonlin. Dyn.*, 7, 317–333.

Carboni, B. and Lacarbonara, W. (2015). Dynamic response of nonlinear oscillators with hysteresis. *Proceedings of the ASME Design Engineering Technical Conference*, August 2–5, Boston, MA.

Carpineto, N., Lacarbonara, W., Vestroni, F. (2013). Hysteretic tuned mass dampers for structural vibration mitigation. *J. Sound Vibr.*, 333, 1302–1318.

Casini, P. and Vestroni, F. (2011). Characterization of bifurcating nonlinear normal modes in piecewise linear mechanical systems. *Int. J. Non-Lin. Mech.*, 46, 142–150.

Casini, P. and Vestroni, F. (2018). Nonlinear resonances of hysteretic oscillators. *Acta Mech.*, 229, 939–952.

Den Hartog, J.P. (1934). *Mechanical Vibrations*, McGraw-Hill, New York.

Gendelman, O.V. (2004). Bifurcations of nonlinear normal modes of linear oscillator with strongly nonlinear damped attachment. *Nonlin. Dyn.*, 37, 115–128.

Haller, G. and Ponsioen, S. (2016). Nonlinear normal modes and spectral submanifolds: Existence. Uniqueness and use in model reduction. *Nonlin. Dyn.*, 86, 1493–1534.

Hassani, V., Tjahjowidodo, T., Do, T.N. (2014). A survey on hysteresis modeling, identification and control. *Mech. Syst. Signal Process.*, 49, 209–233.

Ismail, M., Ikhouane F., Rodellar, J. (2009). The hysteresis Bouc–Wen model, a survey. *Arch. Comput. Methods Eng.*, 16, 161–188.

Jo, H. and Yabuno, H. (2009). Amplitude reduction of primary resonance of nonlinear oscillator by a dynamic vibration absorber using nonlinear coupling. *Nonlin. Dyn.*, 55, 67–78.

Lacarbonara, W. and Vestroni, F. (2003). Nonclassical responses of oscillators with hysteresis. *Nonlin. Dyn.*, 32(3), 235–258.

Laxalde, D., Thouverez, F., Sinou, J.-J. (2006). Dynamics of a linear oscillator connected to a small strongly non-linear hysteretic absorber. *Int. J. Non-Lin. Mech.*, 41, 969–978.

Leine, R. and Nijmeijer, H. (2007). *Dynamics and Bifurcations of Non-Smooth Mechanical Systems*. Springer, Berlin, Heidelberg.

Masiani, R., Capecchi, D., Vestroni, F. (2002). Resonant and coupled response of hysteretic two-degree-of-freedom systems using harmonic balance method. *Int. J. Non-Lin. Mech.*, 37, 1421–1434.

Muravskii, G.B. (2005). On description of hysteretic behaviour of materials. *Int. J. Solids Struct.*, 42(9–10), 2625–2644.

Nayfeh, A.H. and Balachandran, B. (1995). *Applied Nonlinear Dynamics: Analytical, Computational, and Experimental Methods*. Wiley, New York.

Nayfeh, A.H. and Mook, D.T. (1979). *Nonlinear Oscillations*. Wiley, New York.

Ni, Y.Q., Ko, J.M., Wong, C.W. (1998). Identification of non-linear hysteretic isolators from periodic vibration tests. *J. Sound Vibr.*, 217, 747–756.

Pak, C.H. (1989). On the stability behaviour of bifurcated normal modes in coupled nonlinear systems. *J. Applied Mech.*, 56, 155–161.

Rand, R.H., Pak, C.H., Vakakis, A.F. (1992). Bifurcation of nonlinear normal modes in a class of two degree of freedom systems. *Acta Mech.*, 3, 129–145.

Rosenberg, R.M. (1962). On normal vibrations of a general class of nonlinear dual-mode systems. *J. Appl. Mech.*, 29, 7–14.

Shaw, S.W. and Pierre, C. (1991). Nonlinear normal modes and invariant manifolds. *J. Sound Vibr.*, 150(1), 170–173.

Vakakis, A.F. (1992). Non-similar normal oscillations in a strongly non-linear discrete system, *J. Sound Vibr.*, 158(2), 341–361.

Vakakis, A.F. (1997). Non-linear normal modes and their applications in vibration theory: An overview. *Mech. Syst. Signal. Process.*, 11, 3–22.

Vakakis, A.F. (2017). Intentional utilization of strong nonlinearity in structural dynamics. *Procedia Eng.*, 199, 70–77.

Vakakis, A.F., Manevitch, L., Gendelman, O., Bergman, L. (2003). Dynamics of linear discrete systems connected to local, essentially non-linear attachments. *J. Sound Vibr.*, 264(3), 559–577.

Vestroni, F. and Casini, P. (2020). Mitigation of structural vibrations by hysteretic oscillators in internal resonance. *Nonlin. Dyn.*, 99, 505–518.

Vestroni, F. and Noori, M. (2002). Hysteresis in mechanical systems: Modeling and dynamic response. *Int. J. Non-Lin. Mech.*, 37(8), 1261–1262.

Vestroni, F., Luongo, A., Paolone, A. (2008). A perturbation method for evaluating nonlinear normal modes of a piecewise linear 2-DOF system. *Nonlin. Dyn.*, 54, 379–393.

Wen, Y.K. (1976). Method of random vibration of hysteretic systems. *ASCE J. Eng. Mech.*, 102(2), 249–263.

Wiercigroch, M. and de Kraker, B. (2000). *Applied Nonlinear Dynamics and Chaos of Mechanical Systems with Discontinuities.* World Scientific Publishing Co. Pte. Ltd., Singapore.

9

Bridging Waves on a Membrane: An Approach to Preserving Wave Patterns

We introduce a novel metamaterial intended to "bridge" a gap between two membranes using a periodic array of strings, with the aim of identically reproducing an incident wave form on the other side of the void. This involves both homogenization and an exact Fourier series to treat the connection between the two materials. Two bridges are considered. For the first, the bridging effect is both broadband and tunable but there is no way of using this scheme to successfully bridge waves at multiple incident angles. While the second bridge is broadband, tunable and independent of the incident wave angle, the construction is more involved than the first bridge. However the simplicity of the bridging scheme gives scope for further development and extension to other media.

9.1. Introduction

Recently, there has been a significant increase in the amount of work produced with the aim of redirecting or suppressing waves in a wide array of systems. This follows a long tradition of attempting to control wave propagation, including the use of mirrors and basic lenses to focus and redirect optical wavelengths of electromagnetic radiation. Following the discovery of microwaves in the late 19th century, there were attempts to control the newly discovered electromagnetic waves using waveguides and differently shaped antennae (Ramsay 1958).

These waveguides included wave gratings, periodic structures classically originating in optics (Draper 1874) used in elasticity to filter and control wave

Chapter written by Peter WOOTTON and Julius KAPLUNOV.

propagation (Williamson and Smith 1973; Glass and Maradudin 1981). More recently, the effect that "structured interfaces" and periodic connections between continuous materials have on wave motion has been investigated (Bertoldi *et al.* 2007a, b). These structures are capable of exhibiting behavior such as wave filtering Brun *et al.* (2010a) and polarization (Brun *et al.* 2010b). Particular attention is given to structured interfaces and gratings that are intended to enhance transmission, using the resonance of successive internal reflections (Haslinger *et al.* 2011, 2013).

The study of waveguides has been complemented by recent advancements in the field of metamaterials. These materials are specially designed structures engineered in order to exhibit material parameters, effects and wave behavior that was not physically possible with continuous materials. Developments in optics and electromagnetism (Pendry *et al.* 1999; Fischer *et al.* 2011) have been followed by developments in many other fields (Kadic *et al.* 2013) including acoustics (Liu *et al.* 2000; Li and Chan 2004; Movchan and Guenneau 2004), linear bulk elasticity (Milton and Willis 2007; Zhou *et al.* 2012), flexural plate waves (Xiao *et al.* 2012; Colombi *et al.* 2014; Williams *et al.* 2015) and elastic surface waves (Brûlé *et al.* 2014; Ege *et al.* 2018; Wootton *et al.* 2020).

One such effect, referred to as "transformation cloaking", has been produced for the vertical displacement of a thin elastic membrane (Colquitt *et al.* 2013). This scheme proposes a metamaterial that can be embedded in a membrane to redirect waves around a specified region, leaving the region undisturbed. However this relies on the use of a metamaterial with anisotropic material parameters. In this chapter, we propose an alternative method to guiding waves around an inclusion using a metamaterial constructed only with standard isotropic materials, based on the example of waves on a thin elastic membrane. This "bridging" method is distinct from cloaking, and uses periodic inclusions to carry wave energy across a void in a continuous material. A similar system has been previously considered, using a flexural array to carry wave motion through a void in a fluid, with some added "waveshift" effects (Porter 2018).

Previous efforts to solve multiscale problems have included matched asymptotics (Tuck 1971), the so-called "Arlequin method" (Bauman *et al.* 2008; Kolpakov *et al.* 2018), and homogenization (see, for example, Colquitt *et al.* (2015) and references therein). While the Arlequin method matches the work done in the system, the treatment of the boundaries in this chapter involves either homogenization or Fourier series expansions to match the displacement and conserve tensions along the boundary. For more complicated systems, such treatments may not be possible, or will not give an appropriate degree of accuracy.

We begin by introducing an elastic membrane that carries wave energy through out-of-plane motion. This material has been chosen for multiple reasons. First, the governing equations have a simple form that makes manipulation relatively straightforward. Second, cloaking has already been achieved for a membrane (Colquitt *et al.* 2013), so the "bridging" result can potentially be compared with the membrane cloak. Finally, the governing equations for the out-of-plane motion of a string under tension are a clear 1D analogue for the governing equations for a membrane, so this gives an obvious choice for the bridging material.

This problem is established in section 9.2, with governing equations for the given materials and boundary conditions. A homogenization scheme is introduced to distribute the periodic loading from the strings. The resultant wave amplitudes from this scheme are given in section 9.3, and a consideration of the phase changes is given in section 9.4. The homogenized results are verified by using a Fourier series method in section 9.5. A more sophisticated bridge is then introduced in section 9.6, based on a previously studied periodic structure of strings (Martinsson and Movchan 2003). The advantages and disadvantages of each of these schemes are then discussed in section 9.7.

9.2. Problem statement

On defining a Cartesian coordinate system (x, y), we introduce two membranes separated by an infinite rectangular void of thickness d, given by the two domains $y < 0$ and $y > d$, for $-\infty < x < \infty$. On these domains, the out-of-plane displacement, u_m, for a membrane with mass per unit area m and internal in-plane tension T, which is assumed to be uniform in all directions, is given by the solutions:

$$m\frac{\partial^2 u_m}{\partial t^2} = T\left(\frac{\partial^2 u_m}{\partial x^2} + \frac{\partial^2 u_m}{\partial y^2}\right). \qquad [9.1]$$

At the edges $y = 0, d$ of the rectangular membrane, the vertical force per unit length along the boundary, P, is given by:

$$P = T\frac{\partial u_m}{\partial y}. \qquad [9.2]$$

A harmonic forcing of the angular frequency ω gives the following solutions:

$$u_m = A_m e^{i[k_m(x\sin\theta + y\cos\theta)\, \pm\, \omega t]}, \qquad [9.3]$$

where the amplitude A_m is an arbitrary constant and θ denotes the wave propagation angle from the y axis. The frequency ω and the wavenumber k_m are related by the membrane wave speed, c_m, by the dispersion relation:

$$\frac{\omega}{k_m} = c_m = \sqrt{\frac{T}{m}}, \qquad [9.4]$$

so for $k_m > 0$ and $-\pi/2 < \theta \leq \pi/2$ the choice of sign in the exponent determines the direction of wave propagation.

We next consider the behavior of infinitely thin strings intended to "bridge" the gap and carry the wave motion from the membrane. Since this string is only carrying the wave motion from the membrane, we assume that only the out-of-plane motion occurs. We then represent the horizontal plane by the same coordinate system (x, y), such that the string motion is perpendicular to both the x and y axes. Since there are infinitely many strings, we denote the vertical displacement of the nth string by u_{sn}. In order to control the phase difference of the waves in the strings, we allow the strings to be held at an angle to the boundaries. Then, if ϑ is the angle between the string and the y axis, we introduce a new variable ζ to represent the distance along the string, where $\zeta = x \sin\vartheta + y \cos\vartheta$. For these coordinates, the equation of motion for vertical displacement is given by:

$$\mu \frac{\partial^2 u_{sn}}{\partial t^2} = Q \frac{\partial^2 u_{sn}}{\partial \zeta^2}, \qquad [9.5]$$

where μ and Q are the mass per unit length and in-plane force holding the strings, respectively. At the ends of the string, the vertical force, F_n, is given by:

$$F_n = Q \frac{\partial u_{sn}}{\partial \zeta}. \qquad [9.6]$$

This leads to harmonic wave solutions along the string of the form:

$$u_{sn} = A_{sn} e^{i[k_s \zeta \,\pm\, \omega t]} \qquad [9.7]$$

where the amplitude A_{sn} is an arbitrary constant arising from the boundary conditions and k_s is determined by the string dispersion relation:

$$\frac{\omega}{k_s} = c_s = \sqrt{\frac{Q}{\mu}}, \qquad [9.8]$$

where c_s is the wave speed in the string. Again, for $k_s > 0$, the choice of sign in the exponent determines the direction of wave propagation. To use these strings to connect the void between the two halves of the membrane, we define the string ends to be at $y = 0$ and $y = d$, as shown in Figure 9.1. This initially appears to be similar to a wave grating; however, in the current treatment, the aim is to reproduce wave patterns, hence avoiding diffraction patterns or short wavelength behavior. This requires the incident wavelength, given by $\lambda = 2\pi/k_m$, to be much greater than the string separation, l. While not a grating in the traditional sense, this system is similar in appearance and is capable of exhibiting similar properties. Most notably, it is capable of internal reflections, as seen in Haslinger *et al.* (2011, 2013).

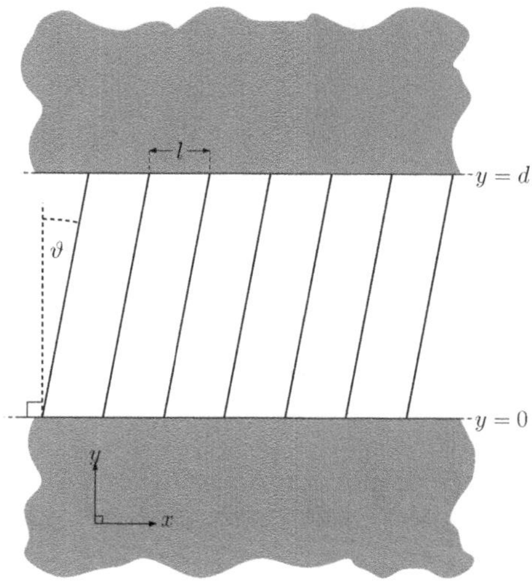

Figure 9.1. *Schematic of an infinite elastic membrane with periodically inserted strings with separation* l*, bridging a void between* $y = 0$ *and* $y = d$

The nth string occupies $x = nl + y \tan \vartheta$, $0 < y < d$, where $n = 0, \pm 1, \pm 2...$. Since these strings do not overlap, we introduce a homogenized displacement in the bridge u_s that satisfies the string equation of motion [9.5] on the domain $-\infty < x < \infty$, $0 < y < d$, such that u_s is equal to u_{sn} at all points where the homogenized and discrete domains overlap.

Since the connections at the boundaries are fixed, conserve displacements such that at the connections,

$$u_s = u_m. \qquad [9.9]$$

If the connections between the strings and membranes are ideal, then at the boundaries the Dirac delta function can be used to model the point connection. However, such point actions are incompatible with the exact treatment of a continuous boundary (Slepyan 1967) so instead, below, we introduce a distributed delta function, δ_h, to represent a real-world forcing along the boundaries. This real-world forcing must be periodic, continuous and differentiable, and it will also be assumed that the tension in the membrane must be finite and decay exponentially with distance away from any junction with a string. As an example, one such function is the Jacobi ϑ_3 function (Whittaker and Watson 1996), which is defined by

$$\delta_h(x) = \tfrac{1}{l}\, \vartheta_3\left(\tfrac{\pi x}{l},\, e^{\frac{-\pi^2 h^2}{4l^2 \ln(2)}}\right) = \frac{2}{h}\sqrt{\ln(2)} \sum_{n=-\infty}^{\infty} e^{-4\ln(2)\frac{(x+nl)^2}{h^2}}, \qquad [9.10]$$

where the half-width h is the distance from the center of the string for which the tension is halved.

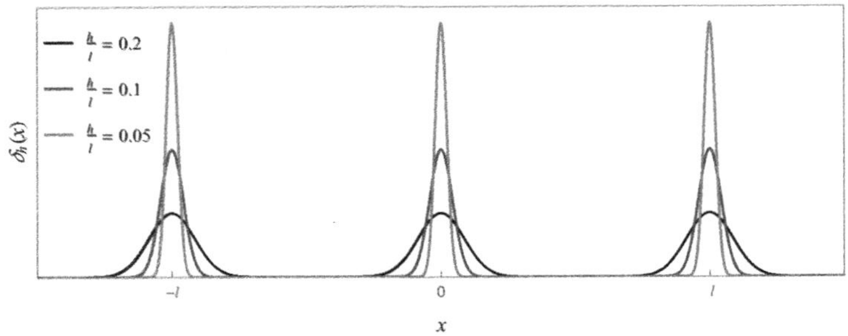

Figure 9.2. *Comparisons of the distributed delta function $\delta_h(x)$ for different values of the half-thickness h, where l is the separation between successive strings*

This half-width should be related to the thickness of the strings, so that it can be determined for different physical systems. As required, this distributed function will tend to a Dirac delta function at $x = nl$, in the limit of $h/l \to 0$. Figure 9.2 shows how this function behaves for decreasing values of h/l, quickly converging to the delta function. An exact treatment of the problem using this distributed function will be discussed in section 9.5, but for a homogenized boundary an approximate treatment will suffice.

For infinitely many evenly spaced strings, the matching of tensions at the boundaries $y = 0, d$ gives:

$$P = \delta_h(x) \sum_{n=-\infty}^{\infty} F_n \qquad [9.11]$$

which gives the boundary condition

$$T\frac{\partial u_m}{\partial y} = Q\frac{\partial u_s}{\partial \zeta}\delta_h(x) \qquad [9.12]$$

Taking the integral of [9.11] over a single unit cell gives

$$\int_{(n-\frac{1}{2})l}^{(n+\frac{1}{2})l} P\, dx = \int_{(n-\frac{1}{2})l}^{(n+\frac{1}{2})l} \delta_h(x) \sum_{n=-\infty}^{\infty} F_n\, dx, \qquad [9.13]$$

where h is assumed to be sufficiently small for δ_h to behave like the delta function, and if the separation of the strings is assumed to be much less than the wavelength of the incident wave, then the stress at the boundary from the incident waves will be approximately uniform over a single unit cell. Taking this integral and distributing across the entire boundary gives the homogenized boundary condition:

$$Pl \approx \sum_{n=-\infty}^{\infty} F_n \qquad [9.14]$$

which becomes, on substitution:

$$T\frac{\partial u_m}{\partial y} = \frac{Q}{l}\frac{\partial u_s}{\partial \zeta}. \qquad [9.15]$$

9.3. Homogenized bridge

For a simple treatment of the bridge, we assume the homogenized boundary condition [9.15]. We then consider an incident wave and analyze the effect that the matching has at the wave form, with the aim of perfectly reproducing the incident wave pattern on the other side of the void.

Suppose that an arbitrary incident plane wave is propagating in the positive y direction from $-\infty$, at some angle θ to the boundary normal. This wave will have the form [9.3]:

$$u_m^{(i)} = A_i e^{i[k_m(x\sin\theta + y\cos\theta) - \omega t]}, \qquad [9.16]$$

where A_i is a constant amplitude and k_m and ω satisfy the membrane dispersion relation [9.4]. This wave will first be incident on the boundary at $y = 0$ and, for the nth string, will transmit waves of the form:

$$u_{sn}^{(t)} = A_t e^{i(k_s\zeta - \omega t)} e^{i\chi nl}, \qquad n = 0, \pm 1, \pm 2... \qquad [9.17]$$

where k_s and ω satisfy the string dispersion relation [9.8] and χ is a constant caused by a variation in amplitudes between successive strings along the boundary. Matching the form of the displacement at the boundary,

$$e^{i(k_m x\sin\theta - \omega t)} = e^{i(k_s nl\sin\vartheta + \chi nl - \omega t)}, \qquad [9.18]$$

from which, it follows that at $x = nl$, $\chi = k_m \sin\theta - k_s \sin\vartheta$. Substituting this into the discrete string displacement, $nl = x - y\tan\theta = x_0$ and thus the homogenized displacement becomes:

$$u_s^{(t)} = A_t e^{i(k_s\zeta - \omega t)} e^{i\chi x_0}. \qquad [9.19]$$

Since the displacement along the boundary must be continuous and it is assumed that there is only one transmitted wave direction, it equally follows that the reflected wave has the form:

$$u_m^{(r)} = A_r e^{i[k_m(x\sin\theta - y\cos\theta)-\omega t]}, \qquad [9.20]$$

where the overall displacement in the membrane is given by:

$$u_m = u_m^{(i)} + u_m^{(r)}. \qquad [9.21]$$

As expected, the angle of incidence is equal to the angle of reflection. On substituting these assumed wave solutions into the boundary condition [9.15], we have

$$A_i - A_r = A_t\gamma_\theta, \qquad [9.22]$$

where a bridge function $\gamma_\theta(\theta)$ and bridging parameter γ are defined in terms of system parameters and angle of incidence:

$$\gamma_\theta(\theta) = \gamma\sec\theta, \qquad \gamma = \frac{Qc_m}{Tlc_s}. \qquad [9.23]$$

Moreover, substituting the assumed waves into the continuity condition [9.9] yields:

$$A_i + A_r = A_t \qquad [9.24]$$

and thus the transmission and reflection coefficients are:

$$A_r = \frac{1-\gamma_\theta}{1+\gamma_\theta}A_i, \qquad A_t = \frac{2}{1+\gamma_\theta}A_i. \qquad [9.25]$$

The main motivation of this system is in perfectly transmitting the wave energy and reproducing the incident wave pattern. Since the wave pattern and wave energy do not depend on changes in phase, an initial consideration of the second boundary only needs to consider changes in amplitude. Repeating the same procedure at the second boundary gives assumed incident, reflection and transmission amplitudes, respectively, of the form:

$$\begin{aligned} u_s^{(i)} &= B_i\, e^{i(k_s\zeta-\omega t)}e^{i\chi x_0}, \\ u_s^{(r)} &= B_r\, e^{-i(k_s\zeta+\omega t)}e^{i\chi x_0}, \\ u_m^{(t)} &= B_t\, e^{i[k_m(x\sin\theta+y\cos\theta)-\omega t]}. \end{aligned} \qquad [9.26]$$

Since these waves are fully periodic, and currently only the amplitudes are being considered, the coordinate system can be arbitrarily shifted so that the "top" boundary is redefined as $y = 0$. Substitution into the boundary conditions [9.15] and [9.9] yields:

$$B_i - B_r = \frac{1}{\gamma_\theta}B_t, \qquad B_i + B_r = B_t. \qquad [9.27]$$

Hence, the reflection and transmission coefficients from the second boundary are:

$$B_r = \frac{\gamma_\theta - 1}{\gamma_\theta + 1} B_i, \qquad B_t = \frac{2\gamma_\theta}{\gamma_\theta + 1} B_i. \qquad [9.28]$$

We define the overall transmission through the system as being given by the ratio of the transmission from the $y = d$ boundary to the initial amplitude incident at $y = 0$. This overall transmission T_1 is given by:

$$T_1(\gamma_\theta) = \frac{B_t}{A_i} = \frac{4\gamma_\theta}{(1+\gamma_\theta)^2}, \qquad [9.29]$$

as shown in Figure 9.3.

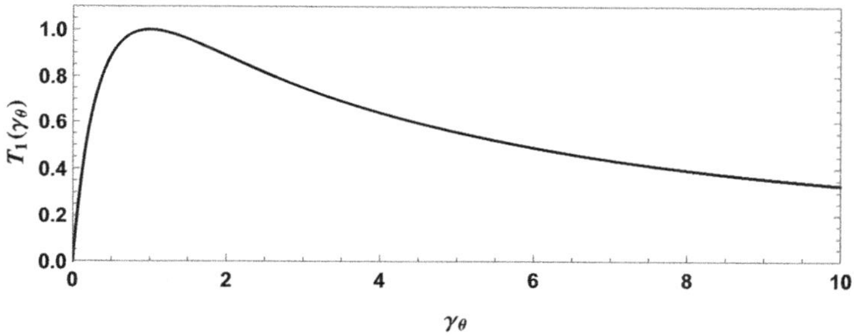

Figure 9.3. *Ratio of the amplitude of the overall transmission to the initial wave amplitude [9.29] against the bridging parameter γ_θ [9.23]*

This transmission has a maximum of $T_1 = 1$ at $\gamma_\theta = 1$, with any other γ_θ values causing a transmission loss. It is interesting to note that the angle of the strings ϑ does not affect the amplitude of the waves transmitted and reflected from this system.

Since γ_θ varies with θ, it is not possible to construct a bridge, such that $\gamma_\theta = 1$ for all possible incident angles. Hence, this system cannot be used to simultaneously bridge a range of waves with different incidences without losing transmission; instead, it is limited to perfectly bridging waves from one particular angle. To find the average transmission over all incident angles, we evaluate the integral:

$$\tilde{T}_1(\gamma) = \frac{1}{\pi}\int_{-\pi/2}^{\pi/2} T_1 d\theta = \frac{1}{\pi}\int_{-\pi/2}^{\pi/2} \frac{4\gamma\cos\theta}{(\cos\theta+\gamma)^2}\, d\theta, \qquad [9.30]$$

$$= \frac{8\gamma}{\pi(1-\gamma^2)}\left(2\frac{\tanh^{-1}\left(\sqrt{\frac{1-\gamma}{1+\gamma}}\right)}{\sqrt{1-\gamma^2}} - 1\right) \qquad [9.31]$$

which has a global maximum $\tilde{T}_1 \approx 0.891$ for $\gamma = 0.612871...$, compared to $\tilde{T}_1 \approx 0.849$ for a bridge with $\gamma = 1$, aimed at perfectly transmitting only normally incident waves. For some constant "tuned" value of γ, transmission has a dependence on the angle:

$$T_1(\theta) = \frac{4\gamma\cos\theta}{(\gamma + \cos\theta)^2} \qquad [9.32]$$

which is shown for $\gamma = 1$ and $\gamma = 0.612871...$ in Figure 9.4.

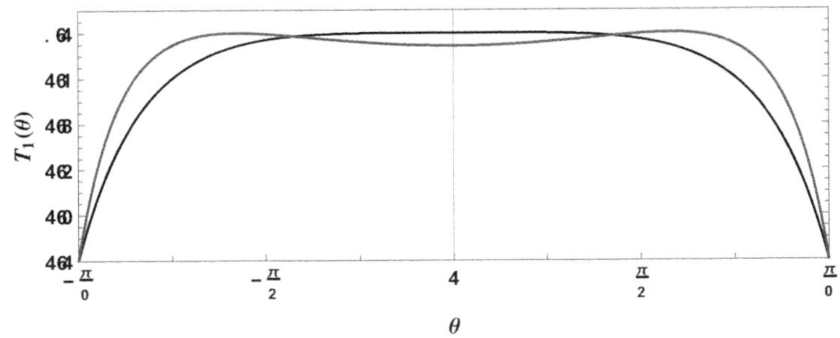

Figure 9.4. *Ratio of the amplitude of the overall transmission to the initial wave amplitude [9.32] for different incident angles θ on a string bridge. The two plot lines show different choices of parameter tuning, with normal incidence tuning $\gamma = 1$ denoted in gray and maximum average transmission tuning $\gamma = 0.612871...$ denoted in black*

Both of the choices of parameter effectively transfer the wave amplitude from one side of the void to the other. While the maximized average tuning transmits the wave more effectively over a wider range of angles, the normal incidence tuning is more effective for small incident angles. Depending on the choice of the system, either of these tuning choices may be more effective at bridging an incoming wave.

Since we have shown that changes in the amplitude of the wave only depend on system parameters and the angle of incidence, and not on the string angle, we now consider the effect of adding a phase change to the waves traveling through the system. For this purpose, we define two different phase shifts, a reflection phase shift, φ_s, and a transmission phase shift, φ_t. The reflection phase shift denotes the phase shift undergone by a wave traveling from one end of the string to another. This is referred to as the "reflection" phase shift, since a wave reflected at the $y = d$ boundary will be out of phase with the incident wave by a factor of $2\varphi_s$ when it returns to the $y = 0$ boundary. The transmission phase shift will instead concern the $y = d$ boundary, representing the difference in phase of waves produced by the bridge and the phase of the corresponding wave produced by a system where the entire space is filled by the elastic membrane.

We begin by considering an arbitrary point of the $y = 0$ boundary, and set this to be the new origin of the coordinate system. Disregarding the constant variation of phase with time, we define the phase at the origin to be 0. For a wave traveling along a string, the change in phase will simply be:

$$\varphi_s = k_s d(\cos\vartheta + \sin\vartheta\tan\vartheta) = \frac{\omega d}{c_s\cos\vartheta}, \qquad [9.33]$$

which does not depend on the angle of the incident wave. Thus, by only knowing the frequency of an incoming wave, it is possible to tune the phase changes and control wave interference. In the long wave limit, the width of the gap will be much smaller than the wavelength, i.e. $k_m d \ll 1$. With the exception of $\vartheta = \pm\pi/2$, at this limit, the reflection phase difference will tend to 0.

We next attempt to determine the transmission phase shift. An unbridged wave propagating from the origin will have a phase shift φ_0 of

$$\varphi_0 = k_m d(\cos\theta + \sin\theta\tan\theta), = \frac{\omega d}{c_m\cos\theta}. \qquad [9.34]$$

The corresponding "originator" point on the $y = 0$ boundary for a bridged wave will be $(d(\tan\theta - \tan\vartheta), 0)$. Since the change in phase along a string has already been calculated, the overall phase difference will be:

$$\begin{aligned}\varphi_t &= k_m d\sin\theta(\tan\theta - \tan\vartheta) + \varphi_s - \varphi_0\\ &= \frac{\omega d}{c_m}\left(\frac{1}{\cos\vartheta}\left(\frac{c_m}{c_s} - \sin\theta\sin\vartheta\right) - \frac{1}{\cos\theta}\right),\end{aligned} \qquad [9.35]$$

which has the trivial solution of $\varphi_t = 0$ for $c_m = c_s$, when $\vartheta = \theta$, i.e. when the wave speeds match and both the incident waves and strings have the same angle as the boundary. It also shows an infinite phase difference as the string angle $\vartheta \to \pm\pi/2$, representing the infinite increase in the string length. When the strings are held perpendicular to the boundary, this simplifies to:

$$\varphi_t = \frac{\omega d}{c_m}\left(\frac{c_m}{c_s} - \frac{1}{\cos\theta}\right). \qquad [9.36]$$

Again, in the long wave limit $k_m d \ll 1$, for an incident wave not parallel to the boundary, this gives $\varphi_t \to 0$.

9.4. Internal reflections

Up until now, the treatment of this system has only considered the first transmission resulting from one incident wave peak. However, when the wave on the strings is incident on a boundary, a part of the amplitude is reflected back into the strings to be retransmitted later. From the primary reflection and transmission coefficients given above [9.25] and [9.28], these successive transmissions and reflections can be calculated as:

$$R_m = \begin{cases} \dfrac{1-\gamma_\theta}{1+\gamma_\theta}, & m = 1 \\ \dfrac{4\gamma_\theta}{(1+\gamma_\theta)^2}\left(\dfrac{\gamma_\theta - 1}{\gamma_\theta + 1}\right)^{(2m-3)}, & m \geq 2 \end{cases} \qquad [9.37]$$

and

$$T_m = \frac{4\gamma_\theta}{(1+\gamma_\theta)^2}\left(\frac{\gamma_\theta - 1}{\gamma_\theta + 1}\right)^{(2m-2)} \qquad [9.38]$$

where $m = 1, 2, 3...$ and R_m and T_m denote the amplitudes of the mth transmissions from the $y = 0$ and $y = d$ boundaries, respectively. These successive wave amplitudes are plotted in Figure 9.5 against the bridge parameter γ_θ. As expected, at $\gamma_\theta = 1$, all of the coefficients, except T_1, are equal to 0 as all of the wave energy is taken by the first transmission.

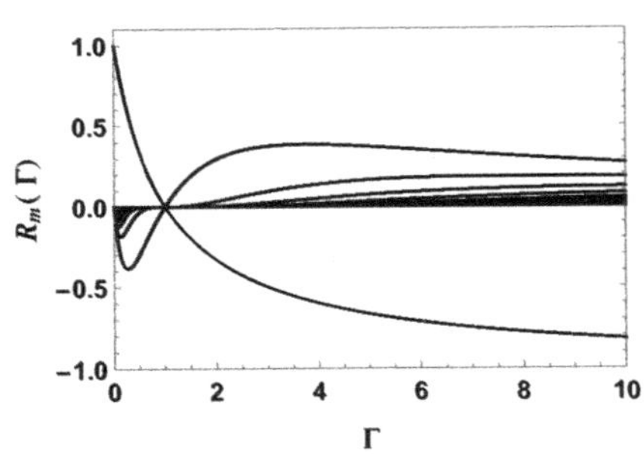

Reflection ratios, R_m

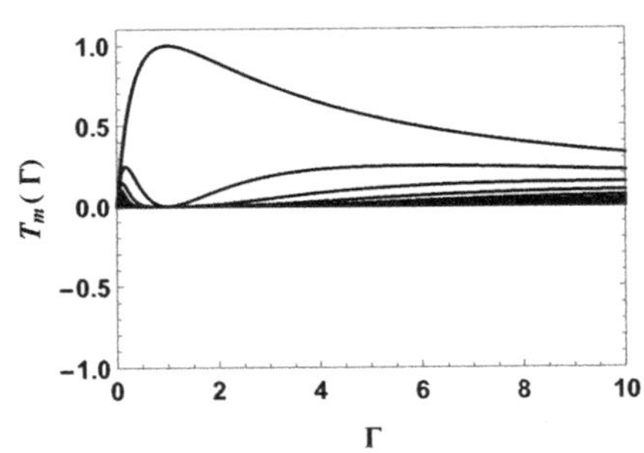

Transmission ratios, T_m

Figure 9.5. *Ratios of the amplitudes of successive reflections [9.37] and transmissions [9.38] to the initial wave amplitude against the bridging parameter γ_θ [9.23]. For both the reflections and the transmissions, the amplitude decreases as m increases*

Accounting for internal reflections, the incident wave produces an infinite number of subsequent reflections and transmissions into the membrane that will interfere depending on their phase difference, to produce one overall transmission and reflection as $t \to \infty$. Since the phase difference undergone by the wave traveling along the string has been previously defined as φ_s, we use this definition again to produce the summation of overall transmissions:

$$R = \frac{1-\gamma_\theta}{1+\gamma_\theta} + \sum_{m=2}^{\infty} \frac{4\gamma_\theta}{(1+\gamma_\theta)^2} \left(\frac{\gamma_\theta - 1}{\gamma_\theta + 1}\right)^{(2m-3)} e^{i2(m-1)\varphi_s}, \qquad [9.39]$$

and

$$T = \sum_{m=1}^{\infty} \frac{4\gamma_\theta}{(1+\gamma_\theta)^2} \left(\frac{\gamma_\theta - 1}{\gamma_\theta + 1}\right)^{2(m-1)} e^{i(2m-1)\varphi_s}, \qquad [9.40]$$

where R and T denote the complex amplitude of the overall reflection and transmission, respectively. These have real amplitudes of

$$|R| = \frac{1-\gamma_\theta^2}{4\gamma_\theta} \left(\frac{1}{2} + \left(\frac{1-\gamma_\theta^2}{4\gamma_\theta}\right)^2 \sin^2 2\varphi_s\right)^{-1/2} \sin 2\varphi_s, \qquad [9.41]$$

and

$$|T| = \left(1 + \left(\left(\frac{1+\gamma_\theta^2}{4\gamma_\theta}\right)^2 - 1\right) \sin^2 \varphi_s\right)^{-1/2}, \qquad [9.42]$$

which are plotted in Figure 9.6 against γ_θ and φ_s.

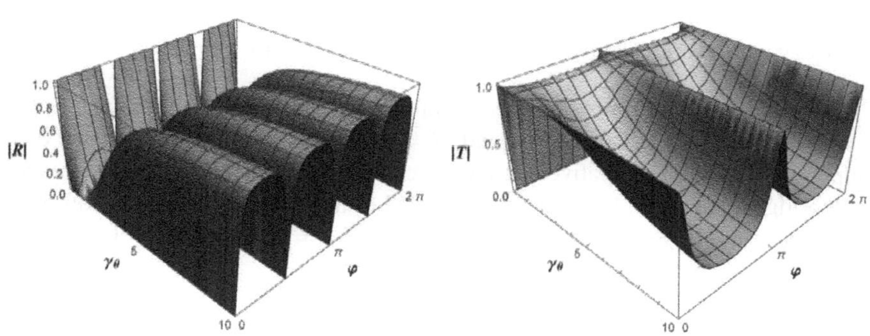

Overall reflection ratio $|R|$ Overall transmission ratio $|T|$

Figure 9.6. *Ratios of the amplitudes of the overall reflection [9.41] and transmission [9.42] to the initial wave amplitude against the parameter γ_θ and the added phase difference φ_s. For a color version of this figure, see www.iste.co.uk/challamel/mechanics2.zip*

From Figure 9.6, it is clear that the overall transmission is only perfect for either $\gamma_\theta = 1$, or when the phase difference φ_s is an integer multiple of π. Outside of these conditions, the overall transmission decreases and the reflection increases. These two cases will be investigated individually.

When $\gamma_\theta = 1$, it is not necessary to consider φ_s as there are no successive reflections and transmissions. In the long wave limit, assuming that neither the incident wave nor the strings are parallel to the boundary, both φ_s and φ_t tend to 0, guaranteeing the perfect matching of the wave.

If $\theta \to \pm\pi/2$, then the long wave assumption is no longer valid. When considering the first transmission, T_0, perfect amplitude matching requires $\gamma_\theta = 1$ so:

$$\cos\theta = \gamma, \tag{9.43}$$

and to perfectly match the wave, the overall phase change needs to be $\varphi_t = 2m\pi,\ m = 0, \pm 1, \pm 2....$ This leads to the relation for the string angle to be:

$$\left(2m\pi\frac{c_m}{\omega d}\gamma + 1\right)\cos\vartheta = \gamma\left(\frac{c_m}{c_s} - \sqrt{1-\gamma^2}\sin\vartheta\right), \tag{9.44}$$

and if it is assumed that the strings are not held parallel to the boundary, $\cos\vartheta \neq 0$. So, as both $\cos\theta, \omega d \to 0$, the resulting wave matches if the incident wavelength, the incident wave angle and the size of the void satisfy

$$\frac{d}{\lambda\cos\theta} = 0, 1, 2... \tag{9.45}$$

Interestingly, this condition is independent of the properties of the bridge itself, so this bridge construction cannot arbitrarily bridge any void for any incident wave.

In the case of an infinite number of incident wavefronts, when $\varphi_s = j\pi,\ j = 0, \pm 1, \pm 2...$, $T = 1$ and $R = 0$. The overall wave, like each of the component waves, has a relative phase difference of 0. Hence, the phase difference of the unbridged wave will be given by φ_t [9.35]. Since perfect matching requires the bridged and unbridged waves to be identical, $\varphi_t = 2m\pi,\ m = 0, \pm 1, \pm 2....$ Matching φ_s from [9.33] gives:

$$\cos\vartheta = \frac{\omega d}{c_s r\pi}, \tag{9.46}$$

which in the long wave limit gives $\vartheta = 0, j \neq 0$. Then:

$$2m\pi = \frac{\omega d}{c_m}\left(\frac{c_m}{c_s} - \frac{1}{\cos\theta}\right), \tag{9.47}$$

which gives the angle requirement:

$$\cos\theta = \frac{c_s}{c_m - \frac{m\lambda}{d}c_s}, \tag{9.48}$$

and which in the long wave limit will give different results depending on m. For $m = 0$, this gives the angle requirement:

$$\cos\theta = \frac{c_s}{c_m} \qquad [9.49]$$

and, for $m > 0$, gives the same requirement as for a single wave front [9.45].

Therefore, for this system, whether a wave can be perfectly bridged or not depends on the properties of the void and the incident wave. While a bridge exists for all angles of incidence, which can ideally transfer the wave amplitude, for shallow incidence waves, the phase change cannot be controlled by changing bridge parameters. This is true when both neglecting and accounting for internal reflection.

The system also produces no reflections for $\varphi_s = (j + 1/2)\pi$, but, unlike the other cases for no reflection, there is still a loss in transmission. For this phase difference, the overall transmission is the same as the first wave transmission from the system [9.29], so $T = T_1$. For this phase difference, all the R_n terms cancel, and so do the T_n terms, except T_1. This represents a wave suppression, where the wave pattern is replicated with no reflections, but the amplitude is reduced.

Also note that the transmission for R and T is only valid, accounting for infinite internal reflections. For a finite number of internal reflections, only a bridge with system parameters $\gamma_\theta = 1$ will give perfect transmission.

9.5. Discrete bridge

Complementary to the homogenization approach presented above, this problem can also be treated using a Fourier series approach. The purpose of this section is to show that the previous homogenization formulation is valid for this system, given that the periodic scale of the boundary is much smaller than the characteristic length scale of the wave, the wavelength λ.

Since the forcing is periodic, we introduce a Fourier series representation for the distributed delta function with coefficients E_j:

$$\delta_h(x) = \sum_{j=-\infty}^{\infty} \frac{E_j}{l} e^{-ij\pi x/l}, \qquad [9.50]$$

where the previously defined example for $\delta_h(x)$ [9.10] yields the coefficients:

$$E_j = \int_{-l/2}^{l/2} \tfrac{1}{l}\,\vartheta_3\left(\tfrac{\pi x}{l},\, e^{\frac{-\pi^2 h^2}{4l^2 \ln(2)}}\right) e^{-i\,j\pi x/l} dx. \qquad [9.51]$$

Therefore, the tension boundary condition [9.12] takes the form:

$$T\frac{\partial u_m}{\partial y} = Q\frac{\partial u_{sn}}{\partial \zeta}\sum_{j=-\infty}^{\infty}\frac{E_j}{l}e^{-i\,j\pi x/l}. \qquad [9.52]$$

Regardless of the choice of parameters, for this function, $E_0 = 1$. Note that for the forcing to conserve forces along the boundary, the $j = 0$ integral should be equal to 1 and so, $E_0 = 1$ in general. Furthermore, in the limit $h/l \to 0$, this gives coefficients $E_j = 1, j = 0, \pm 1, \pm 2....$, which, as required, is the Fourier series representation of the periodic delta function. This will yield the boundary condition

$$T\frac{\partial u_m}{\partial y} = Q\frac{\partial u_{sn}}{\partial \zeta}\sum_{j=-\infty}^{\infty}\frac{1}{l}e^{i\,2j\pi x/l}, \qquad [9.53]$$

which clearly diverges at $x = nl$. Thus, in order to balance both sides of this equation for a finite and non-zero displacement in the membrane, the displacement in the strings needs to be zero. Therefore, the homogenization procedure presented above is not valid in this limit.

However, it can be supposed that this non-physical result is a consequence of using the delta function, a non-physical representation of point loading that disregards material deformations along the boundary. To undertake the short-scale analysis, it is necessary to use the distributed delta as a more realistic coupling function.

We will follow the same technique before assuming a solution form and substituting into the new boundary conditions. The transmitted waves are independent of the boundary change so they will also remain in the same form as before [9.19]. However, it is clear that due to the summation in the boundary condition, the waves in the membrane must also form an infinite summation. We will assume the same incident wave form as before [9.16] and thus, the reflected waves must take the form:

$$u_m^{(r)} = \sum_{j=-\infty}^{\infty} A_{r,j}e^{i[k_m(x\sin\theta - \kappa_j y\cos\theta)-\omega t]}e^{i\,2j\pi x/l}, \qquad [9.54]$$

where to satisfy the membrane governing equation [9.1],

$$\kappa_j = \sqrt{1-\left(\frac{j\lambda}{l\cos\theta}\right)^2}, \qquad [9.55]$$

and as before, λ is the wavelength of the incident wave and θ is the angle of incidence. Substituting these waves into the stress boundary condition [9.15] yields:

$$\left(A_i - \sum_{j=-\infty}^{\infty}\kappa_j A_{r,j}e^{i\,2j\pi x/l}\right) = \gamma_\theta A_t\sum_{j=-\infty}^{\infty}E_j e^{i\,2j\pi x/l}. \qquad [9.56]$$

Each of these summations has the form of a Fourier series and since the terms of a Fourier series are unique, we match each term of the summation to obtain:

$$A_i - A_{r,0} = E_0\gamma_\theta A_t, \qquad \kappa_j A_{r,j} = E_j\gamma_\theta A_t,\ j \neq 0. \tag{9.57}$$

We introduce the parameter ϵ to represent the higher-order terms of the summation:

$$\epsilon = \sum_{j=-\infty}^{-1} \frac{E_j}{\kappa_j E_0} + \sum_{j=1}^{\infty} \frac{E_j}{\kappa_j E_0}. \tag{9.58}$$

Combining the continuity of displacement [9.9] with [9.57]:

$$A_i + A_{r,0} = (1 + E_0\gamma_\theta\epsilon)A_t. \tag{9.59}$$

These give the reflection and transmission coefficients of

$$\frac{A_{r,0}}{A_i} = \frac{1 - E_0\gamma_\theta(1-\epsilon)}{1 + E_0\gamma_\theta(1+\epsilon)}, \qquad \frac{A_t}{A_i} = \frac{2}{1 + E_0\gamma_\theta(1+\epsilon)}. \tag{9.60}$$

By the same process at the $y = d$ boundary, we assume the same incident and reflected wave forms [9.26], and as with the reflection at the $y = 0$, we introduce a summation of transmissions

$$u_m^{(t)} = \sum_{j=-\infty}^{\infty} B_{t,j}\, e^{i[k_m(x\sin\theta + \kappa_j y\cos\theta) - \omega t]} e^{i\,2j\pi x/l}. \tag{9.61}$$

Substituting these wave forms into the boundary conditions yields:

$$B_i - B_r = \frac{B_{t,0}}{E_0\gamma_\theta}, \qquad B_i + B_r = (1+\epsilon)B_{t,0}, \tag{9.62}$$

so then, the coefficients of reflection and transmission are:

$$\frac{B_r}{B_i} = \frac{E_0\gamma_\theta(1+\epsilon) - 1}{E_0\gamma_\theta(1+\epsilon) + 1}, \qquad \frac{B_{t,0}}{B_i} = \frac{2E_0\gamma_\theta}{E_0\gamma_\theta(1+\epsilon) + 1}, \tag{9.63}$$

which leads to the overall transmission through the system as being:

$$T_1(\gamma_\theta) = \frac{B_{t,0}}{A_i} = \frac{4E_0\gamma_\theta}{(1 + E_0\gamma_\theta(1+\epsilon))^2}. \tag{9.64}$$

Each of the above expressions is identical to the homogenized case if $= 0$ and $E_0 = 1$. Thus, when ϵ is small, the transmission will be perfect for $E_0\gamma_\theta = 1$.

In order to verify this solution, we must first determine whether this series will diverge, in order to find out whether this is a problem with the system or a particular feature of using delta function forcing. Hence, the behavior of ϵ must be investigated in a general case.

In the particular case when the forcing is symmetric, the values of E_j must be given by $E_j = E_{-j}$. In this case, the expression for ϵ becomes:

$$\epsilon = \sum_{j=1}^{\infty} \frac{2E_j}{\kappa_j E_0} = \sum_{j=1}^{\infty} \epsilon_j. \qquad [9.65]$$

In the limit $j \to \infty$, each term $_j$ is given by:

$$\epsilon_j \to 2i \frac{l \cos\theta}{E_0 \lambda} \frac{E_j}{j}, \qquad [9.66]$$

and since it is necessary that at the boundary the Fourier series converge, it follows that the sum $\sum_{j=1}^{\infty} E_j$ must converge. From this, it can also be shown that the sum $\sum_{j=1}^{\infty} \frac{E_j}{j}$ converges and thus, ϵ will have a finite value. Therefore, a divergent summation only occurs when using a non-physical divergent boundary condition, such as the delta function.

Also, if $\lambda \cos\theta \gg l$, the summation will have all imaginary terms, so when $\epsilon \neq 0$, in general, $T_1 \neq 1$ since there will be some imaginary component. Hence, for full amplitude transmission, the argument of the overall transmission ratio must be equal to 1, with the imaginary part causing a phase shift in the wave.

In the case $\epsilon = \epsilon_{re} + i\epsilon_{im}$, where ϵ_{re} and ϵ_{im} are both real,

$$\left| \frac{4E_0\gamma_\theta}{(1 + E_0\gamma_\theta(1 + \epsilon_{re} + i\epsilon_{im}))^2} \right| = 1. \qquad [9.67]$$

When ϵ has no real part, $\epsilon_{re} = 0$. Since the bridge parameters are all real and positive, the above condition can be simplified to:

$$\left|(1 + E_0\gamma_\theta)^2 - E_0^2\gamma_\theta^2\epsilon_{im}^2 - 2i(1 + E_0\gamma_\theta)E_0\gamma_\theta\epsilon_{im}\right| = \frac{\left((1 + E_0\gamma_\theta)^2 - E_0^2\gamma_\theta^2\epsilon_{im}^2\right)^2 + 4(1 + E_0\gamma_\theta)^2 E_0^2\gamma_\theta^2\epsilon_{im}^2}{4E_0\gamma_\theta} \qquad [9.68]$$

which has solutions:

$$E_0\gamma_\theta = \frac{3 \pm 2\sqrt{2 + i\epsilon_{im}} + i\epsilon_{im}}{\epsilon_{im}^2 - 1 - 2i\epsilon_{im}}, \quad \frac{1 \pm 2(-1)^{3/4}\sqrt{\epsilon_{im}} - i\epsilon_{im}}{1 + 2i\epsilon_{im} - \epsilon_{im}^2}. \qquad [9.69]$$

However, these solutions only give real positive values of $E_0\gamma_\theta$ for $\epsilon_{im} = 0$. Thus, even though a sufficiently small value of ϵ will lead to diminishingly small transmission losses, there is no way to perfectly reproduce all the aspects of the wave pattern, even taking into account the coupling between the two materials.

Since perfect transmission is not possible by solely selecting bridge parameters, it is again possible to consider the effect of the summation of successive reflections and transmissions from the system as before. Like the simple case of homogenization [9.37] and [9.38], the Fourier series solutions [9.60], [9.63] and [9.64] give the reflection and transmission coefficients of

$$R_m = \begin{cases} \dfrac{1 - E_0\gamma_\theta(1-\epsilon)}{1 + E_0\gamma_\theta(1+\epsilon)}, & m = 1 \\ \dfrac{4E_0\gamma_\theta}{(1 + E_0\gamma_\theta(1+\epsilon))^2}\left(\dfrac{E_0\gamma_\theta(1+\epsilon) - 1}{E_0\gamma_\theta(1+\epsilon) + 1}\right)^{(2m-3)}, & m \geq 2 \end{cases} \quad [9.70]$$

and

$$T_m = \frac{4E_0\gamma_\theta}{(1 + E_0\gamma_\theta(1+\epsilon))^2}\left(\frac{E_0\gamma_\theta(1+\epsilon) - 1}{E_0\gamma_\theta(1+\epsilon) + 1}\right)^{(2m-2)} \quad [9.71]$$

for which the summations of R_m and T_m are again a geometric series, which gives the homogenized result if $\epsilon = 0$. Since in the long wave limit, $\varphi_s \to 0$, the total reflected and transmitted amplitudes, R and T, respectively, are

$$R = \frac{\epsilon}{1+\epsilon}, \quad T = \frac{1}{1+\epsilon}. \quad [9.72]$$

Unlike the homogenized case, these reflection and transmission totals do not tend to a single value. They are, however, still independent of the system parameters and instead, only depend on the nature of the coupling chosen. To recover the homogenized solution, when $\epsilon \to 0$, $R \to 0$ and $T \to 1$. In a more general case, the sum of the transmitted and reflected wave amplitudes is 1, which conserves energy in the system. Furthermore, if the phase of the transmission and reflection is disregarded and we assume that all of the higher summation terms are imaginary so $\epsilon = i\epsilon_{im}$, then the wave amplitudes are:

$$|R| = \frac{\epsilon_{im}}{\sqrt{1+\epsilon_{im}^2}}, \quad |T| = \frac{1}{\sqrt{1+\epsilon_{im}^2}} \quad [9.73]$$

which again, are only equal to 0 and 1, respectively, when $\epsilon_{im} = 0$. For any other value of ϵ_{im}, the transmission decreases and reflection increases.

Therefore, the perfect matching is fundamentally limited by the matching between the materials, and a "good" matching can only be produced for a small value of ϵ. It is therefore necessary to examine more realistic models for the boundary to see whether this condition can be met.

To determine how valid the above approach is, we use the previously defined distributed delta function [9.51] to investigate, which can represent forcing along the boundary in a real physical system.

To find the values for ϵ of the above $\delta_h(x)$, the sum of a numerical Fourier series approximation can be made. For demonstration, a range of ϵ values for a normally incident wave with this type of boundary forcing are given in Table 9.1, for different peak widths and incident wavelengths to an appropriate level of accuracy.

		l/λ			
	ϵ	1×10^{-1}	1×10^{-2}	1×10^{-3}	1×10^{-4}
h/l	0.5	0.0854 i	$0.850\,\mathrm{i}\times 10^{-2}$	$0.850\,\mathrm{i}\times 10^{-3}$	$0.850\,\mathrm{i}\times 10^{-4}$
	0.1	0.3930 i	$3.919\,\mathrm{i}\times 10^{-2}$	$3.919\,\mathrm{i}\times 10^{-3}$	$3.919\,\mathrm{i}\times 10^{-4}$
	0.05	0.5312 i	$5.301\,\mathrm{i}\times 10^{-2}$	$5.300\,\mathrm{i}\times 10^{-3}$	$5.300\,\mathrm{i}\times 10^{-4}$
	0.01	0.8530 i	$8.518\,\mathrm{i}\times 10^{-2}$	$8.518\,\mathrm{i}\times 10^{-3}$	$8.518\,\mathrm{i}\times 10^{-4}$

Table 9.1. *Values of ϵ [9.65] for a Jacobi theta forcing [9.51] on a bridge with string separation l and varying incident wavelengths λ and forcing half-widths h*

Table 9.1 shows, as previously discussed, that a decrease in h/l causes ϵ to increase, as δ_h becomes closer to the exact delta function. However, as indicated in the table, for small values of l/λ, the value of ϵ is small even for reasonably "thin" strings. This shows that, when using a long wavelength assumption where $l \ll \lambda$, the lower-order terms will be negligible regardless of the string thickness and matching at the boundary. Therefore, the homogenization technique developed earlier reproduces the results of the exact solution, while also being much simpler to solve.

9.6. Net bridge

We next introduce a net of strings, as described in Martinsson and Movchan (2003) as a "membrane-like lattice". This square lattice consists of an infinite array of strings, which we will insert into the void between membranes, as shown in Figure 9.7. While this structure allows for x-direction wave propagation in the bridge, the size of the "holes" are much smaller, meaning that the maximum size of inclusion that can be concealed by the bridge is correspondingly much smaller.

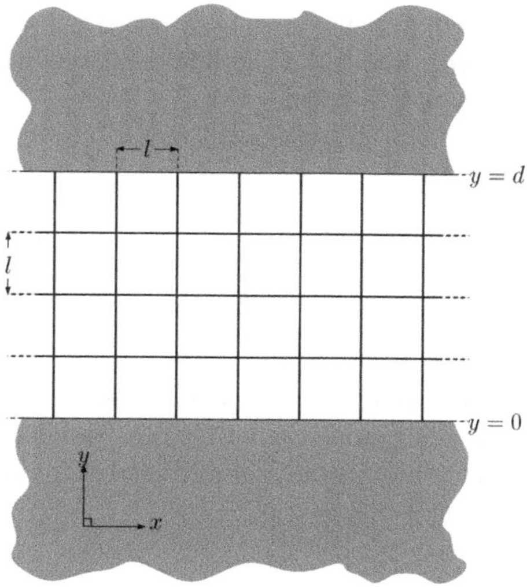

Figure 9.7. *Schematic of an infinite elastic membrane with a square lattice of strings with separation l, bridging a void between $y = 0$ and $y = d$*

Due to the connections between the strings along the x axis, the wave propagation can be 2D in this structure. The dispersion of waves in such a net has already been studied, and for a harmonic wave of amplitude A_n being excited by a constant angular frequency ω, the out-of-plane displacement waves in the net have the form:

$$u_n = A_n e^{i(k_x x + k_y y - \omega t)}, \tag{9.74}$$

where k_x and k_y are the wavenumbers in the x and y directions for waves in the net and in the notation of this chapter:

$$\cos(k_x l) + \cos(k_y l) = 2\cos\left(\frac{\omega l}{c_s}\right). \tag{9.75}$$

Using the same long wave assumption as before, if $\omega \ll c_s l$, then $k_x,\ k_y \ll l$ and thus, using the Taylor expansion of the cosine function yields:

$$k_x^2 + k_y^2 = \sqrt{2}\frac{\omega^2}{c_s^2}, \tag{9.76}$$

which is similar to the dispersion relation for a membrane [9.4]. We define a new wave speed c_n and wavenumber k_n as:

$$c_n = \frac{c_s}{\sqrt[4]{2}} = \frac{\omega}{k_n}. \tag{9.77}$$

For $c_m = c_n$, this net will have the same wave speed as the membrane. Following the same procedure as before yields the altered boundary conditions for forces and displacements, respectively, at $y = 0$ and $y = d$:

$$T\frac{\partial u_m}{\partial y} = Q\frac{\partial u_n}{\partial y}\delta_h(x), \qquad [9.78]$$

and

$$u_m = u_n. \qquad [9.79]$$

Since we have already made use of a long wave assumption and shown the effectiveness of the homogenization scheme for this boundary, we only consider a homogenized boundary condition for this bridge. As before, the homogenization of the distributed delta function yields:

$$T\frac{\partial u_m}{\partial y} = \frac{Q}{l}\frac{\partial u_n}{\partial y} \qquad [9.80]$$

and on assuming waves in the membrane and net, respectively, of the form

$$\begin{aligned} u_m &= A_i e^{i[k_m(x\sin\theta + y\cos\theta) - \omega t]} + A_r e^{i[k_m(x\sin\theta - y\cos\theta) - \omega t]}, \\ u_n &= A_t e^{i(k_x x + k_y y - \omega t)}, \end{aligned} \qquad [9.81]$$

for which, substitution into the boundary conditions at $y = 0$ yields:

$$A_r = \frac{1-\gamma_n}{1+\gamma_n}A_i, \qquad A_t = \frac{2}{1+\gamma_n}A_i, \qquad [9.82]$$

where in this case, the bridging function $\gamma_n(\theta)$ is given by:

$$\gamma_n(\theta) = \frac{Q}{Tl}\sqrt{1 - \left(1 - \frac{c_m^2}{c_n^2}\right)\sec^2\theta}. \qquad [9.83]$$

Similarly, taking the assumed wave forms

$$\begin{aligned} u_n &= B_i e^{i(k_x x + k_y y - \omega t)} + B_r e^{i(k_x x - k_y y - \omega t)}, \\ u_m &= B_t e^{i[k_m(x\sin\theta + y\cos\theta) - \omega t]}, \end{aligned} \qquad [9.84]$$

and substituting into the boundary $y = d$ yields:

$$B_r = \frac{\gamma_n - 1}{\gamma_n + 1}B_i, \qquad B_t = \frac{2\gamma_n}{\gamma_n + 1}B_i, \qquad [9.85]$$

for which the coefficients obtained from this system are identical to [9.25] and [9.28], with γ_θ replaced by γ_n and thus, the previous analysis also applies here, with the overall first wave transmission coefficient given by:

$$T_1(\gamma_n) = \frac{B_t}{A_i} = \frac{4\gamma_n}{(1+\gamma_n)^2}, \tag{9.86}$$

which has a maximum $T_1(\gamma_n) = 1$ at $\gamma_n = 1$. Unlike the previous string bridging problem, it is possible to ensure that perfect bridging can be obtained in this system regardless of the incident angle since, when $c_m = c_n$, there is no angular dependence on the value of γ_n. Furthermore, for this material tuning, the natural matching of tensions $Tl = Q$ gives $\gamma_n = 1$ and there is no phase change introduced, leading to a perfect bridging.

This bridging system also allows for tunable wave filtering based on the incident angle. While the previous string bridging scheme had a decrease in transmission from angles away from the maximum, there was always some transmission from every incident wave angle. However, for this bridge system, there are no propagating waves for

$$\sin\theta > \frac{c_m}{c_n} \tag{9.87}$$

so if $c_n > c_m$, there are always some waves that are not transmitted. To filter any waves with an incident angle greater than a previously defined filtering angle, θ_f, requires

$$\frac{c_m}{c_n} = \sin\theta_f \tag{9.88}$$

and to ensure a maximum at some given angle, θ_{max}, requires $\gamma_n = 1$ at this angle. Therefore, on substitution, the ratio of tensions must be given by:

$$\frac{Tl}{Q} = \sqrt{1 - \frac{\cos\theta_f}{\cos\theta_{max}}} \tag{9.89}$$

so by altering the tension and wave speed in the strings proportional to the tension and the wave speed in the membrane, any incident wave can be fully bridged and any range of incident angles greater than a specified angle can be fully filtered. There is, however, no way to filter a range of angles less than a given angle.

Furthermore, the previous discussion on the effect of successive transmissions and reflections remains valid, where in this system, if the propagation angle of the waves in the net is θ_n, the reflection phase change φ_n is given by:

$$\begin{aligned}\varphi_n &= d(k_n(\cos\theta_n + \sin\theta_n\tan\theta_n) - k_m\sin\theta\tan\theta)\\ &= \frac{\omega d}{c_m}\left(\frac{c_m}{c_n}\left(1 - \frac{c_n^2}{c_m^2}\sin^2\theta\right)^{-1/2} - \sin\theta\tan\theta\right).\end{aligned} \tag{9.90}$$

This phase change is much more involved than the previous phase changes due to the 2D wave propagation in the bridge. However, as before, when φ_n is an integer multiple of π, wave superposition leads to the overall reflected wave tending to a relative amplitude of 0 and the transmitted wave tending to a relative amplitude of 1.

However, as before, this phase change depends on both the wave frequency and the incident wave angle. Therefore, it is not possible to produce perfect transmission by internal reflections, which is either broadband or angle independent.

9.7. Concluding remarks

For the bridge system discussed above, it has been demonstrated that two different periodic arrays of strings can carry wave behavior across a void in an elastic membrane. Not only is it possible to select material parameters for the strings to perfectly transmit a wave pattern incident from a single angle, but also, if the phase of the wave is correctly tuned, internal reflections on the bridge can interfere and for an infinite number of wavefronts, either the same wave pattern or a suppressed wave can be transmitted across the void with no reflection. Importantly, all of the required material parameters are isotropic constants. This is unlike the scheme for membrane cloaking, where the cloak stiffness and density are both spatially dependent and anisotropic.

It has also been shown that a homogenization approach can be used to determine the wave transmission and reflection across the void caused by the periodic insertion of strings. Furthermore, it has been shown using a Fourier series to represent the forcing at the boundary that the transmission and reflection coefficients determined by this approach are accurate for long wavelength forcing. Thus, the homogenization is not only practical to apply and use, but also, the results are close to the exact solution to a good degree of accuracy. As a result, this homogenization can be applied to other more complicated schemes where the exact solution cannot be so easily obtained.

The proposed scheme does have limitations. While the results for the string bridge are broadband in the long wave limit, this bridge cannot ideally reproduce the wave pattern for all incident angles. Furthermore, while the net bridge produces a bridging scheme that can be both broadband and ideally reproduce the incident wave pattern for all incident angles, it strictly requires a matching of wave speeds $c_m = c_n$, which may not be possible in a physical system. The size of possible inclusions that can be concealed is also smaller than for the string bridge. While the "holes" in the string bridge have area $d \times l$, the net bridge has significantly smaller holes of area l^2.

The net bridge can also produce a filtering effect, preventing the transmission of waves with an incidence angle greater than a critical angle, determined by system parameters. This filtering effect also allows for the perfect transmission of one incident angle, which can be changed by tuning the net parameters. However, there is no way to perfectly transmit over a range of angles while filtering another. There is also no way to filter out small incident angle waves while allowing larger incident angle waves to be transmitted.

The simplicity of this bridging setup does, however, give scope for further development. The same type of metamaterial can also be proposed for other continuous media to extend the same bridging ideas to other systems. Notably, many other systems have the same Helmholtz governing equations, including the asymptotic leading order motion along the surface of a linearly elastic half plane (Kaplunov and Prikazchikov 2017; Fu *et al.* 2020), so the results obtained can potentially be extended to other systems.

9.8. Acknowledgments

JK acknowledges the support from the Russian Science Foundation (grant no. 20-11-20133). PW is grateful to Keele University ACORN funding for supporting his PhD studies.

9.9. References

Bauman, P.T., Dhia, H.B., Elkhodja, N., Oden, J.T., Prudhomme, S. (2008). On the application of the Arlequin method to the coupling of particle and continuum models. *Comput. Mech.*, 42(4), 511–530.

Bertoldi, K., Bigoni, D., Drugan, W. (2007a). Structural interfaces in linear elasticity. Part I: Nonlocality and gradient approximations. *J. Mech. Phys. Solids*, 55(1), 1–34.

Bertoldi, K., Bigoni, D., Drugan, W. (2007b). Structural interfaces in linear elasticity. Part II: Effective properties and neutrality. *J. Mech. Phys. Solids*, 55(1), 35–63.

Brûlé, S., Javelaud, E., Enoch, S., Guenneau, S. (2014). Experiments on seismic metamaterials: Molding surface waves. *Phys. Rev. Lett.*, 112(13), 133901.

Brun, M., Guenneau, S., Movchan, A.B., Bigoni, D. (2010a). Dynamics of structural interfaces: Filtering and focussing effects for elastic waves. *J. Mech. Phys. Solids*, 58(9), 1212–1224.

Brun, M., Movchan, A., Movchan, N. (2010b). Shear polarisation of elastic waves by a structured interface. *Continuum Mech. Therm.*, 22(6–8), 663–677.

Colombi, A., Roux, P., Rupin, M. (2014). Sub-wavelength energy trapping of elastic waves in a metamaterial. *J. Acoust. Soc. Am.*, 136(2), EL192–EL198.

Colquitt, D., Jones, I., Movchan, N., Movchan, A., Brun, M., McPhedran, R. (2013). Making waves round a structured cloak: Lattices, negative refraction and fringes. *Proc. Royal Soc. A*, 469, 20130218.

Colquitt, D., Craster, R., Makwana, M. (2015). High frequency homogenisation for elastic lattices. *Quart. J. Mech. Appl. Math.*, 68(2), 203–230.

Draper, H. (1874). On diffraction spectrum photography, and the determination of the wave-lengths of the ultraviolet rays. *Nature*, 9, 224–226.

Ege, N., Erbaş, B., Kaplunov, J., Wootton, P. (2018). Approximate analysis of surface wave-structure interaction. *J. Mech. Mater. Struct.*, 13(3), 297–309.

Fischer, J., Ergin, T., Wegener, M. (2011). Three-dimensional polarization-independent visible-frequency carpet invisibility cloak. *Opt. Lett.*, 36(11), 2059–2061.

Fu, Y., Kaplunov, J., Prikazchikov, D. (2020). Reduced model for the surface dynamics of a generally anisotropic elastic half-space. *Proc. Roy. Soc. A*, 476(2234), 20190590.

Glass, N. and Maradudin, A. (1981). Shear surface elastic waves on large amplitude gratings. *Electronics Letters*, 17(21), 773–774.

Haslinger, S., Movchan, N., Movchan, A., McPhedran, R. (2011). Transmission, trapping and filtering of waves in periodically constrained elastic plates. *Proc. Roy. Soc. A: Math. Phys. Eng. Sci.*, 468(2137), 76–93.

Haslinger, S., McPhedran, R., Movchan, N., Movchan, A. (2013). Structured interfaces for flexural waves – Trapped modes and transmission resonances. In *Journal of Physics: Conference Series*. IOP Publishing, Bristol.

Kadic, M., Bückmann, T., Schittny, R., Wegener, M. (2013). Metamaterials beyond electromagnetism. *Rep. Prog. Phys.*, 76(12), 126501.

Kaplunov, J. and Prikazchikov, D.A. (2017). *Asymptotic Theory for Rayleigh and Rayleigh-Type Waves*. Elsevier, Amsterdam.

Kolpakov, A.G., Andrianov, I.V., Prikazchikov, D.A. (2018). Asymptotic strategy for matching homogenized structures. conductivity problem. *Quart. J. Mech. Appl. Math.*, 71(4), 519–535.

Li, J. and Chan, C.T. (2004). Double-negative acoustic metamaterial. *Phys. Rev. E*, 70(5), 055602.

Liu, Z., Zhang, X., Mao, Y., Zhu, Y., Yang, Z., Chan, C.T., Sheng, P. (2000). Locally resonant sonic materials. *Science*, 289(5485), 1734–1736.

Martinsson, P. and Movchan, A. (2003). Vibrations of lattice structures and phononic band gaps. *Quart. J. Mech. Appl. Math.*, 56(1), 45–64.

Milton, G.W. and Willis, J.R. (2007). On modifications of Newton's second law and linear continuum elastodynamics. *Proc. Royal Soc. A*, 463(2079), 855–880.

Movchan, A. and Guenneau, S. (2004). Split-ring resonators and localized modes. *Phys. Rev. B*, 70(12), 125116.

Pendry, J.B., Holden, A.J., Robbins, D., Stewart, W. (1999). Magnetism from conductors and enhanced nonlinear phenomena. *IEEE Trans. Microwave Theory Tech.*, 47(11), 2075–2084.

Porter, R. (2018). Plate arrays as a water-wave metamaterial. *Proc. 33rd Int. Workshop on Water Waves and Floating Bodies*, pp. 165–169.

Ramsay, J.F. (1958). Microwave antenna and waveguide techniques before 1900. *Proc. IRE*, 46(2), 405–415.

Slepyan, L. (1967). The strain wave in a bar with vibration-isolated masses. *Mech. Solids*, 2, 57–64.

Tuck, E. (1971). Transmission of water waves through small apertures. *J. Fluid Mech.*, 49(1), 65–74.

Whittaker, E.T. and Watson, G.N. (1996). *A Course of Modern Analysis*. Cambridge University Press, Cambridge.

Williams, E.G., Roux, P., Rupin, M., Kuperman, W. (2015). Theory of multiresonant metamaterials for A_0 Lamb waves. *Phys. Rev. B*, 91(10), 104307.

Williamson, R.C. and Smith, H.I. (1973). The use of surface-elastic-wave reflection gratings in large time-bandwidth pulse-compression filters. *IEEE Trans. Microw. Theory Tech.*, 21(4), 195–205.

Wootton, P.T., Kaplunov, J., Prikazchikov, D. (2020). A second-order asymptotic model for Rayleigh waves on a linearly elastic half plane. *IMA J. Appl. Math.*, 85(1), 113–131. Available at: https://doi.org/10.1093/imamat/hxz037.

Xiao, Y., Wen, J., Wen, X. (2012). Flexural wave band gaps in locally resonant thin plates with periodically attached spring–mass resonators. *J. Phys. D*, 45(19), 195401.

Zhou, X., Liu, X., Hu, G. (2012). Elastic metamaterials with local resonances: An overview. *Theor. Appl. Mech. Lett.*, 2(4).

10

Dynamic Soil Stiffness of Foundations Supported by Layered Half-Space

In the soil–structure interaction (SSI) analysis, dynamic soil stiffness, which is an essential element in analyzing the SSI effect, characterizes the response of the soil base under the foundation. A semi-analytical method is developed to generate dynamic soil stiffness for rigid or flexible foundations supported by layered half-space. The flexibility function under a point load is derived analytically, in terms of the soil properties, based on the wave propagation function in the wave number domain. Green's influence function is then formulated under three-dimensional loads using the Fourier series and Bessel transform pair in the frequency domain. In order to model the unbounded soil, the boundary element method (BEM) is used at the foundation surface to determine the dynamic soil stiffness. By applying the BEM, the proposed method is applicable to foundations of arbitrary shapes. Thus, for flexible foundations with $6m$ degrees of freedom, the $6m\times 6m$ dynamic soil stiffness matrix can be determined. Because each soil layer is treated as a single element and only the soil–foundation interface is discretized into subdisks, only a limited number of elements are involved in the analysis, which guarantees the efficiency of the proposed method. Numerical examples are performed for different foundation shapes to demonstrate the accuracy of the proposed method.

10.1. Introduction

The soil–structure interaction (SSI) effect plays an important role in the seismic analysis of large structures, such as nuclear power plants. An efficient approach for SSI analysis is the substructure method, which considers the structure and soil

Chapter written by Yang ZHOU and Wei-Chau XIE.

separately. In the substructure method, the structure is considered to be supported by a generalized spring. This generalized spring is determined by both the structure and soil properties (Jiang 2016). The seismic input is the three-dimensional free-field motion at the soil base, which is applied at the end of the generalized spring. Based on structural modal information and soil stiffness, a transfer function is derived in the frequency domain to convert the free-field motion into foundation level input response spectra (FLIRS). Therefore, the soil–structure system under the excitation of free-field motion is considered as a fixed-base structure under the modified free-field motion FLIRS. Since the soil stiffness varies with the excitation frequency and the SSI analysis is performed in the frequency domain, it is necessary to generate the frequency-dependent stiffness of the soil deposit that determines the response of the soil base under the foundations.

A flexible foundation can be modeled using m nodes. If the foundation is rigid, m is equal to 1. For each node, there are three translational and three rotational degrees of freedom. Hence, the dynamic soil stiffness is a $6m\times 6m$ frequency-dependent matrix. This matrix can be decomposed into $m\times m$ sub-matrices. The sub-matrix $\mathbf{S}_{ij}$ are the amplitudes of the six-dimensional reaction forces at node i when the six-dimensional displacement at node j is equal to $e^{i\omega t}$, where ω is the excitation frequency and t is time.

Since the 1960s, numerous studies have focused on the generation of dynamic soil stiffness. An integral equation approach was first proposed to develop the analytical solutions of the vibration of rigid circular foundations supported by elastic half-space (Luco and Westmann 1971). This method was extended to half-space with multiple horizontal layers in Luco (1974). The dynamic stiffness for embedment foundation with arbitrary shapes was generated for horizontally layered half-space by a semi-analytical method (Luco and Apsel 1983).

Rizzo (1967) developed the boundary element method (BEM) for boundary value problems of elastostatics. In this approach, only the domain boundaries are discretized and the number of nodes is reduced. The BEM is well adapted to simulate unbounded domains. Based on the BEM, the dynamic response of the three-dimensional foundation–soil–foundation interaction on the layered soil site was studied by Karabalis and Mohammadi (1991). Estorff and Kausel (1989) presented a time domain formulation of the BEM. This approach can also be formulated in the frequency domain (Wolf and Darbre 1984; Wolf 1985).

Research into dynamic soil stiffness has mainly focused on rigid foundations. The objective of this study is to generate dynamic soil stiffness for both rigid and flexible foundations by a semi-analytical method, based on Green's function by Wolf (1985). The soil is modeled as a layered half-space characterized by soil properties, including layer thickness, density, shear wave velocity, Poisson's ratio and damping ratio. The soil properties are assumed to be linear. Starting from the wave propagation function, the relevant displacements under three-dimensional point loads are derived in the wave number domain rigorously. The constitutive relation is established in the frequency domain using the Fourier series and Bessel transform pair. The BEM is then used to determine the total stiffness matrix of the soil base under rigid or flexible foundations of arbitrary shapes. For a horizontally layered soil site, a procedure for generating frequency-dependent dynamic soil stiffness is summarized as follows:

1) Given the soil properties of each layer, the dynamic stiffness matrix is established for each layer under point loads.

2) The dynamic stiffness matrix is assembled to calculate the flexibility function.

3) The foundation is discretized into small subdisks, i.e. circular elements. Through considering the loaded subdisk one by one, Green's function is obtained based on the flexibility function.

4) The BEM is applied to generate frequency-dependent dynamic soil stiffness.

10.2. Generation of dynamic soil stiffness

10.2.1. *Dynamic stiffness matrix under point loads*

The analytical dynamic stiffness matrix of a single soil layer derived by Wolf (1985) is used in this research. Since the directions of the wave propagation are always assumed to be in a vertical plane, for example the x–z plane or the y–z plane. The P-wave and S-wave in the x–z plane are considered. During the propagation of waves, the displacement in the frequency domain can be expressed as

$$u(x,z) = l_x A_{\mathrm{P}} \exp\left[\frac{\mathrm{i}\omega}{c_{\mathrm{P}}^*}(-l_x x - l_z z)\right] + l_x B_{\mathrm{P}} \exp\left[\frac{\mathrm{i}\omega}{c_{\mathrm{P}}^*}(-l_x x + l_z z)\right]$$
$$+ m_z A_{\mathrm{SV}} \exp\left[\frac{\mathrm{i}\omega}{c_{\mathrm{S}}^*}(-m_x x - m_z z)\right] + m_z B_{\mathrm{SV}} \exp\left[\frac{\mathrm{i}\omega}{c_{\mathrm{S}}^*}(-m_x x + m_z z)\right], \quad [10.1]$$

$$v(x,z) = A_{\mathrm{SH}} \exp\left[\frac{\mathrm{i}\omega}{c_{\mathrm{S}}^*}(-m_x x - m_z z)\right] + B_{\mathrm{SH}} \exp\left[\frac{\mathrm{i}\omega}{c_{\mathrm{S}}^*}(-m_x x + m_z z)\right], \quad [10.2]$$

$$w(x,z) = l_z A_{\mathrm{P}} \exp\left[\frac{\mathrm{i}\omega}{c_{\mathrm{P}}^*}(-l_x x - l_z z)\right] + l_z B_{\mathrm{P}} \exp\left[\frac{\mathrm{i}\omega}{c_{\mathrm{P}}^*}(-l_x x + l_z z)\right]$$
$$+ m_x A_{\mathrm{SV}} \exp\left[\frac{\mathrm{i}\omega}{c_{\mathrm{S}}^*}(-m_x x - m_z z)\right] + m_x B_{\mathrm{SV}} \exp\left[\frac{\mathrm{i}\omega}{c_{\mathrm{S}}^*}(-m_x x + m_z z)\right], \quad [10.3]$$

where A and B are the amplitudes of the incoming and outgoing waves, respectively. The subscripts P, SH and SV represent the P-wave, SH-wave and SV-wave, respectively. The direction of the P-wave is the same as the propagation direction, and x and z are the coordinates of any point. l_x and l_z, m_x and m_z are the directional cosines of the directions of wave propagation of the P-wave and S-wave, respectively. We define the complex P-wave velocity $c_{\mathrm{P}}^* = c_{\mathrm{P}}\sqrt{(1+2\mathrm{i}\zeta_{\mathrm{P}})}$ and the complex S-wave velocity $c_{\mathrm{S}}^* = c_{\mathrm{S}}\sqrt{(1+2\mathrm{i}\zeta_{\mathrm{S}})}$, where c_{P} and c_{S} are the P-wave and S-wave velocities, respectively, and ζ_{P} and ζ_{S} are the damping ratios for the P-wave and S-wave, respectively.

A horizontal layer with thickness d, density ρ and complex shear modulus $G^* = G(1+2\mathrm{i}\zeta_{\mathrm{S}})$ is studied. The dynamic stiffness of this layer under three-dimensional point loads with frequency ω is determined. The displacements u, v and w, stresses τ_{xz}, τ_{yz} and σ_z, and external forces P, Q and R are shown in Figure 10.1. The origin is located at the top of the layer and the z axis is pointing downwards.

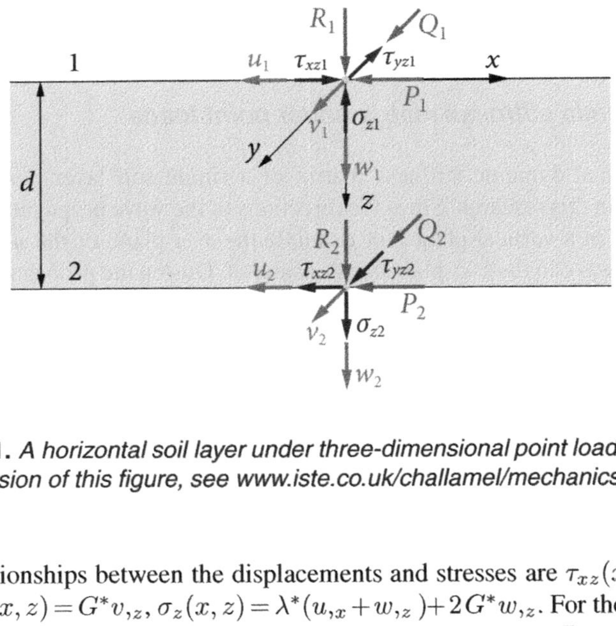

Figure 10.1. *A horizontal soil layer under three-dimensional point loads. For a color version of this figure, see www.iste.co.uk/challamel/mechanics2.zip*

The relationships between the displacements and stresses are $\tau_{xz}(x,z) = G^*(u_{,z} + w_{,x})$, $\tau_{yz}(x,z) = G^* v_{,z}$, $\sigma_z(x,z) = \lambda^*(u_{,x} + w_{,z}) + 2G^* w_{,z}$. For the top surface 1 and the bottom surface 2, the external loads are denoted as $P_1 = -\tau_{xz1}$, $Q_1 = -\tau_{yz1}$, $R_1 = -\sigma_{z1}$, $P_2 = \tau_{xz2}$, $Q_2 = \tau_{yz2}$ and $R_2 = \sigma_{z2}$. Thus, all displacements and external loads are the functions of amplitudes A_{P}, B_{P}, A_{SH}, B_{SH}, A_{SV} and B_{SV}. By eliminating all A and B, the external loads, $P_1, P_2, Q_1, Q_2, R_1, R_2$, can be expressed by displacements, $u_1, u_2, v_1, v_2, w_1, w_2$. Since the out-of-plane motion v and the in-plane motions u and w are decoupled, the out-of-plane motion and in-plane motions can be studied separately.

For the out-of-plane motion, introducing the wave number $k = \omega m_x / c_S^*$ and parameter $\alpha = \sqrt{m_x^{-2} - 1}$ leads to the dynamic stiffness matrix $\mathbf{K}_{\text{out}}$ of the out-of-plane motion in a single soil layer

$$\begin{Bmatrix} Q_1 \\ Q_2 \end{Bmatrix} = \mathbf{K}_{\text{out}} \begin{Bmatrix} v_1 \\ v_2 \end{Bmatrix}, \quad \mathbf{K}_{\text{out}} = \frac{\alpha k G^*}{\sin \alpha k d} \begin{bmatrix} \cos \alpha k d & -1 \\ -1 & \cos \alpha k d \end{bmatrix}, \qquad [10.4]$$

For in-plane motions, letting $l_x / c_P^* = m_x / c_S^*$, which allows the analysis to concentrate on the variation with z, and introducing $\beta = \sqrt{l_x^{-2} - 1}$ results in

$$\{P_1, \mathrm{i}R_1, P_2, \mathrm{i}R_2\}^T = \mathbf{K}_{\text{in}} \{u_1, \mathrm{i}w_1, u_2, \mathrm{i}w_2\}^T, \quad \mathbf{K}_{\text{in}} = (1 + \alpha^2 k G^*) D^{-1} \hat{\mathbf{K}}_{\text{in}}, \qquad [10.5]$$

in which $\mathbf{K}_{\text{in}}$ is the dynamic stiffness matrix of the in-plane motion in a single soil layer, i is added before R_1, R_2, w_1 and w_2 to keep $\mathbf{K}_{\text{in}}$ symmetric, and the elements are

$$[\hat{\mathbf{K}}_{\text{in}}]_{11} = [\hat{\mathbf{K}}_{\text{in}}]_{33} = \alpha^{-1} \cos \beta k d \sin \alpha k d + \beta \sin \beta k d \cos \alpha k d$$

$$[\hat{\mathbf{K}}_{\text{in}}]_{22} = [\hat{\mathbf{K}}_{\text{in}}]_{44} = \beta^{-1} \sin \beta k d \cos \alpha k d + \alpha \cos \beta k d \sin \alpha k d$$

$$\begin{aligned} [\hat{\mathbf{K}}_{\text{in}}]_{12} &= -[\hat{\mathbf{K}}_{\text{in}}]_{21} = -[\hat{\mathbf{K}}_{\text{in}}]_{34} = [\hat{\mathbf{K}}_{\text{in}}]_{43} \\ &= \mathrm{i}\frac{3 - \alpha^2}{1 + \alpha^2}(1 - \cos \beta k d \cos \alpha k d) + \mathrm{i}\frac{1 + 2\alpha^2\beta^2 - \alpha^2}{\alpha\beta(1 + \alpha^2)} \sin \beta k d \sin \alpha k d \end{aligned}$$

$$[\hat{\mathbf{K}}_{\text{in}}]_{13} = [\hat{\mathbf{K}}_{\text{in}}]_{31} = -\beta \sin \beta k d - \alpha^{-1} \sin \alpha k d$$

$$[\hat{\mathbf{K}}_{\text{in}}]_{14} = -[\hat{\mathbf{K}}_{\text{in}}]_{41} = [\hat{\mathbf{K}}_{\text{in}}]_{23} = -[\hat{\mathbf{K}}_{\text{in}}]_{32} = \mathrm{i} \cos \beta k d - \mathrm{i} \cos \alpha k d$$

$$[\hat{\mathbf{K}}_{\text{in}}]_{24} = [\hat{\mathbf{K}}_{\text{in}}]_{42} = -\beta^{-1} \sin \beta k d - \alpha \sin \alpha k d$$

$$D = 2(1 - \cos \beta k d \cos \alpha k d) + [\alpha\beta + (\alpha\beta)^{-1} \sin \beta k d \sin \alpha k d].$$

For half-space, which means that d approaches infinity, there are only waves propagating into it, and no waves are coming from half-space. Thus, the amplitudes of incoming waves A in equations [10.1]–[10.3] are equal to 0, and P_2, Q_2, R_2, u_2, v_2 and w_2 do not exist. Equations [10.4] and [10.5] become

$$Q_1 = \mathbf{K}_{\text{out}}^{\text{HF}} v_1, \quad \mathbf{K}_{\text{out}}^{\text{HF}} = [\mathrm{i}\alpha k G^*], \qquad [10.6]$$

$$\begin{Bmatrix} P_1 \\ \mathrm{i}R_1 \end{Bmatrix} = \mathbf{K}_{\text{in}}^{\text{HF}} \begin{Bmatrix} u_1 \\ \mathrm{i}w_1 \end{Bmatrix}, \quad \mathbf{K}_{\text{in}}^{\text{HF}} = \frac{kG^*}{1 + \alpha\beta} \begin{bmatrix} \mathrm{i}\beta(1 + \alpha^2) & 1 + 2\alpha\beta - \alpha^2 \\ 1 + 2\alpha\beta - \alpha^2 & \mathrm{i}\alpha(1 + \alpha^2) \end{bmatrix}. \qquad [10.7]$$

where the superscript HF represents half-space.

10.2.2. *Formulation of the flexibility function*

A site with n horizontal layers on a half-space is considered. Three-dimensional forces with frequency ω are applied at the origin, as shown in Figure 10.2.

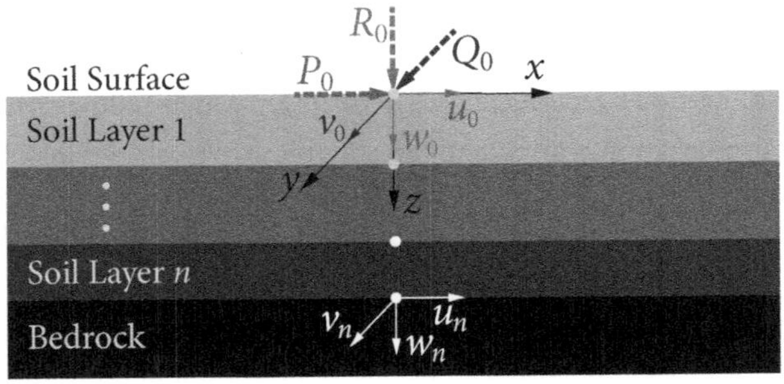

Figure 10.2. *Layered half-space under three-dimensional forces. For a color version of this figure, see www.iste.co.uk/challamel/mechanics2.zip*

10.2.2.1. *Flexibility function of the out-of-plane motion*

From equation [10.4], the equilibrium equations for the ith layer are expressed as

$$\{Q_{1,i}, Q_{2,i}\}^T = \mathbf{K}_{\text{out},i} \{v_{1,i}, v_{2,i}\}^T, \quad i = 1, 2, \cdots, n-1. \qquad [10.8]$$

When i is equal to n, i.e. the top of the bedrock half-space, the equilibrium equation is $Q_{1,n} = \mathbf{K}_{\text{out},n}\, v_{1,n}$. Because all the forces, excluding the external forces applied at the surface, are internal forces, this means that the forces at any interface follow $Q_{2,i} + Q_{1,i+1} = 0$. Then, the total equilibrium equations for the out-of-plane motion is obtained by assembling the equilibrium equations of each layer as $\mathbf{L}_{\text{out}} = \mathbf{S}_{\text{out}}\boldsymbol{\Delta}_{\text{out}}$, where $\mathbf{L}_{\text{out}} = \{Q_1, Q_2, \cdots, Q_n\}^T = \{Q_1, 0, \cdots, 0\}^T$ is the vector of external loads, $\boldsymbol{\Delta}_{\text{out}} = \{v_1, v_2, \cdots, v_n\}^T$ is the vector of displacements in the y direction and $\mathbf{S}_{\text{out}}$ is an $n \times n$ matrix assembled by $\mathbf{K}_{\text{out},i}$ and $\mathbf{K}_{\text{out},n}$. For two adjacent layers, $\mathbf{K}_{\text{out},i}$ are partly overlapped. Partitioning the equilibrium equation as

$$\begin{Bmatrix} \mathbf{Q}_1 \\ \mathbf{Q}_{2\sim n} \end{Bmatrix} = \begin{bmatrix} \mathbf{S}_{\text{out},1,1} & \mathbf{S}_{\text{out},1,2\sim n} \\ \mathbf{S}_{\text{out},2\sim n,1} & \mathbf{S}_{\text{out},2\sim n,2\sim n} \end{bmatrix} \begin{Bmatrix} \mathbf{v}_1 \\ \mathbf{v}_{2\sim n} \end{Bmatrix}$$

and eliminating $\mathbf{v}_{2\sim n}$ gives $\mathbf{v}_1 = \mathbf{F}_{\text{out}}\mathbf{Q}_1$, where $\mathbf{F}_{\text{out}}$ is the flexibility function of the out-of-plane motion given by

$$\mathbf{F}_{\text{out}} = \mathbf{F}_{vv} = (\mathbf{S}_{\text{out},1,1} - \mathbf{S}_{\text{out},1,2\sim n}\mathbf{S}^{-1}_{\text{out},2\sim n,2\sim n}\mathbf{S}_{\text{out},2\sim n,1})^{-1}. \qquad [10.9]$$

10.2.2.2. *Flexibility function of the in-plane motion*

Similar to the out-of-plane case, the total equilibrium equations for in-plane motions are obtained by assembling the equilibrium equations of each layer as $\mathbf{L}_{\text{in}} = \mathbf{S}_{\text{in}}\boldsymbol{\Delta}_{\text{in}}$, where $\mathbf{L}_{\text{in}} = \{P_1, R_1; 0, 0; \cdots; 0, 0\}^T$ and $\boldsymbol{\Delta}_{\text{in}} = \{u_1, w_1; u_2, w_2; \cdots; u_n, w_n\}^T$ are the $2n\times1$ vectors of external loads and displacements in the x and z directions, respectively, and $\mathbf{S}_{\text{in}}$ is a $2n\times2n$ matrix assembled by $\mathbf{K}^i_{\text{in}}$ and $\mathbf{K}^n_{\text{in}}$. For two adjacent layers, $\mathbf{K}^i_{\text{in}}$ are partly overlapped. The surface displacements and external loads can be expressed as

$$\{u_1, w_1\}^T = \mathbf{F}_{\text{in}}\{P_1, R_1\}^T, \qquad [10.10]$$

in which $\mathbf{F}_{\text{in}}$ is the flexibility function of the in-plane motion

$$\mathbf{F}_{\text{in}} = \begin{bmatrix} F_{uu} & F_{uw} \\ F_{wu} & F_{ww} \end{bmatrix} = (\mathbf{S}_{\text{in},1\sim2,1\sim2} - \mathbf{S}_{\text{in},1\sim2,3\sim2n}\mathbf{S}^{-1}_{\text{in},3\sim2n,3\sim2n}\mathbf{S}_{\text{in},3\sim2n,1\sim2})^{-1}. \qquad [10.11]$$

10.2.3. *Formulation of Green's influence function*

Based on the flexibility function, Green's influence function is formulated to obtain the displacement of any points when a small circular disk is loaded (Wolf 1985). In cylindrical coordinates, the load and displacement can be expanded in both the frequency domain and the wave number domain by the Bessel transform pair (Kausel and Roësset 1981)

$$\bar{\mathbf{f}}(k,n) = a_n \int_{r=0}^{\infty} r\,\mathbf{C}_n(kr) \int_{\theta=0}^{\pi} \mathbf{D}_{n\theta}\,\mathbf{f}(r,\theta)\,\mathrm{d}\theta\,\mathrm{d}r, \qquad [10.12]$$

$$\mathbf{f}(r,\theta) = \sum_{r=0}^{\infty} \mathbf{D}_{n\theta} \int_{k=0}^{\infty} k\,\mathbf{C}_n(kr)\,\bar{\mathbf{f}}(k,n)\,\mathrm{d}k, \qquad [10.13]$$

where k is the wave number and n is the index of the Fourier series. a_n is equal to $1/2\pi$ $(n=0)$, or $1/\pi$ $(n\neq0)$. $\mathbf{f}(r,\theta)$ and $\bar{\mathbf{f}}(k,n)$ are the values in the frequency domain and the wave number domain, respectively; they are 3×1 vectors, and the three elements are the values in the two horizontal directions and vertical direction. r and θ are the coordinates in cylindrical coordinates. $\mathbf{C}_n(kr)$ is given by

$$\mathbf{C}_n(kr) = \begin{bmatrix} \frac{1}{k}J_n(kr)_{,r} & \frac{n}{kr}J_n(kr) & 0 \\ \frac{n}{kr}J_n(kr) & \frac{1}{k}J_n(kr)_{,r} & 0 \\ 0 & 0 & -J_n(kr) \end{bmatrix}, \qquad [10.14]$$

in which $J_n(kr)$ is the Bessel function of order n of the first kind. $\mathbf{D}_{n\theta}$ is given by

$$\mathbf{D}_{n\theta} = \begin{bmatrix} \cos n\theta & 0 & 0 \\ 0 & -\sin n\theta & 0 \\ 0 & 0 & \cos n\theta \end{bmatrix} \quad \text{or} \quad \begin{bmatrix} \sin n\theta & 0 & 0 \\ 0 & \cos n\theta & 0 \\ 0 & 0 & \sin n\theta \end{bmatrix}. \qquad [10.15]$$

for the symmetric case and the antisymmetric case, respectively.

10.2.3.1. *Green's influence function under a vertical load*

For a vertical uniform load r_0 in the z-direction, p_z can be considered as symmetric and expressed by the Fourier series with the index of 0 as $p_z(r,\theta) = r_0$. According to equation [10.12], p_z can be transformed in the k domain as

$$p_z(k) = \frac{1}{2\pi}\int_{r=0}^{a} r\left[-J_0(kr)\right]\int_{\theta=0}^{2\pi} r_0 \cos\theta \, \mathrm{d}\theta \, \mathrm{d}r = -\frac{a}{k}J_1(ka)r_0. \qquad [10.16]$$

Then, the displacements u in the radial direction and w in the vertical direction, in the frequency domain, are given as, based on equation [10.13],

$$\begin{Bmatrix} u(r,\theta) \\ w(r,\theta) \end{Bmatrix} = \begin{bmatrix} \cos\theta & 0 \\ 0 & \cos\theta \end{bmatrix} \int_{k=0}^{\infty} k \begin{bmatrix} \frac{1}{r}J_0(kr)_{,r} & 0 \\ 0 & -J_0(kr) \end{bmatrix} \begin{Bmatrix} u(k) \\ w(k) \end{Bmatrix} \mathrm{d}k. \qquad [10.17]$$

Substituting the flexibility functions, [10.9] and [10.11], and equation [10.16] into equation [10.17], and applying the identity of Bessel functions leads to

$$\begin{Bmatrix} u(r,\theta) \\ w(r,\theta) \end{Bmatrix} = \begin{Bmatrix} G_{uw} \\ G_{ww} \end{Bmatrix} r_0 = a r_0 \int_{k=0}^{\infty} J_1(ka) \begin{Bmatrix} F_{uw}(k)J_1(kr) \\ F_{ww}(k)J_0(kr) \end{Bmatrix} \mathrm{d}k, \qquad [10.18]$$

where G_{uw} and G_{ww} are Green's function in the radial and vertical directions under the vertical load.

10.2.3.2. *Green's influence function under a horizontal load*

For a horizontal uniform load p_0 in the x-direction, it can be considered as symmetric and expressed by the Fourier series with the index of 1 as $p_r(r,\theta) = p_0\cos\theta$, $p_\theta(r,\theta) = -p_0\sin\theta$. From equation [10.12], they can be expressed in the k domain as

$$\begin{Bmatrix} p_r(k) \\ p_\theta(k) \end{Bmatrix} = \frac{p_0}{k}\int_{r=0}^{a} \begin{bmatrix} rJ_1(kr)_{,r} + J_1(kr) \\ rJ_1(kr)_{,r} + J_1(kr) \end{bmatrix} \mathrm{d}r. \qquad [10.19]$$

Using identities of Bessel functions, equation [10.19] can be rewritten as

$$p_r(k) = p_\theta(k) = \frac{p_0}{k}\int_{r=0}^{a} \left(rJ_1(kr)_{,r} + J_1(kr)\right)\mathrm{d}r = \frac{p_0 a}{k}J_1(ka). \qquad [10.20]$$

Green's function in the radial, angular and vertical directions G_{uu}, G_{vu} and G_{wu} under the horizontal load in the x-direction are obtained as

$$\{G_{uu}, G_{vu}, G_{wu}\}^T = a\,\mathbf{D}_\theta \int_{k=0}^{\infty} J_1(ka)\,\mathbf{J}\,\{F_{uu}(k), F_{vv}(k), F_{wu}(k)\}^T\, \mathrm{d}k, \quad [10.21]$$

$$\mathbf{J} = \begin{bmatrix} \frac{1}{2}J_0(kr) - \frac{1}{2}J_2(kr) & \frac{1}{2}J_0(kr) + \frac{1}{2}J_2(kr) & 0 \\ \frac{1}{2}J_0(kr) + \frac{1}{2}J_2(kr) & \frac{1}{2}J_0(kr) - \frac{1}{2}J_2(kr) & 0 \\ 0 & 0 & -J_1(kr) \end{bmatrix}.$$

The horizontal load q_0 in the y-direction can be considered as an antisymmetric load in the cylindrical coordinates, and $\mathbf{D}_\theta$ for the antisymmetric case in [10.15] is used.

10.2.3.3. *Green's influence function under three-dimensional loads*

When three-dimensional loads, i.e. p_0, q_0 and r_0, are considered simultaneously, the displacements at any point, for example, a point i, in the cylindrical coordinates are

$$\begin{Bmatrix} u_i(r,\theta) \\ v_i(r,\theta) \\ w_i(r,\theta) \end{Bmatrix} = \mathbf{G}_i^{\mathrm{c}} \begin{Bmatrix} p_0 \\ q_0 \\ r_0 \end{Bmatrix}, \quad \mathbf{G}_i^{\mathrm{c}} = \begin{bmatrix} G_{uu,i} & G_{uv,i} & G_{uw,i} \\ G_{vu,i} & G_{vv,i} & 0 \\ G_{wu,i} & G_{wv,i} & G_{ww,i} \end{bmatrix}, \quad [10.22]$$

where the superscript c represents cylindrical coordinates. The displacements in the x, y and z directions in Cartesian coordinates, $\delta_{1,i}$, $\delta_{2,i}$ and $\delta_{3,i}$, can be expressed as

$$\begin{Bmatrix} \delta_{1,i}(x,y) \\ \delta_{2,i}(x,y) \\ \delta_{3,i}(x,y) \end{Bmatrix} = \mathbf{T}_i \begin{Bmatrix} u_i(r,\theta) \\ v_i(r,\theta) \\ w_i(r,\theta) \end{Bmatrix} = \mathbf{T}_i\,\mathbf{G}_i^{\mathrm{c}} \begin{Bmatrix} p_0 \\ q_0 \\ r_0 \end{Bmatrix}, \quad \mathbf{T}_i = \begin{bmatrix} \cos\theta_i & -\sin\theta_i & 0 \\ \sin\theta_i & \cos\theta_i & 0 \\ 0 & 0 & 1 \end{bmatrix}, \quad [10.23]$$

in which θ^i is the coordinate of point i in the cylindrical coordinates with the origin at the center of the disk. Green's function in the Cartesian coordinates is given by

$$\mathbf{G}_i = \mathbf{T}_i \mathbf{G}_i^{\mathrm{c}}. \quad [10.24]$$

10.2.4. *Total dynamic soil stiffness by the boundary element method*

The total dynamic stiffness of the soil deposit can be obtained based on the boundary conditions. Since Green's influence function is available, only the interface between the foundation and the soil base needs to be considered according to the BEM. The interface, rather than the entire soil space, is meshed into a limited number of elements.

10.2.4.1. *Dynamic soil stiffness for a rigid foundation*

A rigid foundation can be simplified as one node with 6 degrees of freedom, i.e. three translational components and three rotational components. A rigid foundation with arbitrary shape supported by a horizontally layered half-space is shown in Figure 10.3. The foundation is discretized into n uniformly distributed subdisks of radius a. The total area of subdisks is approximately the same as the original foundation. a is small enough that the displacement within each subdisk can be considered the same. Since 6 points in a wavelength are needed to address each wave (Wolf 1985), a should be no more than $c_S/6f_{max}$, in which f_{max} is the maximum frequency under consideration.

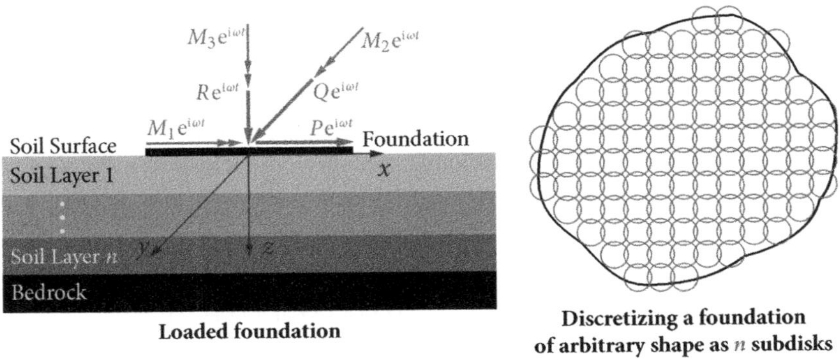

Figure 10.3. *Foundation-site model and discretization of foundation. For a color version of this figure, see www.iste.co.uk/challamel/mechanics2.zip*

The six-dimensional external loads $\mathbf{P}_R$, where the subscript R represents rigid foundations, applied on the foundation and the displacements $\mathbf{u}_R$ caused by the load vector are

$$\mathbf{P}_R = \{P, Q, R, M_1, M_2, M_3\}^T, \quad \mathbf{u}_R = \{u_1, u_2, u_3, u_4, u_5, u_6\}^T, \qquad [10.25]$$

where P, Q and R are the translational loads and M_1, M_2 and M_3 are the moments in the x, y and z directions, respectively. u_1, u_2 and u_3 are the translational displacements and u_4, u_5 and u_6 are the rotations in the x, y and z directions, respectively.

Consider the ith subdisk, $i = 1, 2, \ldots, n$. When the three-dimensional uniform loadings p_i, q_i and r_i are applied on this disk, the displacements of the jth disk are $\{\delta_{1,ji}, \delta_{2,ji}, \delta_{3,ji}\}^T = \mathbf{G}_{ji}\{p_i, q_i, r_i\}^T$, where $\mathbf{G}_{ji}$ is a 3×3 matrix given by equation [10.24]. By loading the disks one by one, the total displacements of each disk can be obtained as $\boldsymbol{\delta} = \mathbf{G}\mathbf{p}$, where $\boldsymbol{\delta}$ and $\mathbf{p}$ are the $3n \times 1$ translational displacement vector and the load vector of all nodes, respectively, as

$$\boldsymbol{\delta} = \left\{\delta_{1,1}, \delta_{2,1}, \delta_{3,1}; \cdots; \delta_{1,n}, \delta_{2,n}, \delta_{3,n}\right\}^T, \quad \mathbf{p} = \left\{p_1, q_1, r_1; \cdots; p_n, q_n, r_n\right\}^T,$$

and $\mathbf{G}$ is the $3n \times 3n$ total Green's function matrix as $\mathbf{G} = [\mathbf{G}_{ji}]$. The relationship of $\mathbf{u}_\mathrm{R}$ and $\boldsymbol{\delta}$ is $\boldsymbol{\delta} = \mathbf{N}_\mathrm{R}\mathbf{u}_\mathrm{R}$, where $\mathbf{N}_\mathrm{R}$ is a $3n \times 6$ matrix given by

$$\mathbf{N}_\mathrm{R} = \left\{\mathbf{N}_1, \mathbf{N}_2, \cdots, \mathbf{N}_n\right\}^T, \quad \mathbf{N}_i = \begin{bmatrix} 1 & 0 & 0 & 0 & 0 & -y_i \\ 0 & 1 & 0 & 0 & 0 & x_i \\ 0 & 0 & 1 & y_i & x_i & 0 \end{bmatrix},$$

where x_i and y_i are the coordinates of the center of the ith disk. For the external loads, we have $\mathbf{P}_\mathrm{R} = \mathbf{N}_\mathrm{R}^T\mathbf{p}$. Hence, $\mathbf{P}_\mathrm{R} = \mathbf{N}_\mathrm{R}^T\mathbf{G}^{-1}\boldsymbol{\delta} = \mathbf{N}_\mathrm{R}^T\mathbf{G}^{-1}\mathbf{N}_\mathrm{R}\mathbf{u}_\mathrm{R} = \mathbf{S}_\mathrm{R}\mathbf{u}_\mathrm{R}$, where $\mathbf{S}_\mathrm{R}$ is the 6×6 dynamic stiffness matrix of the soil base. $\mathbf{S}_\mathrm{R}$ is determined by the frequency of excitation forces, the geometry of the foundation, soil properties and the thickness of each layer.

10.2.4.2. *Dynamic soil stiffness for a flexible foundation*

A flexible foundation can be simplified as m nodes with $6m$ degrees of freedom, as shown in Figure 10.4. m nodes are selected according to the structure characteristics. The whole flexible foundation is then discretized into n subdisks with radius a. n subdisks are uniformly distributed in the foundation and between m foundation nodes. The total area of subdisks is approximately the same as the original foundation.

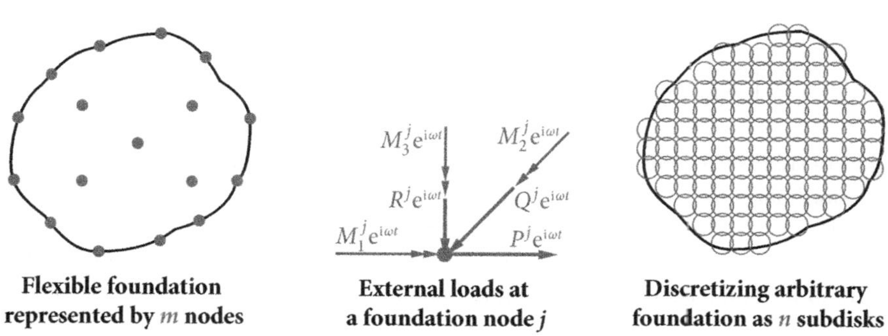

Figure 10.4. *Flexible foundation nodes and discretization of foundation. For a color version of this figure, see www.iste.co.uk/challamel/mechanics2.zip*

The external load vector $\mathbf{P}_\mathrm{F}$, the subscript F representing flexible foundations, applied on the foundation and the displacement vector $\mathbf{u}_\mathrm{F}$ caused by the load vector are $\mathbf{P}_{\mathrm{F}}=\left\{\mathbf{P}^1,\mathbf{P}^2,\cdots,\mathbf{P}^m\right\}^T$, $\mathbf{P}^j=\left\{P^j,Q^j,R^j,M_1^j,M_2^j,M_3^j\right\}^T$, $\mathbf{u}_{\mathrm{F}}=\left\{\mathbf{u}^1,\mathbf{u}^2,\cdots,\mathbf{u}^m\right\}^T$, $\mathbf{u}^j=\left\{u_1^j,u_2^j,\cdots,u_6^j\right\}^T$, where the superscript j is the foundation node number, P^j, Q^j and R^j are the translational loads and M_1^j, M_2^j and M_3^j are the moments in the x, y and z directions on the jth foundation member, respectively. u_1^j, u_2^j and u_3^j are the translational displacements and u_4^j, u_5^j and u_6^j are the rotations in the x, y and z directions on the jth foundation member, respectively.

Similar to the rigid foundation, the relationship of $\mathbf{u}_{\mathrm{F}}$ and $\boldsymbol{\delta}$ is $\boldsymbol{\delta}=\mathbf{N}_{\mathrm{F}}\,\mathbf{u}_{\mathrm{F}}$, where $\mathbf{N}_{\mathrm{F}}$ is a $3n\times 6m$ matrix. When the foundation is considered as flexible and represented by m foundation nodes, $\mathbf{N}_{\mathrm{F}}$ can be obtained from the stiffness and damping matrices of foundations. If external loads $\mathbf{P}\mathrm{e}^{\mathrm{i}\omega t}$ are applied on m foundation nodes, the displacements are governed by

$$\begin{bmatrix}\mathbf{M}_m & \mathbf{0}\\ \mathbf{0} & \mathbf{0}\end{bmatrix}\begin{Bmatrix}\ddot{\mathbf{u}}_m(t)\\ \ddot{\mathbf{u}}_n(t)\end{Bmatrix}+\begin{bmatrix}\mathbf{C}_{mm} & \mathbf{C}_{mn}\\ \mathbf{C}_{nm} & \mathbf{C}_{nn}\end{bmatrix}\begin{Bmatrix}\dot{\mathbf{u}}_m(t)\\ \dot{\mathbf{u}}_n(t)\end{Bmatrix}+\begin{bmatrix}\mathbf{K}_{mm} & \mathbf{K}_{mn}\\ \mathbf{K}_{nm} & \mathbf{K}_{nn}\end{bmatrix}\begin{Bmatrix}\mathbf{u}_m(t)\\ \mathbf{u}_n(t)\end{Bmatrix}=\begin{bmatrix}\mathbf{P}\mathrm{e}^{\mathrm{i}\omega t}\\ 0\end{bmatrix}, \tag{10.26}$$

where the subscripts m and n indicate representative foundation nodes and centers of subdisks. $\mathbf{M}$, $\mathbf{C}$, $\mathbf{K}$ are, respectively, the mass, damping and stiffness matrices. They can be extracted from the finite element model in which all m foundation nodes and n nodes are considered together and connected by finite elements. Introducing $\mathbf{u}_m(t)=\mathbf{u}\,\mathrm{e}^{\mathrm{i}\omega t}$ and $\mathbf{u}_n(t)=\boldsymbol{\delta}\,\mathrm{e}^{\mathrm{i}\omega t}$, the second block row can be written as $\boldsymbol{\delta}=\mathbf{N}_{\mathrm{F}}\mathbf{u}$, where

$$\mathbf{N}_{\mathrm{F}}=\begin{Bmatrix}\mathbf{N}_{11} & \mathbf{N}_{12} & \cdots & \mathbf{N}_{1m}\\ \mathbf{N}_{21} & \mathbf{N}_{22} & \cdots & \mathbf{N}_{2m}\\ \vdots & \vdots & \ddots & \vdots\\ \mathbf{N}_{n1} & \mathbf{N}_{2n} & \cdots & \mathbf{N}_{nm}\end{Bmatrix},\quad \mathbf{N}_{ij}=\begin{bmatrix}1\,0\,0 & 0 & 0 & -(y_i-y_j)\\ 0\,1\,0 & 0 & 0 & x_i-x_j\\ 0\,0\,1 & y_i-y_j & x_i-x_j & 0\end{bmatrix}.$$

if the ith subdisk is within the scope of the jth foundation node, where x_j and y_j are the coordinates of the jth foundation node; otherwise, $\mathbf{N}_{ij}$ is a zero matrix. Considering the external loads, we obtain $\mathbf{P}_{\mathrm{F}}=\mathbf{N}_{\mathrm{F}}^T\mathbf{p}=\mathbf{N}_{\mathrm{F}}^T\mathbf{G}^{-1}\boldsymbol{\delta}_{3n\times 1}$ $=\mathbf{N}_{\mathrm{F}}^T\mathbf{G}^{-1}\mathbf{N}_{\mathrm{F}}\mathbf{u}_{\mathrm{F}}=\mathbf{S}_{\mathrm{F}}\mathbf{u}_{\mathrm{F}}$, where $\mathbf{S}_{\mathrm{F}}$ is the frequency-dependent $6m\times 6m$ stiffness matrix of the soil base under the flexible foundation.

10.3. Numerical examples of the generation of dynamic soil stiffness

In order to verify the proposed method, four numerical examples are performed. The resultant dynamic soil stiffness results are compared. The horizontal, vertical, rocking and torsion stiffness are studied in Zhou (2020); only the horizontal and vertical stiffness are presented in this chapter.

10.3.1. *A rigid square foundation supported by a layer on half-space*

The dynamic soil stiffness of a rigid square foundation with side length of L is considered on a layer with the thickness of L resting on a half-space. Poisson's ratio for both layers is 0.33, and the damping ratios are 0.05 for the top layer and 0.03 for the half-space. The ratios of the shear wave velocities and densities are $c_{s,2}/c_{s,1} = 1.25$ and $\rho_2/\rho_1 = 1.13$. The square foundation is divided into 256 subdisks with the diameter of $L/16$, as shown in Figure 10.5.

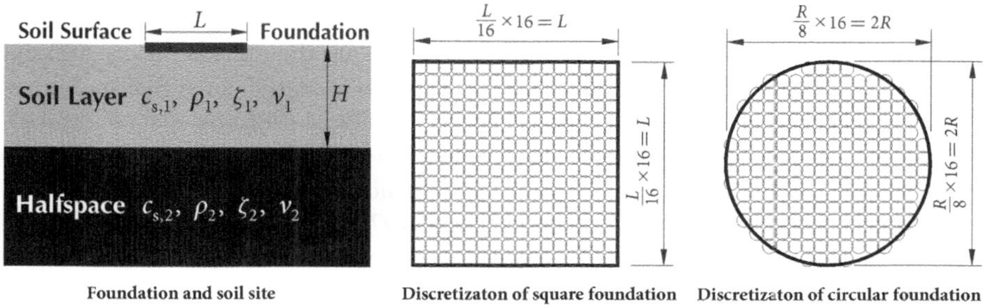

Figure 10.5. *Foundation-site models and discretization of foundations. For a color version of this figure, see www.iste.co.uk/challamel/mechanics2.zip*

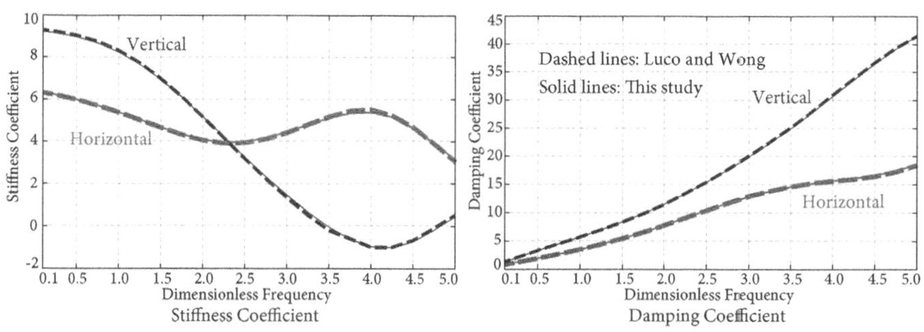

Figure 10.6. *Dynamic soil stiffness of the square foundation on a layer and half-space. For a color version of this figure, see www.iste.co.uk/challamel/mechanics2.zip*

The results are compared to those obtained by Wong and Luco (1985). A dimensionless frequency a_0 is introduced as $a_0 = \omega R/c_s$, where R is the radius or equivalent radius of foundations, and the dynamic stiffness can be expressed as $S = K_s + \mathrm{i}a_0 C_s$, where K_s is the stiffness coefficient and C_s is the damping coefficient. A comparison of dynamic soil stiffness is shown in Figure 10.6. It can be seen that the dynamic soil stiffness results of both methods agree very well.

10.3.2. *A rigid circular foundation supported by a layer on half-space*

The dynamic stiffness of a rigid circular foundation with the radius of R is considered on a layer with the thickness of R resting on a half-space. For both layers, Poisson's ratio is 0.25 and the damping ratio is 0.001. The ratios of the shear wave velocities and densities are $c_{s,1}/c_{s,2} = 0.8$ and $\rho_1/\rho_2 = 0.85$, respectively. The circular foundation is divided into 200 subdisks with the radius of $R/8$, as shown in Figure 10.5.

The decomposed dynamic stiffness based on the dimensionless frequency a_0 is introduced to yield $S = S_0(k + \mathrm{i}\, a_0 c)$, where k and c are the dimensionless stiffness and damping coefficients, respectively. S_0 is equal to a coefficient β that multiplies the static stiffness, and the static stiffness is given as (Wolf 1985) $S_h = 8GR/(2-\nu)$, $S_v = 4GR/(1-\nu)$, $S_{\phi h} = 8GR^3/[3(1-\nu)]$, $S_{\phi v} = 16GR^3/3$, where S_h, S_v, $S_{\phi h}$ and $S_{\phi v}$ are the horizontal, vertical, rocking and torsional stiffness, respectively. The results are compared to those obtained by Luco (1974) in Figure 10.7. The excellent agreement shows the high accuracy of the proposed method.

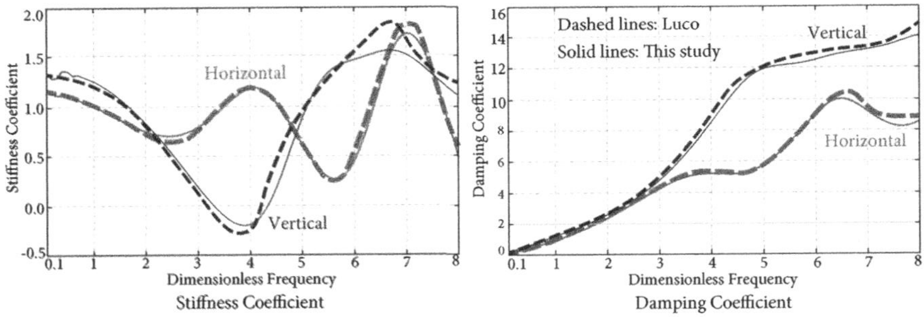

Figure 10.7. *Dynamic soil stiffness of the circular foundation on a layer and half-space. For a color version of this figure, see www.iste.co.uk/challamel/mechanics2.zip*

10.3.3. *A rigid circular foundation supported by half-space and a layer on half-space*

The dynamic stiffness of a rigid circular foundation with the radius of R is considered in two sites. In the first case, the foundation is supported by a homogeneous half-space. Poisson's ratio is $\nu = 0.33$. For the second case, a layer of thickness R rests on a half-space. For both layers, Poisson's ratio is 0.33, the damping ratio is equal to 0.05 and the densities are the same. The ratio of the shear wave velocities is $c_{s,2}/c_{s,1} = 2$. The circular foundation is divided into 200 subdisks, which is the same as the circular foundation, as shown in Figure 10.5.

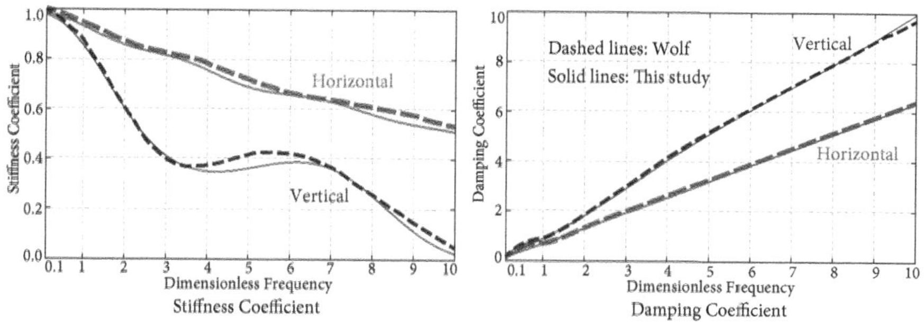

Figure 10.8. *Dynamic soil stiffness of the circular foundation on half-space. For a color version of this figure, see www.iste.co.uk/challamel/mechanics2.zip*

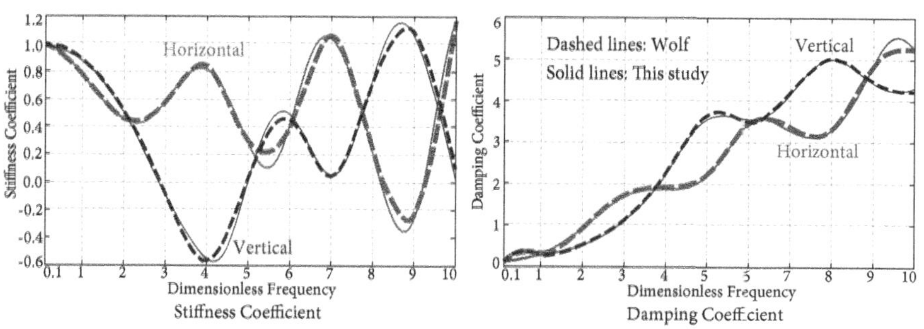

Figure 10.9. *Dynamic soil stiffness of circular foundation on a layer and half-space. For a color version of this figure, see www.iste.co.uk/challamel/mechanics2.zip*

The results are compared to those calculated by Wolf (1985) in Figures 10.8 and 10.9. The coefficients β_h, β_v, $\beta_{\phi h}$ and $\beta_{\phi v}$ are 1.01, 1.02, 1.03 and 1.00 for the half-space case, respectively, while they are 1.32, 1.82, 1.19 and 1.04 for a layer on the half-space. It is seen that the dynamic stiffness results agree well.

10.4. Numerical examples of the generation of FRS

Floor response spectra (FRS) of a typical reactor building in a nuclear power plant are generated based on the dynamic soil stiffness given by the proposed method and the commercial software ACS SASSI to examine the proposed method in the application.

There are three steps to generate FRS considering the SSI: (1) the soil stiffness is determined by the proposed method or ACS SASSI; (2) SSI analysis is performed based on the soil stiffness to develop the modified free-field motion FLIRS (Jiang 2016); and (3) FRS are then generated in a fixed-base model by a direct method (Jiang *et al.* 2015) under the excitation of FLIRS. The two sets of soil stiffness, FLIRS and FRS, are compared and discussed.

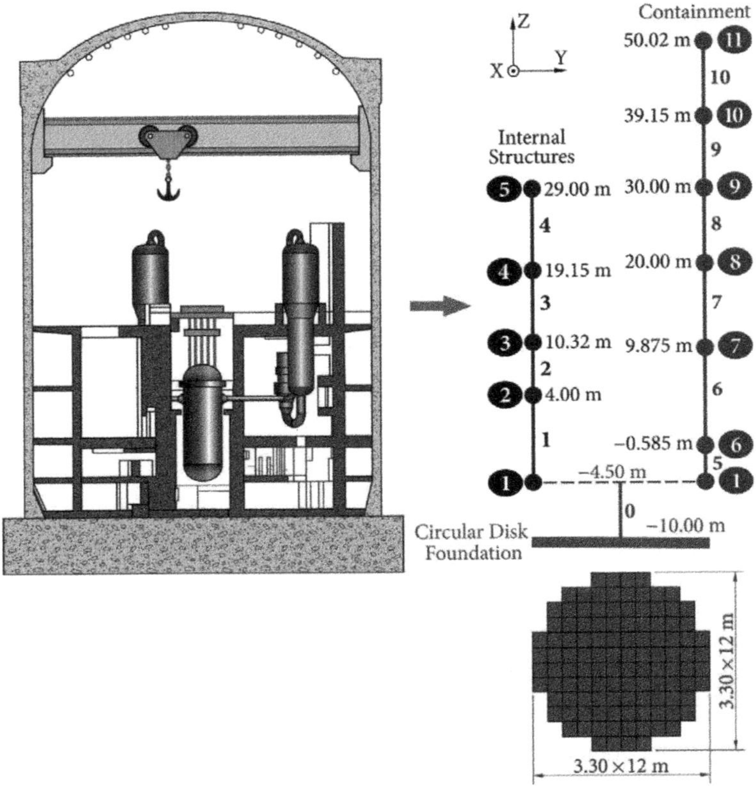

Figure 10.10. *Primary and secondary systems in a nuclear power plant. For a color version of this figure, see www.iste.co.uk/challamel/mechanics2.zip*

Model information: the selected reactor building consists of a containment and an internal structure supported by a circular disk foundation of radius 19.8 m, as shown in Figure 10.10. Using the commercial finite element software ANSYS, the building is modeled as a lumped-parameter system with a foundation at the bottom. The foundation is considered as rigid in this example. Hence, the foundation mass is lumped at the center. There are 112 plates used to represent the foundation, which results in 128 nodes. Detailed information of the finite element model is described in Li *et al.* (2005).

Soil property: the underlying site consists of three infinite soil layers 1, 2 and 3 resting on a homogeneous half-space (layer 4). The shear wave velocity V_S, the thickness of the layer H, the damping ratio ζ, Poisson's ratio ν and the unit weight γ are $V_{s,1}=2100$ m/s, $V_{s,2}=2150$ m/s, $V_{s,3}=2200$ m/s, $V_{s,4}=2100$ m/s, $H_1=H_2=60$ m, $H_3=30$ m, and $\zeta_i=1\%$, $\nu_i=0.3$, $\gamma_i=25.89$ kN/m^3, for $i=1,2,3,4$.

Foundation input response spectra: the R.G. 1.60 response spectra (USNRC 1973) are assumed as the foundation input response spectra (FIRS) obtained from a site response analysis. The peak ground accelerations are anchored to 0.3g and 0.2 g for the horizontal and vertical directions, respectively. A total of 30 sets of tri-directional time histories, which are compatible with the target FIRS, are generated by the Hilbert–Huang transform method (Ni *et al.* 2011, 2013), and are used to perform SSI analysis by ACS SASSI for comparison (see Jiang *et al.* (2015) for details).

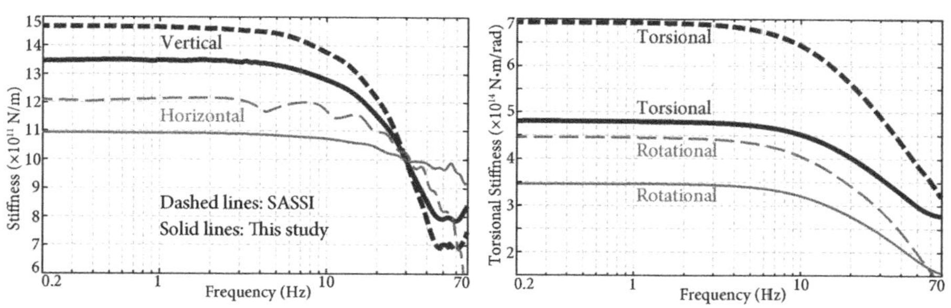

Figure 10.11. *Dynamic soil stiffness of a reactor building. For a color version of this figure, see www.iste.co.uk/challamel/mechanics2.zip*

Soil stiffness: the soil stiffness is determined using the proposed method, as shown in Figure 10.11, along with the results obtained using ACS SASSI. In the proposed method, each square foundation plate is discretized into four uniformly distributed subdisks of radius 0.931 m to calculate the soil stiffness. It can be observed that the difference between horizontal and vertical soil stiffness is small, while the difference between rocking and torsional soil stiffness is significant. For a half-space case, the soil stiffness at the low frequency domain is approximately equal to the static frequency-independent soil stiffness (Luco 1974; Wolf 1985). If the vertical static uniform load q_s is applied at the foundation placed on the soil surface, the additional stresses at the bottoms of the top three layers are $0.14\,q_s$, $0.039\,q_s$ and $0.026\,q_s$, respectively, indicating that the top three layers play the dominant role in the static soil stiffness. Since the V_s of the top three layers are 2100 m/s, 2150 m/s and 2200 m/s, respectively, the soil stiffness in the low frequency domain is reasonable to be larger than half-space with a V_s of 2100 m/s, and smaller than half-space with a V_s of 2200 m/s. The static stiffness of half-space is calculated by equations in ASCE (2000). The rocking and torsional soil stiffness at 0.2 Hz and static values are given in Table 10.1. It can be seen that the rocking and torsional soil stiffness by the proposed method at 0.2 Hz are between the values of two half-space cases, where ACS SASSI results are noticeably larger than the half-space case with a V_s of 2200 m/s, which are incorrect and the errors in soil stiffness will influence the results of SSI analysis.

Case	Rocking stiffness ($\times 10^{14}$N · m/rad)	Torsional stiffness ($\times 10^{14}$N · m/rad)
Half-space ($V_s = 2100$m/s)	3.39	4.75
Half-space ($V_s = 2200$m/s)	3.72	5.21
This study (at 0.2 Hz)	3.50	4.80
ACS SASSI (at 0.2 Hz)	4.46	6.93

Table 10.1. *Rocking and torsional stiffness in the reactor building case*

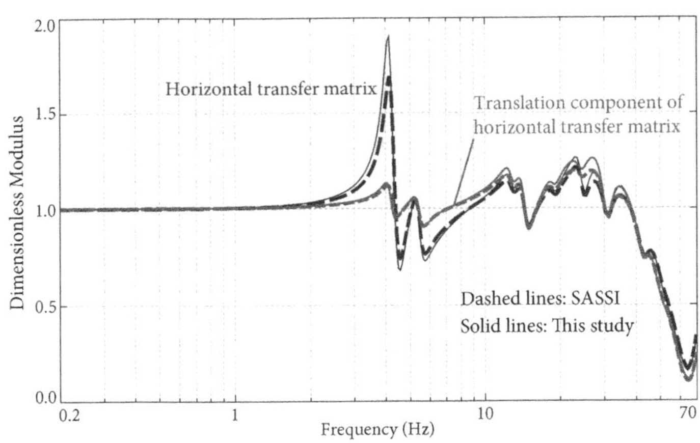

Figure 10.12. *Horizontal transfer matrix of a reactor building*

Developing foundation level input response spectra: the translational component of the horizontal transfer matrix is shown in Figure 10.12. It represents the contribution of translational motions to the determination of FLIRS. The main difference between two sets ranges from 10 Hz to 40 Hz. The horizontal transfer function, which consists of the influence of both translational and rotational components, and characterizes the relationship from FIRS to FLIRS, is also shown in Figure 10.12. The shapes and values of two sets of transfer matrix are almost the same.

Compared to the noticeable difference between peak values at 4.1 Hz, the difference resulting from the translation component between 10 Hz and 40 Hz is not significant. Hence, the main difference is caused by the rotational components influenced by the rotational and torsional stiffness. FLIRS are then developed, as shown in Figure 10.13. It shows that the difference between horizontal FLIRS at approximately 4 Hz and vertical FLIRS at approximately 12 Hz cannot be neglected, which is consistent with the trend of the transfer function. Due to the accurate rocking and torsional soil stiffness obtained by the proposed method, the effect of the SSI is appropriately addressed in this study. Otherwise, the two sets of FLIRS match reasonably well.

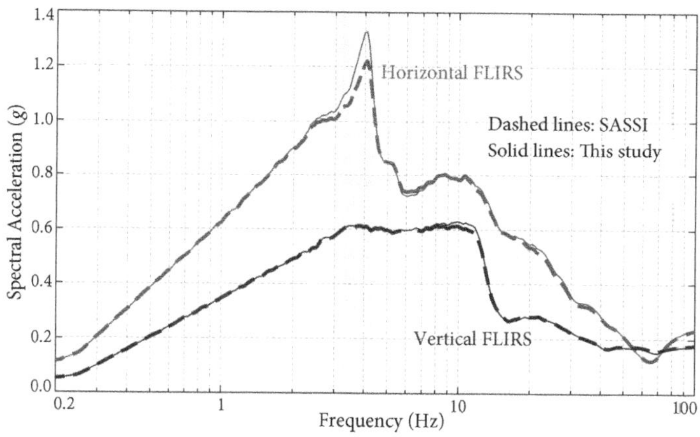

Figure 10.13. *FLIRS of a reactor building*

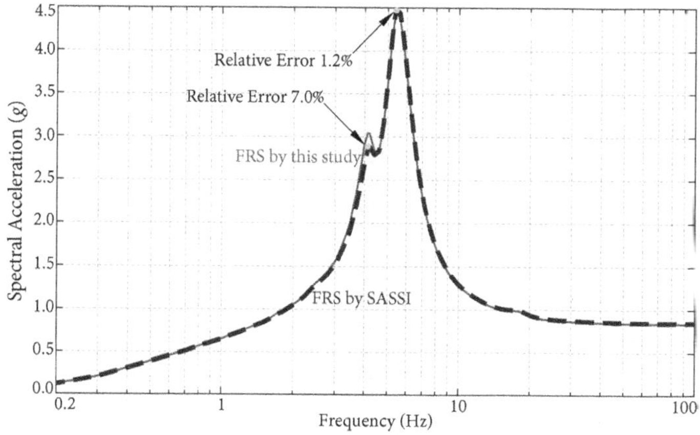

Figure 10.14. *Comparison of FRS at node 4 at the reactor building. For a color version of this figure, see www.iste.co.uk/challamel/mechanics2.zip*

Generating floor response spectra: FRS at node 4 are generated by the direct method (Jiang *et al.* 2015), as shown in Figure 10.14. The average FRS obtained from 30 time history analyses using SASSI is also presented for comparison. It can be seen that both results agree well with a relative error less than 1.2%, except at the first peak (approximately 4 Hz), which is caused by the difference between FLIRS. Hence, it is essential to have accurate dynamic soil stiffness in the SSI analysis.

10.5. Conclusion

A semi-analytical method is developed for accurate and efficient generation of dynamic soil stiffness of both rigid and flexible foundations. Numerical examples are presented to validate the proposed method. Some features and conclusions are summarized as follows:

1) The dynamic soil stiffness varies significantly with excitation frequencies, which cannot be neglected in the SSI analysis.

2) The proposed dynamic soil stiffness is accurate according to the analytical and rigorous formulation of Green's influence function for the given soil properties.

3) Rigid or flexible foundations with arbitrary shapes can be adapted in the proposed method to provide the $6m \times 6m$ dynamic soil stiffness matrix for the SSI analysis.

4) The proposed method is efficient, based on which a very limited number of soil layer elements is involved in the formulation of Green's influence function, and only the soil–foundation interface is meshed.

10.6. References

ASCE (2000). *Seismic Analysis of Safety-related Nuclear Structures and Commentary.* American Society of Civil Engineers, ASCE 4-98. Reston, Virginia.

Estorff, O.V. and Kausel, E. (1989). Coupling of boundary and finite elements for soil-structure interaction problems. *Earthquake Engineering and Structural Dynamics*, 18(7), 1065–1075.

Jiang, W. (2016). Direct method of generating floor response spectra. PhD Thesis, University of Waterloo, Ontario.

Jiang, W., Li, B., Xie, W.-C., Pandey, M. (2015). Generate floor response spectra. Part 1: Direct spectra-to-spectra method. *Nuclear Engineering and Design*, 293, 525–546.

Karabalis, D.L. and Mohammadi, M. (1991). *Foundation-Soil-Foundation Dynamics Using a 3-D Frequency Domain BEM*. Springer, Dordrecht.

Kausel, E. and Roësset, J.M. (1981). Stiffness matrices for layered soils. *Bulletin of the Seismological Society of America*, 71(6), 1743–1761.

Li, Z.-X., Li, Z.-C., Shen, W.-X. (2005). Sensitivity analysis for floor response spectra of nuclear reactor buildings. *Nuclear Power Engineering*, 26(1), 44–50.

Luco, J.E. (1974). Impedance functions for a rigid foundation on a layered medium. *Nuclear Engineering and Design*, 31(2), 204–217.

Luco, J.E. and Apsel, R.J. (1983). On the Green's functions for a layered half-space: Part I. *Bulletin of the Seismological Society of America*, 73(4), 909–929.

Luco, J.E. and Westmann, R.A. (1971). Dynamic response of circular footing. *Journal of the Engineering Mechanics Division*, 97(5), 1381–1395.

Ni, S.-H., Xie, W.-C., Pandey, M.D. (2011). Tri-directional spectrum-compatible earthquake time-histories for nuclear energy facilities. *Nuclear Engineering and Design*, 241(8), 2732–2743.

Ni, S.-H., Xie, W.-C., Pandey, M.D. (2013). Generation of spectrum-compatible earthquake ground motions considering intrinsic spectral variability using Hilbert–Huang transform. *Structural Safety*, 42, 45–53.

Rizzo, F.J. (1967). An integral equation approach to boundary value problems of classical elastostatics. *Quarterly of Applied Mathematics*, 25(1), 83–95.

USNRC (1973). *Design Response Spectra for Seismic Design of Nuclear Power Plants*. U.S. Atomic Energy Commission, USA.

Wolf, J.P. (1985). *Dynamic Soil-Structure Interaction*. Prentice Hall, New Jersey.

Wolf, J.P. and Darbre, G.R. (1984). Dynamic-stiffness matrix of soil by the boundary-element method: Embedded foundation. *Earthquake Engineering and Structural Dynamics*, 12(3), 401–416.

Wong, H.L. and Luco, J.E. (1985). Tables of impedance functions for square foundations on layered media. *Soil Dynamics and Earthquake Engineering*, 4(2), 64–81.

Zhou, Y. (2020). Direct method for floor response spectra considering soil-structure interaction. PhD Thesis, University of Waterloo, Ontario.

List of Authors

Holm ALTENBACH
University of Magdeburg
Germany

Igor V. ANDRIANOV
RWTH Aachen University
Germany

Marcus AßMUS
University of Magdeburg
Germany

Metin AYDOGDU
Trakya University
Turkey

J.R. BANERJEE
City, University of London
UK

Paolo CASINI
Sapienza University of Rome
Italy

Noël CHALLAMEL
University of Southern Brittany
France

Ufuk GUL
Trakya University
Turkey

Sae HOMMA
Kyoto University
Japan

Julius KAPLUNOV
Keele University
UK

Lelya A. KHAJIYEVA
Al-Farabi Kazakh National
University
Kazakhstan

Kotaro KOJIMA
Kyoto Institute of Technology
Japan

Stefano LENCI
Marche Polytechnic University
Italy

Angelo LUONGO
University of L'Aquila
Italy

Giuseppe REGA
Sapienza University of Rome
Italy

Izuru TAKEWAKI
Kyoto University
Japan

Fabrizio VESTRONI
Sapienza University of Rome
Italy

Peter WOOTTON
University of Glasgow
UK

Wei-Chau XIE
University of Waterloo
Canada

Yang ZHOU
University of Waterloo
Canada

Index

Summary of Volume 1

Chapter 6. Buckling and Post-Buckling of Parabolic Arches with Local Damage

Uğurcan EROĞLU, Giuseppe RUTA, Achille PAOLONE and Ekrem TÜFEKCI

Chapter 7. Inelastic Microbuckling of Composites by Wave-Buckling Analogy

Rivka GILAT and Jacob ABOUDI

Chapter 8. Quasi-Bifurcation of Discrete Systems with Unstable Post-Critical Behavior under Impulsive Loads

Mariano P. AMEIJEIRAS and Luis A. GODOY

Summary of Volume 3

Chapter 5. New Approach to the Reliability Verification of Aerospace Structures

Giora MAYMON

Chapter 6. A Review of Interval Field Approaches for Uncertainty Quantification in Numerical Models

Matthias FAES, Maurice IMHOLZ, Dirk VANDEPITTE and David MOENS

Chapter 7. Convex Polytopic Models for the Static Response of Structures with Uncertain-but-bounded Parameters

Zhiping QIU and Nan JIANG

Chapter 8. On the Interval Frequency Response of Cracked Beams with Uncertain Damage

Roberta SANTORO

Chapter 9. Quantum-Inspired Topology Optimization

Xiaojun WANG, Bowen NI and Lei WANG

Chapter 10. Time Delay Vibrations and Almost Sure Stability in Vehicle Dynamics

Walter V. WEDIG

Chapter 11. Order Statistics Approach to Structural Optimization Considering Robustness and Confidence of Responses

Makoto YAMAKAWA and Makoto OHSAKI

Other titles from

ISTE

in

Mechanical Engineering and Solid Mechanics

2021

DAHOO Pierre Richard, POUGNET Philippe, EL HAMI Abdelkhalak
Applications and Metrology at Nanometer Scale 1: Smart Materials, Electromagnetic Waves and Uncertainties
(Reliability of Multiphysical Systems Set – Volume 9)

2020

LEDOUX Michel, EL HAMI ABDELKHALAK
Heat Transfer 1: Conduction

SALENÇON Jean
Elastoplastic Modeling

2019

BAYLE Franck
Reliability of Maintained Systems Subjected to Wear Failure Mechanisms: Theory and Applications
(Reliability of Multiphysical Systems Set – Volume 8)

BEN KAHLA Rabeb, BARKAOUI Abdelwahed, MERZOUKI Tarek
Finite Element Method and Medical Imaging Techniques in Bone Biomechanics
(Mathematical and Mechanical Engineering Set – Volume 8)

IONESCU Ioan R., QUEYREAU Sylvain, PICU Catalin R., SALMAN Oguz Umut
Mechanics and Physics of Solids at Micro- and Nano-Scales

LE VAN Anh, BOUZIDI Rabah
Lagrangian Mechanics: An Advanced Analytical Approach

MICHELITSCH Thomas, PÉREZ RIASCOS Alejandro, COLLET Bernard, NOWAKOWSKI Andrzej, NICOLLEAU Franck
Fractional Dynamics on Networks and Lattices

SALENÇON Jean
Viscoelastic Modeling for Structural Analysis

VÉNIZÉLOS Georges, EL HAMI Abdelkhalak
Movement Equations 5: Dynamics of a Set of Solids
(Non-deformable Solid Mechanics Set – Volume 5)

2018

BOREL Michel, VÉNIZÉLOS Georges
Movement Equations 4: Equilibriums and Small Movements
(Non-deformable Solid Mechanics Set – Volume 4)

FROSSARD Etienne
Granular Geomaterials Dissipative Mechanics: Theory and Applications in Civil Engineering

RADI Bouchaib, EL HAMI Abdelkhalak
Advanced Numerical Methods with Matlab® 1: Function Approximation and System Resolution
(Mathematical and Mechanical Engineering SET – Volume 6)
Advanced Numerical Methods with Matlab® 2: Resolution of Nonlinear, Differential and Partial Differential Equations
(Mathematical and Mechanical Engineering SET – Volume 7)

SALENÇON Jean
Virtual Work Approach to Mechanical Modeling

2017

BOREL Michel, VÉNIZÉLOS Georges
Movement Equations 2: Mathematical and Methodological Supplements (Non-deformable Solid Mechanics Set – Volume 2)
Movement Equations 3: Dynamics and Fundamental Principle (Non-deformable Solid Mechanics Set – Volume 3)

BOUVET Christophe
Mechanics of Aeronautical Solids, Materials and Structures
Mechanics of Aeronautical Composite Materials

BRANCHERIE Delphine, FEISSEL Pierre, BOUVIER Salima, IBRAHIMBEGOVIC Adnan
From Microstructure Investigations to Multiscale Modeling: Bridging the Gap

CHEBEL-MORELLO Brigitte, NICOD Jean-Marc, VARNIER Christophe
From Prognostics and Health Systems Management to Predictive Maintenance 2: Knowledge, Traceability and Decision (Reliability of Multiphysical Systems Set – Volume 7)

EL HAMI Abdelkhalak, RADI Bouchaib
Dynamics of Large Structures and Inverse Problems (Mathematical and Mechanical Engineering Set – Volume 5)
Fluid-Structure Interactions and Uncertainties: Ansys and Fluent Tools (Reliability of Multiphysical Systems Set – Volume 6)

KHARMANDA Ghias, EL HAMI Abdelkhalak
Biomechanics: Optimization, Uncertainties and Reliability (Reliability of Multiphysical Systems Set – Volume 5)

LEDOUX Michel, EL HAMI Abdelkhalak
Compressible Flow Propulsion and Digital Approaches in Fluid Mechanics (Mathematical and Mechanical Engineering Set – Volume 4)
Fluid Mechanics: Analytical Methods (Mathematical and Mechanical Engineering Set – Volume 3)

MORI Yvon
Mechanical Vibrations: Applications to Equipment

2016

BOREL Michel, VÉNIZÉLOS Georges
Movement Equations 1: Location, Kinematics and Kinetics (Non-deformable Solid Mechanics Set – Volume 1)

BOYARD Nicolas
Heat Transfer in Polymer Composite Materials

CARDON Alain, ITMI Mhamed
New Autonomous Systems (Reliability of Multiphysical Systems Set – Volume 1)

DAHOO Pierre Richard, POUGNET Philippe, EL HAMI Abdelkhalak
Nanometer-scale Defect Detection Using Polarized Light (Reliability of Multiphysical Systems Set – Volume 2)

DE SAXCÉ Géry, VALLÉE Claude
Galilean Mechanics and Thermodynamics of Continua

DORMIEUX Luc, KONDO Djimédo
Micromechanics of Fracture and Damage (Micromechanics Set – Volume 1)

EL HAMI Abdelkhalak, RADI Bouchaib
Stochastic Dynamics of Structures (Mathematical and Mechanical Engineering Set – Volume 2)

GOURIVEAU Rafael, MEDJAHER Kamal, ZERHOUNI Noureddine
From Prognostics and Health Systems Management to Predictive Maintenance 1: Monitoring and Prognostics (Reliability of Multiphysical Systems Set – Volume 4)

KHARMANDA Ghias, EL HAMI Abdelkhalak
Reliability in Biomechanics
(Reliability of Multiphysical Systems Set –Volume 3)

MOLIMARD Jérôme
Experimental Mechanics of Solids and Structures

RADI Bouchaib, EL HAMI Abdelkhalak
Material Forming Processes: Simulation, Drawing, Hydroforming and Additive Manufacturing
(Mathematical and Mechanical Engineering Set – Volume 1)

2015

KARLIČIĆ Danilo, MURMU Tony, ADHIKARI Sondipon, MCCARTHY Michael
Non-local Structural Mechanics

SAB Karam, LEBÉE Arthur
Homogenization of Heterogeneous Thin and Thick Plates

2014

ATANACKOVIC M. Teodor, PILIPOVIC Stevan, STANKOVIC Bogoljub, ZORICA Dusan
Fractional Calculus with Applications in Mechanics: Vibrations and Diffusion Processes
Fractional Calculus with Applications in Mechanics: Wave Propagation, Impact and Variational Principles

CIBLAC Thierry, MOREL Jean-Claude
Sustainable Masonry: Stability and Behavior of Structures

ILANKO Sinniah, MONTERRUBIO Luis E., MOCHIDA Yusuke
The Rayleigh–Ritz Method for Structural Analysis

LALANNE Christian
Mechanical Vibration and Shock Analysis – 5-volume series – 3rd edition
Sinusoidal Vibration – Volume 1
Mechanical Shock – Volume 2
Random Vibration – Volume 3
Fatigue Damage – Volume 4
Specification Development – Volume 5

LEMAIRE Maurice
Uncertainty and Mechanics

2013

ADHIKARI Sondipon
Structural Dynamic Analysis with Generalized Damping Models: Analysis

ADHIKARI Sondipon
Structural Dynamic Analysis with Generalized Damping Models: Identification

BAILLY Patrice
Materials and Structures under Shock and Impact

BASTIEN Jérôme, BERNARDIN Frédéric, LAMARQUE Claude-Henri
Non-smooth Deterministic or Stochastic Discrete Dynamical Systems: Applications to Models with Friction or Impact

EL HAMI Abdelkhalak, RADI Bouchaib
Uncertainty and Optimization in Structural Mechanics

KIRILLOV Oleg N., PELINOVSKY Dmitry E.
Nonlinear Physical Systems: Spectral Analysis, Stability and Bifurcations

LUONGO Angelo, ZULLI Daniele
Mathematical Models of Beams and Cables

SALENÇON Jean
Yield Design

2012

DAVIM J. Paulo
Mechanical Engineering Education

DUPEUX Michel, BRACCINI Muriel
Mechanics of Solid Interfaces

ELISHAKOFF Isaac *et al.*
Carbon Nanotubes and Nanosensors: Vibration, Buckling and Ballistic Impact

GRÉDIAC Michel, HILD François
Full-Field Measurements and Identification in Solid Mechanics

GROUS Ammar
Fracture Mechanics – 3-volume series
Analysis of Reliability and Quality Control – Volume 1
Applied Reliability – Volume 2
Applied Quality Control – Volume 3

RECHO Naman
Fracture Mechanics and Crack Growth

2011

KRYSINSKI Tomasz, MALBURET François
Mechanical Instability

SOUSTELLE Michel
An Introduction to Chemical Kinetics

2010

BREITKOPF Piotr, FILOMENO COELHO Rajan
Multidisciplinary Design Optimization in Computational Mechanics

DAVIM J. Paulo
Biotribolgy

PAULTRE Patrick
Dynamics of Structures

SOUSTELLE Michel
Handbook of Heterogenous Kinetics

2009

BERLIOZ Alain, TROMPETTE Philippe
Solid Mechanics using the Finite Element Method

LEMAIRE Maurice
Structural Reliability

2007

GIRARD Alain, ROY Nicolas
Structural Dynamics in Industry

GUINEBRETIÈRE René
X-ray Diffraction by Polycrystalline Materials

KRYSINSKI Tomasz, MALBURET François
Mechanical Vibrations

KUNDU Tribikram
Advanced Ultrasonic Methods for Material and Structure Inspection

SIH George C. *et al.*
Particle and Continuum Aspects of Mesomechanics

Printed and bound by CPI Group (UK) Ltd, Croydon, CR0 4YY

15/09/2026

14962362-0006